T0075092

The Desert Bones

Life of the Past James O. Farlow, editor

THE DESERT BONES

The Paleontology and Paleoecology of Mid-Cretaceous North Africa

JAMALE IJOUIHER

Indiana University Press

This book is a publication of

Indiana University Press
Office of Scholarly Publishing
Herman B Wells Library 350
1320 East 10th Street
Bloomington, Indiana 47405 USA

iupress.org

This book is printed on acid-free paper.

Manufactured in Korea

First printing 2022

Cataloging information is available from the Library of Congress.

ISBN 978-0-253-06331-1 (hdbk.)
ISBN 978-0-253-06332-8 (web PDF)

I want to dedicate this book to my mother, Jacqueline Ijouiher,
and my grandparents, David Brian Evans and Margaret Mary Evans.
I also extend that dedication to my former lecturer and
mentor professor Alan Turner of Liverpool John Moores University.
Great people who are truly missed.
This is for you.

Contents

Acknowledgments *ix*
Preface *xi*
Museum Abbreviations *xiii*

Introduction 1

1. The Paleoenvironments and Stratigraphy of North Africa 7
2. The Flora of North Africa 31
3. The Fauna of North Africa: Invertebrates 59
4. The Fauna of North Africa: Vertebrates (Fish) 101
5. The Fauna of North Africa: Vertebrates (Tetrapoda) 151
6. North African Ecology 211
7. The March of the Oysters 241
8. The Cenomanian Mass Extinction 247

Appendix 257
Literature Cited 263
Index 305

Acknowledgments

The list of people who deserve my gratitude seems endless, and some (like Ernst Stromer) have been dead for decades and never even met me! I thank Thomas Bastelberge, Oliver Demuth, Liz Kennedy, Casey Holliday, Bruno Granier, Walid Makled, Marguerite Miller, Emilie Läng, Omar Mohamed, Thomas Tütken, Olof Moleman, Markus Buhler, Ricardo Pimentel, Guillaume Guinot, Michael Donovan, Jean-François Lhomme, Neil Loneragan, Andrea Cau, Daniela Schwarz-Wings, Didier Dutheil, Mike Taylor, John Adamek, Matt Lamanna, Hilda Silva, Cristiano Dal Sasso, Wagih Hannaa, Mickey Mortimer, Aaron Miller, Lionel Cavin, Hans Larsson, Bradley McFeeters, and Kenneth Lacovara for their assistance in providing me with access to literature and specimens. If I've missed anyone from that list, no slight was ever intended.

Worthy of special mention are the staff of Liverpool John Moores University for their continued patronage and advice. My editor James Farlow and principal reviewer Barbara Grandstaff, who share many of the same interests and sparked many new ideas through our discussions. My thanks are also extended to my former flatmates, whose friendship at university made that period one of the happiest points of my life.

I would also like to thank Joschua Knüppe, Christopher DiPiazza, Andrey Atuchin, and James McKay for contributing artwork and ideas during the writing of this book.

And ultimately, I've got to thank my sister and aunt for their continued love and support.

Preface

I've noted that many paleontologists seem to have the same origin story: a childhood interest in fossils that never ended, and parents who were supportive of these endeavors (once they finally accepted that it was not a phase). In that respect, I'm no different from all the others. People always ask me when my interest in dinosaurs began, but in all honesty, it began so early in my childhood that I can't remember that far back.

I suppose what makes me different is my interest in one particular subject. While I've always been interested in dinosaurs, my fascination with North Africa is a relatively new development. And contrary to what some might be thinking because of my name, it has nothing to do with being of Moroccan descent.

It was in early 2008; I was attending university at the time. My mother had purchased Nothdurft's book *The Lost Dinosaurs of Egypt* for me the previous Christmas, and I was finally getting around to reading it. It was the closest I've ever come to a religious conversion. I stayed up into the night, finishing the book from cover to cover in one go—undoubtedly the highest compliment you can give a book.

I was enthralled: the adventures, the larger-than-life characters who preceded me, the absolute uniqueness of the environment. The more data I collected, the more my love of North African paleontology was fueled. From that moment, I dedicated my life to the paleontology of North Africa, and I've never looked back.

This book is just one further manifestation of my love for this topic. A definitive book on North African paleontology is long overdue. One inspiration was *Jurassic West* by John Foster. No doubt readers will immediately see the similarities in style as I attempt to do for Africa what Foster did for the US.

This project has been a four-year labor of love. It hasn't all been easy to write; some days I'll write three or four pages at a go, and other days I stare at a blank screen, wondering where to begin. I also seem to have developed a habit of writing chapters in no particular order, instead of starting at the beginning; whether other authors use this style or this is just a quirk of my own, I can't say.

Ultimately, my aim with this work is to produce something informal and accessible, yet also to provide a serious scientific discourse that will hopefully become a reference source for the paleontological community for years to come. And even if you don't pay me that ultimate compliment by finishing this book from cover to cover in one go, I hope with all my heart that you will come away having learned something new. As Kenneth Lacovara once said, ancient North Africa "was a huge biome, one that is only just beginning to be explored."

So let's explore it together.

Jamale Ijouiher, July 2022.

Museum Abbreviations

BSP	Bayerische Staatssammlung für Paläontologie und Historische Geologie. Germany.
BSPG	Bayerische Staatssammlung für Paläontologie und Geologie. Germany
CGM	Egyptian Geological Museum. Egypt.
CMN	Canadian Museum of Nature. Canada.
FSAC-KK	Faculte des Sciences Ain Chock, Universite Hassan II. Morocco.
GZG	Geo Wissenschaftliches Zentrum der Universität Göttingen. Germany.
JP	JuraPark. Poland.
MDE	Le Musée des Dinosaures Espéraza-Aude. France.
MHNM	Natural History Museum of Marrakech. Morocco.
MN	Museu Nacional, Rio de Janeiro. Brazil.
MNUFRJ	Museu Nacional of the Universidade Federal do Rio de Janeiro. Brazil.
MPUR NS	Museum of Paleontology, University of Rome. Italy.
MSNM	Museo Civico di Storia Naturale di Milano. Italy.
OLPH	Olphin collection of the Museo Geologico e Paleontologico. Italy
UALVP	University of Alberta Laboratory for Vertebrate Paleontology. Canada.
UCRC	University of California. US.
UCPC	University of Connecticut. US.
UMI	University Moulay Ismail. Morocco.
USNM	National Museum of Natural History, Smithsonian Institution. US.
ZIN	Zoological Institute, Russian Academy of Sciences. Russia.

The Desert Bones

To the eyes of an outsider, the initial impression of ancient Egypt is one of unfamiliarity.

—**JAROMIR MALEK**, *In the Shadow of the Pyramids: Egypt during the Old Kingdom*

Introduction

Egyptologists refer to the age in the third millennium BC as the Old Kingdom, a period when Egyptian civilization, what would be the first of three kingdoms, truly began. However, Egypt was home to kingdoms *much* older than that.

Unfortunately, the geological record is incomplete. If we look at the Bahariya oasis, for example, not only do we know little of the people who dwelled in the region before the Old Kingdom, but much of Egypt's deep history is also missing (El-Sisi et al. 2002). Indeed, this huge gap means that the most recent of these genuinely ancient kingdoms, long before humans arrived in Bahariya, is the basalt flows of the Miocene, a series of rocks formed roughly 16 to 20 million years ago (henceforth abbreviated to mya). Egypt was then a kingdom of arid grasslands, with violent volcanic eruptions racking the landscape as the Gulf of Suez first began to open (El-Sisi et al. 2002).

Going even further back in time, we come to the Radwan Formation of the Oligocene, formed 33–23 mya. Back then, the Nile was far to the west of its present position, and North Africa was seeing the first grasslands expanding at the expense of the forests (El-Sisi et al. 2002). This new biome was home to large numbers of perissodactyl mammals.

As we continue our journey, we arrive at the Hamra, Qazzun, and Naqb Formations of the Eocene, 55–33 mya. Then, much of Egypt's Western Desert was under water, drowned by shallow seas. The only pharaohs to rule this kingdom were the primitive whales, such as *Basilosaurus*, whose remains are found in the Wadi Al-Hitan: the Valley of the Whales.

Now we come to yet another gap in the record, and on the other side, we have left the Cenozoic era behind and arrived at the Mesozoic era: the age of the dinosaurs. First, we come to the Khoman Chalks, formed during the Maastrichtian age right at the very end of the Mesozoic. Going even further back in time, beyond the Campanian El Hefhuf Formation, we come to some of the very oldest rocks preserved at Bahariya: those of the Cenomanian age, which began an estimated 99.0 mya and ended roughly 93.5 mya.

The Mesozoic era is divided into three periods: the Triassic, Jurassic, and Cretaceous, each with its own subdivisions (see fig. 0.1). Contrary to what some think, most dinosaurs did not all coexist at the same time. The Mesozoic lasted for 180 million years, and over that vast period, dinosaurs were continually becoming extinct and being replaced with newly evolved dinosaur species. I think Thomas Holtz Jr. (2012) put it best during an interview in which he pointed out that by the time *Tyrannosaurus*

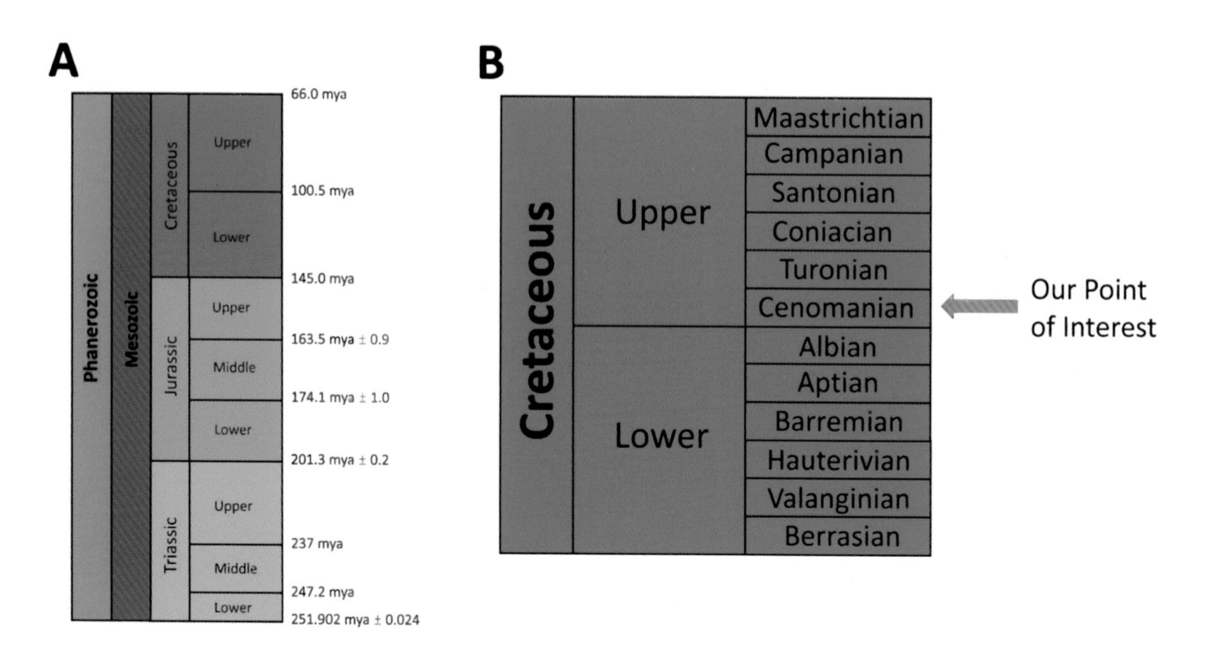

Fig. 0.1. *A*, the geological periods in the Mesozoic era; *B*, a close-up of the Cretaceous period with the Cenomanian age highlighted. Adapted from *The International Commission on Stratigraphy* (2018) with slight modification.

evolved, *Allosaurus* had already been extinct for millions of years; *Allosaurus* was just as far from *Tyrannosaurus* as *Tyrannosaurus* is from us!

The Cenomanian was an age of unprecedented global warming and the highest sea levels observed in the past six hundred million years—approximately 150 meters (492 feet) above present-day levels (Robin et al. 2013). What happened back then can give us a window into the current environmental challenges we face; the miserable end to the Cenomanian may be an omen of what is awaiting us.

The Cenomanian was also an age of unrestricted oceanic currents leading to the widespread distribution of marine species. It would not be until the end of the Cenomanian that local endemism began to rise again (Stilwell and Henderson 2002). On land, this age also saw the extinction of some dinosaur groups and the rise of others that would dominate the planet's ecosystems for the remainder of the Mesozoic.

The Cenomanian is a crucial interval in Mesozoic history, and North Africa contains some of the best known and documented Cenomanian sediments in the world. This is in part because this biome stretched for thousands of miles, from the Gulf of Suez to the Atlantic Ocean. Many fossil vertebrates are also known from partially complete skeletons, making North Africa one of the richest Cenomanian vertebrate assemblages in the world (Cavin et al. 2013).

So one can only wonder why Africa has been continuously overlooked. Russell (1995, 10) considered Africa to be one of the "darkest" of the "lost worlds of the dinosaur era," and with good reason; apart from the well-studied Jurassic fauna of Tendaguru, African dinosaur paleontology has been a mostly stagnant field. While expeditions to North Africa have always met with success, there has been little attempt to build on them until recently (Weishampel et al. 2004).

There is no one reason for this. Partly it has to do with the seemingly constant undercurrent of political unrest in the region (Owen 2012), which perhaps explains why Morocco, one of the more stable countries in the area, also has the most well-established "fossil industry." The exorbitant expense of mounting such expeditions is also a factor; Egypt is hardly the distant land it was in 1914, but it still takes US$60,000 to mount an expedition to Bahariya (Nothdurft et al. 2002).

Many of the best fossils ever found in the region were lost during World War II. From 1912 to 1914, the German paleontologist Ernst Freiherr Stromer von Reichenbach had organized arguably the most successful North African paleontological expedition to date regarding the number of species discovered and the sheer wealth of material. By the start of World War II, his priceless discoveries were housed at the Alte Akademie in Munich, under the control of Karl Beurlen. A Nazi through and through, Beurlen rebuffed repeated requests that the Bavarian state collection be hidden away until the war ended, as other institutions were doing; he even attempted to have Stromer sent to a concentration camp after Stromer begged him to see reason (Nothdurft et al. 2002).

So it would have been interesting to have seen the look on Beurlen's face on the morning of April 25, 1944, when the British scored a direct hit on Munich, including his precious museum. Nothdurft and colleagues (2002) speculate that Air chief Marshal Arthur "Bomber" Harris deliberately targeted the city in a calculated act of revenge for the German destruction of Coventry in 1940.

Regardless of whatever Pyrrhic victory can be envisioned in seeing Beurlen getting his comeuppance, almost the entire Stromer collection was destroyed, most of which we have never been able to replace. Ancient Egyptians were said to have had curses on their tombs (Silverman 2003), so one can't help but wonder whether Egyptian dinosaurs did too. Given the misfortune that befell Stromer, Tutankhamun has nothing on *Spinosaurus*!

Then in the 1950s, dinosaur paleontology entered a period of worldwide decline (Paul 2000). If scientists were no longer interested in the dinosaurs found in their own countries, there was no chance that they would go abroad to look for them.

After all this heartache, North Africa now seems to be experiencing a renaissance. From the 1970s onward, dinosaur paleontology has made an incredible comeback; an estimated 50% of the world's newly discovered dinosaur-bearing formations were only described in the past thirty years (Benton 2015).

In 2008 alone, Mansoura University launched a series of ambitious reconnaissance missions and excavations, which were ongoing until 2010 (O'Connor et al. 2010). More researchers than ever before are now working in the region, and their discoveries are finally getting international attention.

However, it's not all strawberries and cream, to use the old expression. Such renewed interest has opened up North Africa to black-market dealers, with the money from illicit fossil sales helping fund God knows what (Boudad et al. 2014). In 2015, an unnamed Moroccan scientist revealed the existence of a whole network of gangs who smuggle fossils, including a partially complete dinosaur skeleton at one point, out of the country for prices of up to 100,000 dirhams (20minutos.es 2015), which would be the equivalent of GBP£8,215 or US$10,477, going by current exchange rates at the time of writing.

The 1970 UNESCO Convention prohibits the illicit sale of fossils, but enforcing such laws in the third world is difficult. And such laws are controversial; in my experience, many fossil dealers and private collectors are honest people, using legitimate sources, and are happy making their discoveries available to scientists. And those in Morocco who provide these fossils are often just poor rural workers (one can be sure that the 100,000 dirhams above never went into their pockets).

Liston and Long (2015) have even suggested that such stringent laws incite criminal behavior on the part of researchers to circumvent the red tape. However, I vehemently disagree with the suggestion that fossil laws lead to academic theft. On the contrary, cases of academic theft—plagiarism and acquiring specimens from dubious sources—are the result of the paleontological community's continual failure to police itself, not the result of law enforcement (Taylor 2008a, 2008b; Escobar 2015).

One way to help is to encourage the growth of online communities like thefossilforum.com, where collectors can upload and discuss specimens, so privately owned fossils are recorded for public access and don't just disappear into some vault, the traditional complaint leveled against those who buy fossils. While publishing specimens that are not in an accredited institution is traditionally frowned on, the growing ubiquity of the internet, open publishers like *PeerJ*, and online collections like digimorph.org make this inevitable in my mind.

We must also stimulate interest in those countries themselves. Dr. Nizar Ibrahim has a point when he says that precious few fossils are exhibited in their countries of origin (in Clark 2015). And it's not for lack of interest; Ibrahim reports that many people are genuinely interested in their countries' deep history.

The formation of natural history institutions across the region would, in theory, bring the fossil industry under control, as the rural prospectors would still be able to make a living with legitimate outlets for their discoveries (Boudad et al. 2014). However, this is all easier said than done.

For all these issues, this is undoubtedly the greatest period of interest North Africa has seen in a long time. With luck, this will usher in a continuing era of research equal to that seen in Europe, Asia, and the Americas.

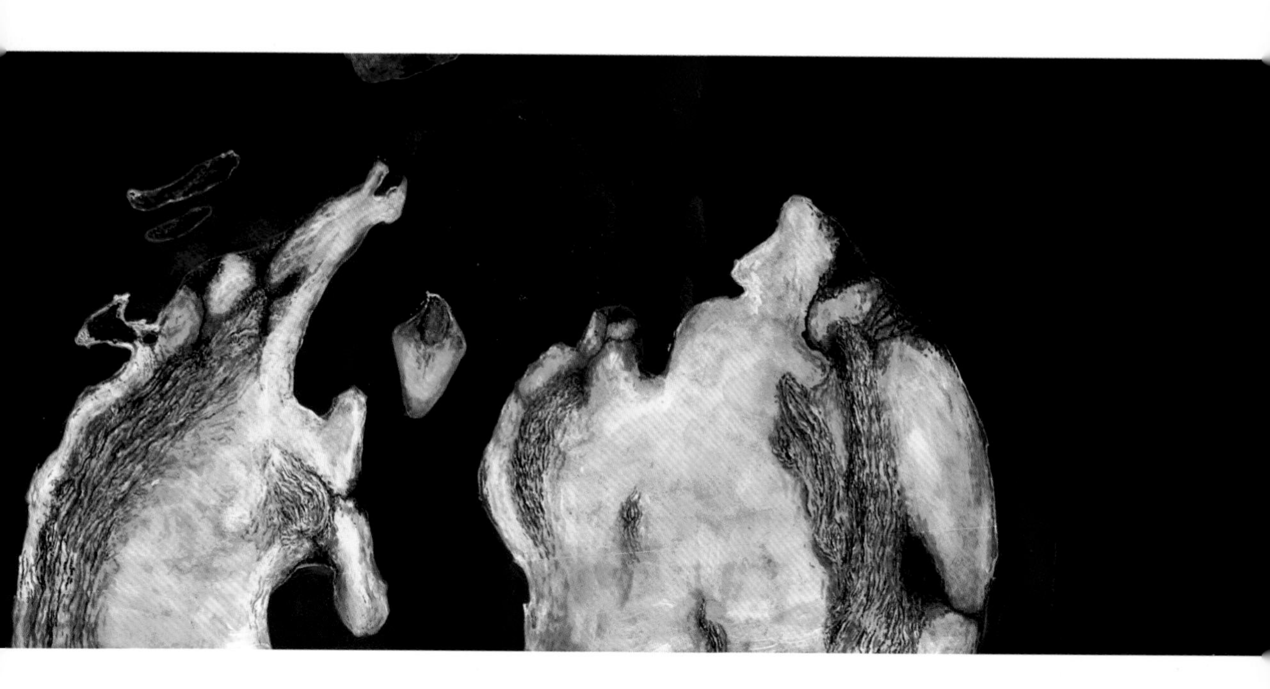

Fig. 1.1. A map of North Africa, circa 93.5 MYA. Adapted from El-Sisi et al. (2002); Lüning et al. (2004); Gertsch et al (2010b); and Contessi (2013b) with slight modification.

The Paleoenvironments and Stratigraphy of North Africa

1

Paleoenvironment

As a boy, I loved the novelization of the TV series *Walking with Dinosaurs*. I purchased my copy at Heathrow Airport when going on vacation and subsequently carried it with me all over the US East Coast. I found the best part to be the series of wonderful maps giving place names to all the ancient locations described within. It made the dinosaurs seem like real creatures living in an actual time and place, which is often the biggest difficulty I've found in communicating the science of paleontology to people.

So what was North Africa like between 99.0 and 93.5 million years ago (mya)? Contrary to what most would expect, the geography was not all that different from what you see today, with Africa, the Americas, and Europe all vaguely recognizable (see fig. 1.1). However, the high sea levels and positions of the continents meant that the world was cut in two: a northern continent named Laurasia and a southern continent, of which Africa was once a part, called Gondwana or Gondwanaland. Gondwana also contained South America, Australia, and Antarctica.

Both these landmasses were remnants of the supercontinent, Pangaea, which split apart in the Jurassic period. By the Cenomanian, however, Gondwana itself was breaking into pieces. While the exact times for the breakup are still debated, it's likely that Africa separated first and Australia last.

Africa, at that time, was southeast of its current position and still close to the remainder of Gondwana. This raises a question: Had Africa completely separated from the remainder of Gondwana by the Cenomanian? Similarities in the faunas of Africa and South America suggest either that travel between the two continents was still possible or that the connection between them had been severed only recently (Candeiro et al. 2011; Medeiros et al. 2014).

A narrow seaway separated Africa's western coastline from South America. That was the southern part of the Atlantic Ocean, only recently formed and still growing. On Africa's northern and eastern coastline was the Tethys Sea. The Tethys was mostly calm and tropical but had a profound impact on the surrounding lands because the collision of warm and cold oceanic currents would have resulted in occasional but violent hurricanes (Nothdurft et al. 2002; Russell and Paesler 2003; Benyoucef et al. 2014b). However, while these currents could be incredibly strong in some regions, in the west the upwelling (wind-driven circulation in which deep water is brought to the surface) was less than 20 cm a day, leaving those areas almost stagnant (Abdelhady 2002).

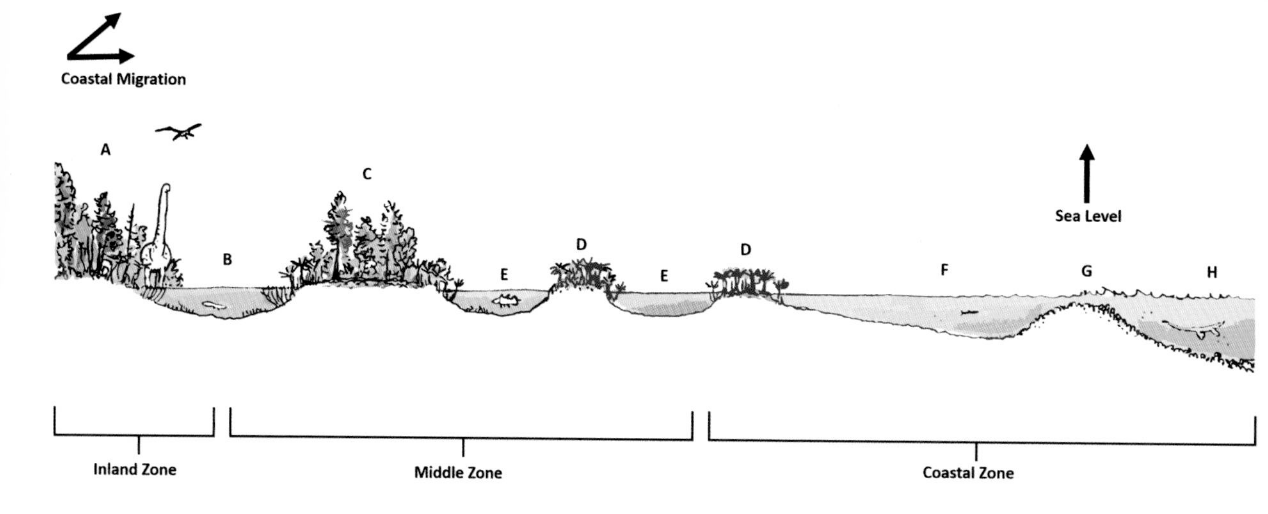

Fig. 1.2. Cross section of a North African Cenomanian river system. *A*, terrestrial gallery forest; *B*, rivers; *C*, marsh/transitional communities; *D*, mangals; *E*, tidal channels; *F*, lagoons; *G*, sandbars/reefs; *H*, epicontinental sea. Despite rising sea levels, the coastlines would have been expanding out into the epicontinental seas as the mangrove vegetation trapped sediment, gradually creating new land. This continued until the end of the Cenomanian, when sea level change finally exceeded sedimentation rates, flooding the river basins. Adapted from Lacovara et al. (2003) with modifications. Artwork by Joschua Knüppe.

Inland, to the south, was the Central Gondwanan Desert (Myers et al. 2011; Mateus et al. 2011; Benson et al. 2012). This desert effectively cut off the northern part of the continent from the south (Rauhut and López-Arbarello 2009). This inhospitable region formed as a result of Africa's geographic position, which placed the central portion of the continent within the planet's arid climactic belt, with local temperatures ranging between 15°C and 30°C (Jacobs et al. 2009; Mateus et al. 2011). There is little evidence of plant life in this region with desert-adapted dinosaurs different from those found in North Africa (Mateus et al. 2011).

The biggest difference was the fact that Africa was divided by the Trans-Saharan Seaway, which cut a swath across the continent, extending from the Tethys in the north to the newly formed Atlantic in the south. The Trans-Saharan Seaway also had a cold-water current, the proto-Benguela current, which flowed northward (Gebhardt 1999).

On the northwestern shore of the Trans-Saharan Seaway were a large lagoon and a series of small islands and carbonate platforms, similar to the Solnhofen islands of Jurassic Germany (Herkat 2003). The exact size and precise location of these still-unnamed islands are unknown. Although it's doubtful that these islands held any large native fauna, many unique species of small animal may have evolved there.

Rising out of the Trans-Saharan Seaway, situated between western and eastern Africa, was the island of Gargal, sometimes spelled as Gargaf (Lüning et al. 2004). Originally an upland region that became an island when the Trans-Saharan Seaway flooded the surrounding lowlands, Gargal would have been one of the largest islands in the world at that point in history.

At the latitude of roughly 10° N, most of North Africa was near the equator at the time, resulting in a tropical climate. This warmth was exacerbated by the higher global temperatures in the Cenomanian (Tarduno et al. 1998; Russell and Paesler 2003; Lacovara et al. 2003; Deaf 2009).

However, precise estimates of the paleotemperatures vary; Russell and Paesler (2003) favor a high-end estimate, while Krassilov and Bacchia (2013) suggest milder conditions. A surface temperature of roughly 34°C, with water temperatures ranging between 21°C and 34°C, now seems like a fair estimate for North Africa at this time (Amiot et al. 2010b; Murray et al. 2013b).

While some areas appear to have had a more humid climate (Schrank 1990; Brenner 1996; Deaf 2009; Gertsch et al. 2010b; Baioumi et al. 2012; Zobaa et al. 2008), these would have been localized and not representative of the climate as a whole. The abundance of xerophytic plant fossils and palynomorphs confirms a hot and arid climate (Bolkhovitina 1953; Krutzsch 1961; Trevisan 1980; Lyon et al. 2001; Zobaa et al. 2008; Deaf 2009; Ibrahim et al. 2009; Bodin et al. 2010; Baioumi et al. 2012).

Contributing to the high temperatures during this period were the global CO_2 levels, which were eleven times higher than today's levels (Bice and Norris 2002; Takashima et al. 2006). The atmosphere at the time was also composed of a mere 10%–11% oxygen, compared to the current level of 21% (Poulsen et al. 2015).

The climate would have remained relatively constant through the year, although it would have endured periodic hurricanes (Russell and Paesler 2003; Benyoucef et al. 2014b) and charcoal layers show that some regions also experienced occasional droughts and forest fires (Nothdurft et al. 2002; Krassilov and Bacchia 2013). There was a brief period of climate change in the early and middle Cenomanian when conditions became wetter, and ocean temperature dropped. The climate then reverted to more arid conditions again (Vullo and Neraudeau 2008; Catuneanu et al. 2006; Gertsch et al. 2010b).

Medeiros and colleagues (2014), when describing similar habitats in Brazil, probably gave the best summation of North Africa at that time: a "coastal forested area . . . surrounded by a dominantly dry environment." In many ways, the script was flipped from the current situation; back then North Africa was a wetland and the interior a desert, whereas today North Africa is desiccated and the continent's interior covered in jungle. The only modern wetland ecosystem on a similar scale is the Amazon basin, which covers roughly 7.5 million km^2 (Goulding et al. 2003).

North Africa also presents us with the first of many paradoxes we'll encounter: a dry region with a seemingly endless supply of water. This was maintained as a result of the area's high water table (Catuneanu et al. 2006), localized humid conditions along parts of the coastline (Deaf 2009); and a low continental gradient, which allowed the Tethys and Trans-Saharan Seas to flood the land and cause the rivers to develop extensive meander systems that slowed down the exit of any rainwater (see fig. 1.2). This created a water-rich environment despite the limited rainfall (Nash 2012).

North Africa was effectively a giant version of the modern Nile and Okavango deltas, a geographic anomaly creating a wetland in an otherwise arid region. However, while the region may appear to be uniform in

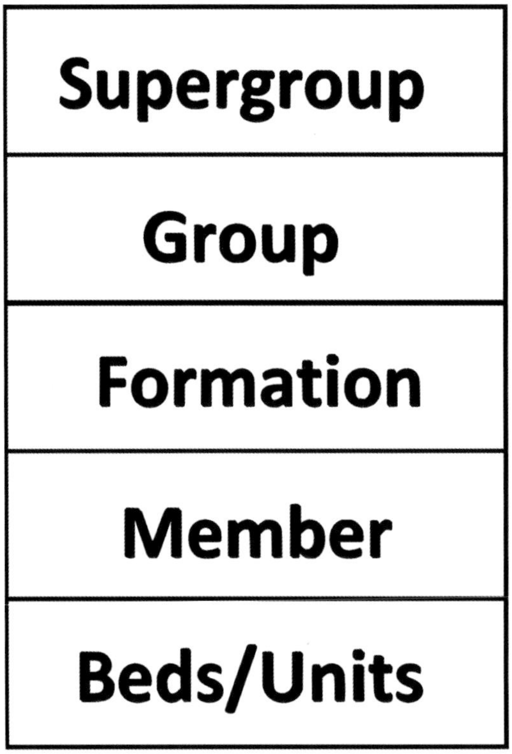

Supergroup

Group

Formation

Member

Beds/Units

Fig. 1.3. The international ranking system used when describing rock formations.

habitat and fauna, a *continual ecological province*, there are differences on closer examination, which will be discussed later.

Stratigraphy

Of course, such a reconstruction of past environments would not be possible without an understanding of stratigraphy and its various subdisciplines. At its simplest, stratigraphy is a study of rock layers (strata), which are themselves separated into different layers, which are identified using an international ranking system (see fig. 1.3).

The ranks used are as follows: supergroup, a collection of associated groups that were formed at the same time and share lithological characteristics; group, a collection of formations, likewise of similar age with similar lithological characteristics; formation, a lithologically distinct sequence of rocks; member, a section of a formation that can be differentiated from the rest on the basis of its unique lithology (not all formations have members); and bed or unit, the lowest rank used and a lithologically unique layer within a member.

These various layers are differentiated from one another because of their composition, fossil content, and other characteristics. It's also possible to correlate these layers with others in different parts of the world (see fig. 1.4). This ability to correlate various rock strata is essential because it gives us a global view of what was happening at various times in the past.

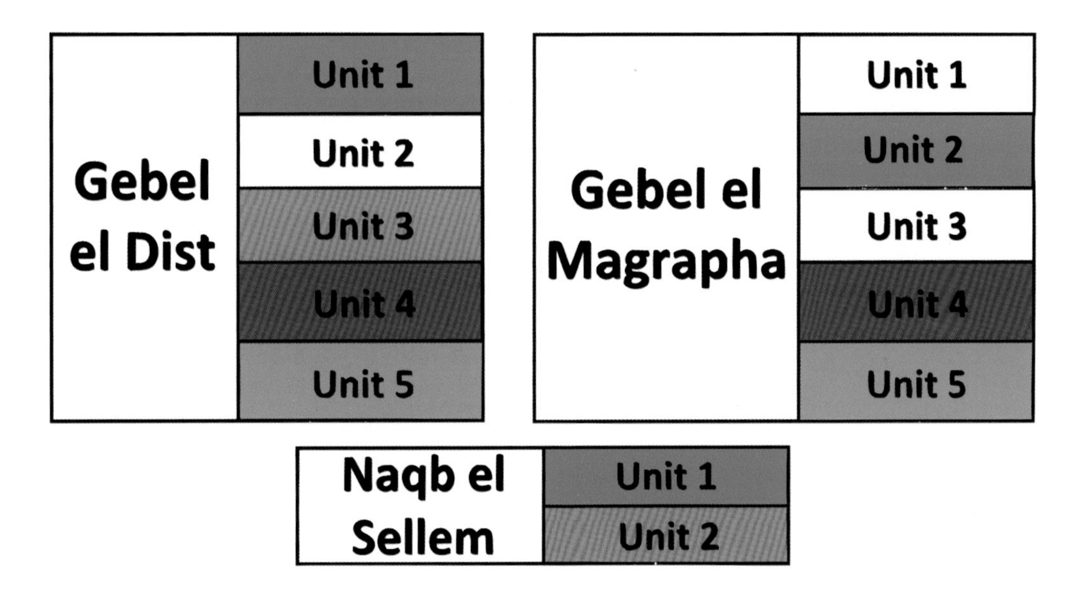

Fig. 1.4. A color-coded stratigraphic chart illustrating how the ages of various beds in different locations can be correlated, using the Bahariya Oasis as an example. It's important to remember that not all these sediments were laid down at the same speed, which is why unit 1 of Gabel el Dist was being deposited at the time that unit 2 of Gabel el Magrapha was being formed. This is worked out depending on the types of rock that make up the sediments of each bed and the fossil content within each. Units without color have yet to be correlated precisely.

Sediments are dated using biostratigraphy, the science of dating lithostratigraphic units via the fossils contained within those rocks. For example, the presence of the trilobite *Paradoxides* in a rock layer proves that those rocks must be between 500 and 509 million years old. In North Africa, ammonites are often used for biostratigraphy.

Most of the sediments in North Africa were eroded from the Neoproterozoic rocks of the Saharan Metacraton (Abdelsalam et al. 2002). A craton is a tectonic region of a continent that has not been deformed to any great degree for extended periods of time (Bates and Jackson, 1980). A metacraton is formed when a craton's margin and interior have been metamorphosed by plate tectonics to form orogenic belts (mountain building) but the craton's original characteristics can still be seen (Liégeois et al. 2013).

The Saharan Metacraton was formed an estimated 3,000–2,000 mya and stretched 5,000,000 km^2 at its largest (Rocci 1965; Kroner 1977; Bertrand and Caby 1978; Tawadros 2001; Tanner and Khalifa 2009). Enhanced weathering in the Cenomanian resulted in the metacraton eroding, with large amounts of iron ore subsequently being injected into the North African ecosystem (Floegel and Hay 2004; Meyer and Kump 2008; Tanner and Khalifa 2009; Baioumy and Boulis 2012).

While the Saharan Metacraton underlies Egypt, Libya, and Nigeria, the sediments of Morocco and Algeria were sourced from the West African Craton (Ennih and Liégeois 2008). The West African Craton was formed from three older cratons that first began to fuse roughly 2,100 mya. The northern portion of Nigeria may also have its separate craton, due to its older sediments (Abdelsalam et al. 2002; Tanner and Khalifa 2009).

It's taken time to work out the stratigraphy of each formation. For instance, the Bahariya Formation has been given multiple names (El

Gezeery and O'Connor 1975), and Martill and colleagues (2011) began documenting the Gara Sbaa region only to have their stratigraphic schematics redrawn before they had even finished (Cavin et al. 2010). Likewise, the multiple outcrops of the Galala Formation have been given different names (El Beialy et al. 2010). For these reasons, we'll be looking at only the best-known formations.

Egypt

Of all the formations in North Africa, Bahariya is the one I love the most. While long since overshadowed by the Kem Kem beds of Morocco, the Bahariya Formation of Egypt should be one of Africa's paleontological hallowed grounds. Bahariya was one of the first to provide scientists with a wealth of partial skeletons, and it continues to be a source of material to this day (Nothdurft et al. 2002).

History has not been kind to Bahariya. Not only have many of the fossils to come out of this formation been destroyed, but few paleontologists have returned in the years since (Nothdurft et al. 2002). The Bahariya Dinosaur Project of 2000 proved the continued potential of this formation, but many of the specimens found remain undescribed (Smith et al. 2001b; Nothdurft et al. 2002). A renewed campaign of exploration and analysis of the existing material languishing in storage is desperately needed.

The Bahariya Formation itself is divided into three members: Gebel Ghorabi, Gabel el Dist, and El Heiz (see fig. 1.5). The Gabel el Dist is the best-studied member of the Bahariya Formation and can be further divided into five units. This gives a complete view of how the biome it records developed over time. Unit 1, composed of bioturbated shales and sandstones, was deposited in a shallow bay with muddy tidal flats (Reineck and Wunderlich 1968; El-Sisi et al. 2002).

The composition of unit 2 is the same (El-Sisi et al. 2002); however, this unit is famed for a high concentration of fossil leaves and dinosaur bones (Stromer 1914, 1936). The depositional environment was a low-energy mangrove swamp, with vegetation extending out into the preexisting bay (Hassouba et al. 1988; Smith et al. 2001b; Lacovara et al. 2003).

I must mention that I'm using the term *mangrove* in its loosest sense: a salt-tolerant coastal forest. Technically *mangrove* only refers to plants of the family Rhizophoraceae, which didn't exist in the Cenomanian. But as most researchers use this term to apply to the ecosystem (Hogarth 2007; Lacovara et al. 2003), not the specific plants, I shall use it herein; especially, as the flora was behaving like an extant rhizophorid community (Lacovara et al. 2003) and would have lived under conditions identical to those of a "true" mangrove swamp.

Unit 3 is formed of sandstone, interspersed with mudstone, and was deposited in a series of wide river channels (El-Sisi et al. 2002). This marks a marine regression; as sea levels dropped, the coastline moved farther away, and the rivers that fed into the mangroves extended into this area. In some areas, unit 3 has eroded the underlying second unit.

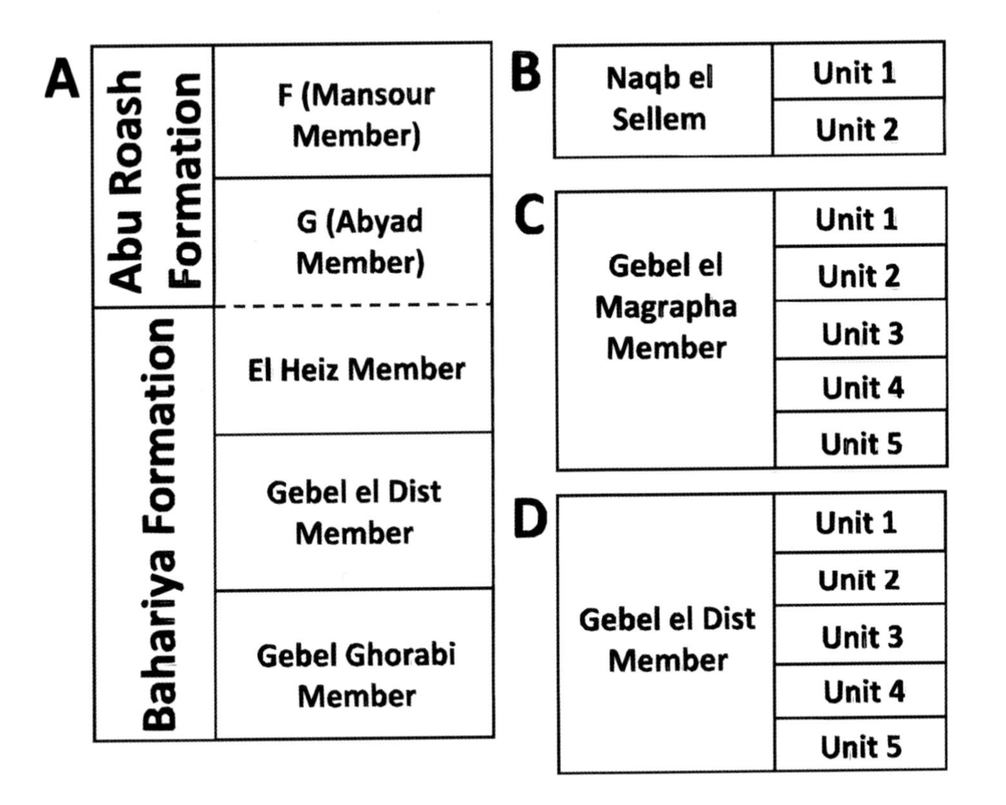

Fig. 1.5. *A*, stratigraphical chart for the Bahariya Formation; and a close-up look at the *B*, Naqb el Sellem; *C*, Gabel el Magrapha, and *D*, Gabel el Dist members. The dotted line indicates the coeval nature of El Heiz and G members. Based on Khaled (1999) and El-Sisi et al. (2002).

Unit 4 marks a return to the depositional environment of unit 1, with shale and sandstone deposited along a series of estuaries and tidal bays (El-Sisi et al. 2002). This unit is rich in shells and bioturbation (Stromer 1914). The formation records multiple marine transgressions, with rising sea levels reflooding the bay and the depositional basin migrating landward (El-Sisi et al. 2002).

Unit 5 is noted for a distinct surface boundary created by subaerial erosion that separates it from unit 4. The surface has lag deposits, limestones, and *Glossifungites* ichnofacies (El-Sisi et al. 2002). Unit 5 is composed of shale, siltstone, mudstones, and sandstones (El-Sisi et al. 2002). This unit is full of fossils (Stromer 1914) and records the ecosystem returning to the mangrove environment of unit 2. It's important to remember that each of these changes in the depositional environment would have taken place over just a few thousand years.

The Bahariya Formation is followed by the Abu Roash Formation, which itself is divided into several members informally designated by the initials A to G (Schlumberger 1995). The terminology is as follows: Ghorab member (A), Rammak member (B), Abu Sennan member (C), Meleiha member (D), Miswag member (E), Mansour member (F), and Abyad member (G). Because the Bahariya and Abu Roash Formations form a continuous sequence through time, parts of which are coeval with each other, they are described together (Beadnell 1902; Norton 1967; Robertson Research International 1982; Schlumberger 1995).

However, as the Abu Roash Formation crosses the Cenomanian/Turonian border, only the Mansour and Abyad members (units F and G) will be considered here. The Abu Roash Formation's Abyad member, the oldest portion of which is probably coeval with the Bahariyan El Heiz member, is made up of limestone, shale, and siltstone (Khaled 1999).

In addition to the Bahariya and Abu Roash Formations, two other, minor, Cenomanian formations are also known from Egypt. The Naqb el Sellem Formation is composed of shales, sandstones, and mudstones; it is divided into two members. This formation was formed along a coastal tidal flat, rich in mangrove vegetation (El-Sisi et al. 2002). The Naqb el Sellem Formation is coeval with the Gebel El Dist and Gebel El Magrafa members.

The Maghrabi Formation is part of the Egyptian Nubia series from the Western Desert. The upper Maghrabi Formation is coeval with the Bahariya Formation and is composed mainly of sandstone, claystone, shales, and siltstone (El-Shayeb et al. 2013). The portion of this formation that outcrops east of the Kharga Oasis is composed mainly of heavily bioturbated mudstone with interbedding sandstone layers (Abuelkheir and Abdelgawad 2017). The depositional environment recorded in this area comprised a series of rivers, tidal flats, and supratidal marshlands (El-Shayeb et al. 2013; Abuelkheir and Abdelgawad 2017). Both are mentioned for the sake of completeness and will not be discussed further.

The environment recorded in this region was an arid coastline covered in mangrove swamps . One of these was located in an embayment referred to as the Bahariya Bight, which stretched for 300 km along the coast (Catuneanu et al. 2006; Grandstaff et al. 2012). The Bahariyan waters were dysoxic/anoxic (El-Soughier et al. 2010; Tahoun et al. 2013; Zobaa et al. 2008), although some areas were suboxic/anoxic (El-Soughier et al. 2011). The rivers and mangrove channels were also deep, with depths of 3 to 6 meters (9 to 19 feet), depending on location (Nothdurft et al. 2002; Baioumy and Boulis 2012).

However, these mangroves were just part of a larger wetland system. Inland, there was a series of vast, braided river networks feeding into the mangroves; the southern part of this network was particularly extensive (Stromer 1914; Werner 1989; Lyon et al. 2001; El-Sisi et al. 2002; Catuneanu et al. 2006; Grandstaff et al. 2012). The Bahariya Oasis also developed a vast lagoon in the south toward the end of the Cenomanian (Said 1962, 1990). Moustafa and Lashin (2012) have shown that the Bahariyan coastline also had a neighboring upland community.

The Galala Formation (see fig. 1.6) is found primarily in Egypt's Eastern Desert and the Sinai Peninsula, to the southeast and east of Bahariya. The East Themed area (a region to the east of the town of Themed) and the outcrop in the Eastern Desert are exceptionally well studied. This formation was named by Abdallah and Adindani (1963), although the first studies on this unit predate them (Abdelhady 2007).

Because this formation covers such a vast area, it has been given multiple names over the years. Ghorab (1961) named the outcrop in the central Sinai region the Raha Formation, while the northern part was named the Halal Formation (Said 1971). Issawi and colleagues (2009) don't consider this to be valid, and El Beialy and colleagues (2010) consider the Halal Formation to be part of the Galala Formation, and this is followed herein (contra Abdelhady 2007). Kora and colleagues (2001) and El Beialy and colleagues (2010) believe that the Raha Formation may be valid, although this name should only be used for the formation in the "central Sinai and the Gulf of Suez" region (El Beialy et al. 2010).

The Galala Formation has three members, the Lower Shale Member, the Middle Siliciclastic/Carbonate Member, and the Upper Carbonate Member. It's possible that the base of the Galala Formation may be of Aptian age (El Beialy et al. 2010), although this formation certainly crosses the Cenomanian boundary at some point.

In general, the Galala Formation is composed mainly of shales and sandstones at its base, with calcareous layers in the middle section that give way to dolomitic rocks near the top of the formation. This formation is also rich in gypsum (El Beialy et al. 2010). Sections of the Galala Formation can also be found in Egypt's Eastern Desert (Nagm and Wilmsen 2012). Notably, this part of the formation possesses an ammonite fauna that is not preserved elsewhere in the formation (Nagm and Wilmsen 2012), providing a complete biostratigraphic record.

Sections of the Galala Formation also outcrop in an area called Wadi Quseib, close to the Gulf of Aqaba, which geologists call the East Themed area. Interestingly, the Galala Formation generally has a diverse macrobenthic fauna except for this one section at Wadi Quseib, which is composed mostly of shale and carbonate. This suggests that the Wadi Quseib section can be correlated with the lower shale member, which is also siliciclastic with a dearth of macrobenthic fauna.

The formation also records three systems tracts: low-stand, transgressive, and high-stand (Khalifa et al. 2003). While marine transgressions were commonplace throughout North Africa at that time, the presence of a high-stand tract shows that the sea reached its maximum possible height in this area, completely covering what is now the Sinai Peninsula (Abdelhady 2007).

Nagm and Wilmsen (2012) identified six facies in the western portion of this formation, in Egypt's Eastern Desert, which would have corresponded to the ancient coastline. Layer one was composed of thick-bedded marl (a type of mudstone). Layer two was made of calcareous packstone (a carbonate rock that is composed of grains between 63 μm and 2 mm in size and a small amount of mud). Layer three was formed from bioturbated wackestone (a carbonate rock that is composed primarily of mud and over 10% grains, most of which are 63 μm in size). Layers four and five were made of rudstone (a carbonate rock similar to wackestone, except the grains are larger than 2 mm), while layer six returned to primarily packstone.

The Galala Formation is followed by the Abu Qada Formation (Ghorab 1961),which is distinguished from the other formations in the Eastern

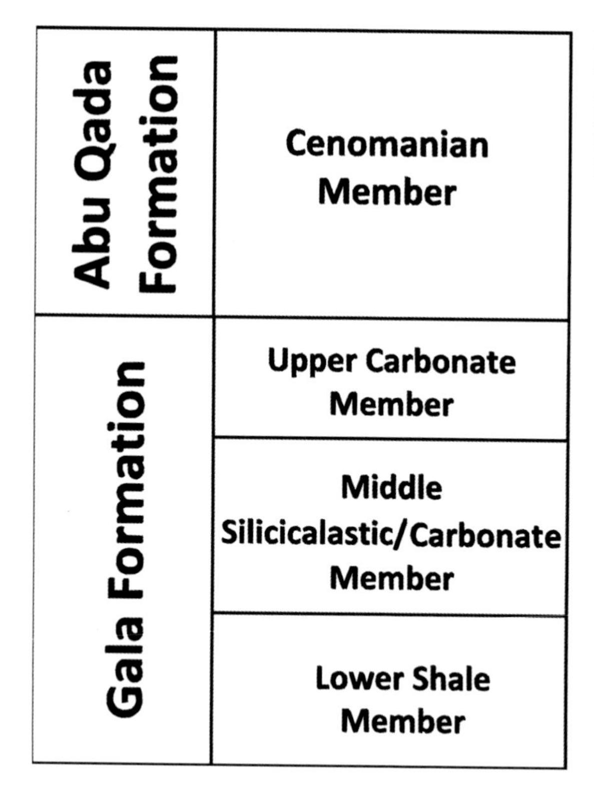

Fig. 1.6. Stratigraphic chart for the Galala and Abu Qada Formations. Based on Issawi et al. (2009) and El Beialy et al. (2010).

Desert by its characteristic green marl and shale layers, interbedded with various types of limestone. The Abu Qada Formation also possesses a large number of fossils, especially ammonites.

Subdivisions of the Abu Qada Formation cannot be identified in the East Themed area with certainty, although in other places the Abu Qada Formation can be divided into three members: lower, middle, and upper. However, these members are of Turonian age, as opposed to the East Themed section of the formation, which is upper Cenomanian and thus postdates the other Abu Qada Formation subsections. For these reasons, the Cenomanian section of the Abu Qada Formation is informally referred to as the Cenomanian member herein.

The environment recorded in this region was a network of large bivalve reefs and marine lagoons along the southeastern shores of the Tethys Sea, with little to no freshwater input (Nagm and Wilmsen 2012). The presence of palynomorphs in the early Cenomanian (El Beialy et al. 2010) suggests that this region was once a freshwater delta that was drowned by rising sea levels (Khalifa et al. 2003; Wilmsen and Nagm 2012).

The reefs within the Galala Sea would have connected, at its western border, with those that surrounded the tip of the Bahariya Bight and then stretched along Gondwana's northern coastline and down into the Trans-Saharan Seaway. Just beyond the edge of the Galala Sea's northern border was another large lagoon with its own group of islands, the Nammoûra

Archipelago. These islands are technically part of the Middle East (what would become Lebanon) and therefore are beyond the scope of this book. They are mentioned herein just for completeness's sake (Vecchia et al. 2001; Vecchia and Chiappe 2002; Elgin and Frey 2011).

The Galala Sea was calm and deep, ranging from 25 to 50 meters (82 to 164 feet) in depth (Ayoub-Hannaa and Fürsich 2012b). Some areas possessed extensive oyster beds and sandbars, indicating some shallow-water regions (Wilmsen and Nagm 2012). These regions became rarer throughout the Cenomanian as sea levels continued to rise, and the coastline moved farther away (Abdelhady 2007). The diversity of the suspension-feeding fauna indicates these waters were eutrophic (Nagm and Wilmsen 2012; Abdelhady 2007). However, the shallow, inner-ramp regions show signs that this region was suffering from ecological difficulties (El Beialy et al. 2010).

Morocco

Of all North African formations, the Kem Kem Formation of Morocco has garnered the most continual attention from researchers. With abundant strata, an established fossil industry, and a location that is both inexpensive to reach (compared to other places in Africa at least) and politically stable, the Kem Kem Formation has become the must-see place for Cenomanian studies.

Sometimes called the Continental Red Beds, this formation has been undergoing stratigraphic study since at least 1931. The unit was initially classified as part of the Continental Intercalaire (Kilian 1931). Later work by Choubert (1948) divided the formation into three units but regarded them as Albian in age. This error in dating is interesting because Choubert was aware of the similarities with Bahariya (Choubert et al. 1952; Cavin et al. 2010) but appears never to have made the connection.

Traditionally, the Kem Kem beds were divided into three formations: the Ifezouane, Aoufous, and Akrabou Formations (Dubar 1948). While these formations were initially considered to be separate from the three units described by Choubert, Lavocat (1948, 1954a, 1954b) later showed that they were the same.

Because the Kem Kem sediments were deposited continuously, without any hiatus where sediments and fossils were not recorded, we have a complete view of how this ecosystem developed over time (see fig. 1.7). What makes it a nightmare to work on is the fact that both the boundary between the formations and the internal stratigraphy within them is poorly defined, making it difficult to get concise dates for the various sections.

As a result, the dating is done mostly by biostratigraphy, as some of the vertebrates are restricted to the Cenomanian age (Arden et al. 2018), and also by comparison with the Bahariya Formation, since the two formations share taxa and the dates for Bahariya are much more concise (El-Sisi et al. 2002; Catuneanu et al. 2006; Arden et al. 2018). Recent work

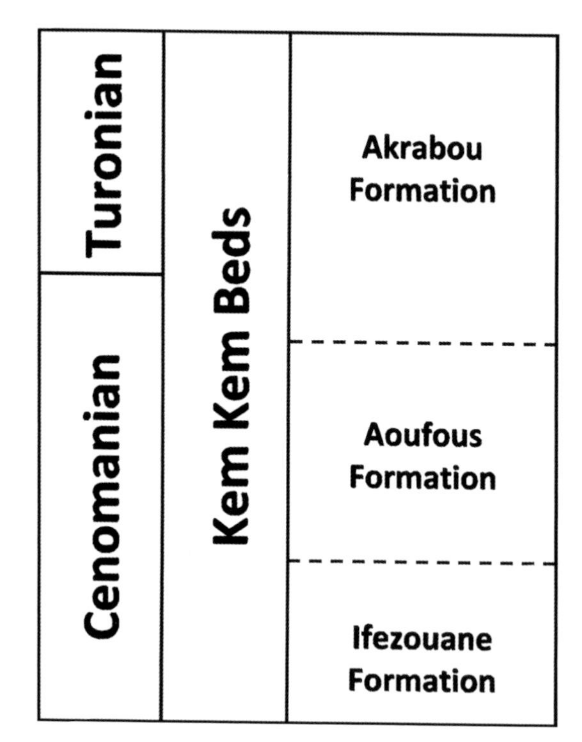

Fig. 1.7. The old stratigraphic chart for the Kem Kem Formation. Based on Calvin et al. (2010).

by Ettachfini and Andreu (2004) and Cavin and colleagues (2010) have given us an even clearer understanding of this formation.

The Ifezouane Formation is composed of sandstones (Cavin et al. 2010). This formation is thicker in the north, 250 meters (820 feet) in some areas, and becomes thinner to the south (Cavin et al. 2010). The Aoufous Formation is more consistent, maintaining an average thickness of about 100 to 200 meters (328 to 656 feet). It is composed mainly of microconglomerates (coarse grains contained in silt or clay) and various sandstones (Cavin et al. 2010). The abundance of gypsum shows that these sediments were deposited in shallow, hypersaline waters.

Because the Akrabou Formation crosses the Cenomanian/Turonian boundary, only the base of the formation will be considered here. Composed of fine-grained sandstones with extensive carbonate preservation (Cavin et al. 2010), the Akrabou Formation records an extreme marine transgression that flooded most of this region. However, the presence of terrestrial plants and animals shows that the sea was still shallow, and the coastline was probably close by (Martill et al. 2011).

The Akrabou Formation is also known for its Konservat Lagerstätte locations (Martill et al. 2011). Lagerstätte refers to a region with anoxic waters that limited scavenging animals and bacterial decay, resulting in excellent fossil preservation.

However, this stratigraphic system has been recently challenged by Ibrahim and colleagues (2020a), who coin a new terminology for this region: Hamadian Supergroup, with the Gara Sbaa and Douira Formations

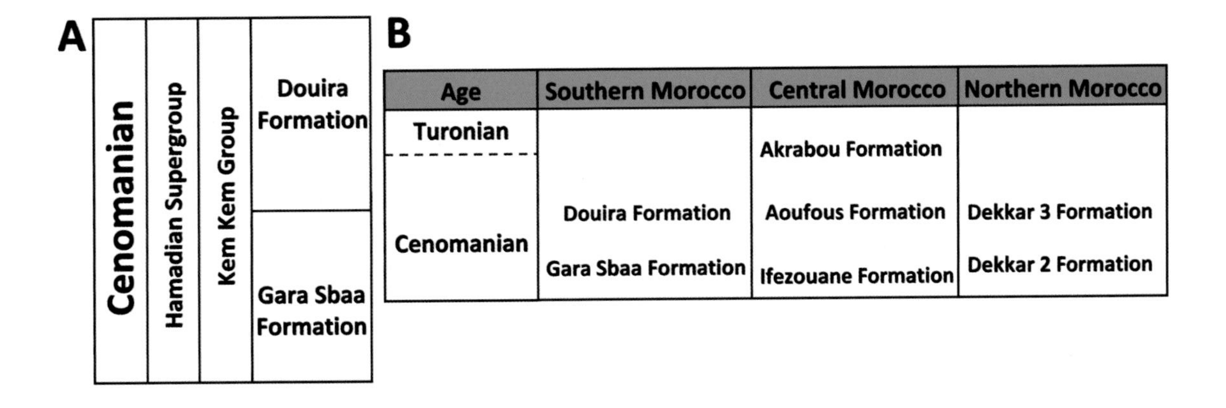

Fig. 1.8. *A*, the newly proposed stratigraphic chart for the Kem Kem Formation; and *B*, a table showing the location of the main Cenomanian formations and how they correlate. Based on Ibrahim et al. (2020a).

replacing the Ifezouane and Aoufous Formations; those names now only apply to the formations in the Tinghir region, although they are clearly correlated with each other (see fig. 1.8). While I personally favor this new stratigraphic system, it remains to be seen whether the majority of workers will accept it.

The Gara Sbaa Formation, based on an outcrop found at Gara Sbaa, has a base composed of coarse sandstone that gives way to more fine-grained sandstones, interspersed with rare layers of siltstone, mudstone, and feldspar clasts. This lower portion of the formation is barren, with no fossils reported yet. The remainder is iron rich and composed of course-grained sandstone, interspersed with rare layers of fine-grained sandstones and paleosols. The beds become thicker toward the top of the formation, with the sandstone intermixing with yellow clay (Ibrahim et al. 2020a).

Like most of the Kem Kem beds, this formation was deposited in an extensive river system and delta. However, desiccation cracks and fossilized ripple marks show that some areas were extremely shallow and occasionally dried out completely. Other areas show signs of brackish water and tidal influences, indicating that the sea was nearby. As time went on, increased erosion in the nearby mountains resulted in many of the river channels shallowing as they were increasingly filled with sediment (Ibrahim et al. 2020a).

The border of the Gara Sbaa and Douira Formations is marked by a sudden marine transgression. While there were repeated marine transgressions throughout the Cenomanian, this one appears to be correlated with the Mid-Cenomanian Event, which would give the boundary a rough date of about 96.0 mya (Kuhnt et al. 2009; Ibrahim et al. 2020a).

The Douira Formation, based on an outcrop near the village of Douira, is composed of fine-grained, red mudstone at its base, which was a result of a marine transgression that temporarily flooded the region. This later gives way to huge fossiliferous limestone layers, topped by claystones or mudstones, which make up 64% of the formation's thickness on average. In some northern locations, there are also thin layers of marl and limestone at the formation's upper boundary (Ibrahim et al. 2020a).

This formation was deposited at a time when marine transgressions were increasing, flooding the land and replacing the river systems with extensive shallow-water coastlines, dominated by tidal flats. Eventually, as sea levels continued to rise, even these were drowned out, resulting an offshore marine environment (Ibrahim et al. 2020a).

Other minor formations are also known from the High Atlas. Part of the Dekkar group, its formations are informally known as Dekkar 1–3. Only Dekkar 3 and the latter part of Dekkar 2 are of Cenomanian age. Still poorly known, these formations were initially laid down on an arid, freshwater floodplain that was gradually drowned by the sea (Haddoumi et al. 2008). This group is mentioned for the sake of completeness and will not be discussed further.

Regardless of how the stratigraphy will ultimately pan out, the environment recorded in Morocco was an arid coastline with carbonate lagoons and inland floodplains (Ettachfini and Andreu 2004). The presence of Konservat Lagerstätten in some areas shows that these coastal waters were becoming hypersaline and anoxic toward the end of the Cenomanian (Martill et al. 2011).

Inland was an extensive network of fast-flowing waterways (Russell 1996; Cavin et al. 2010; Ibrahim et al. 2020a, b). Russell (1996) initially suggests that these rivers flowed westward, but it's now known that they flowed northward (Dutheil 1999a). These river networks are so extensive that they've been nicknamed "the river of giants."

The Kem Kem Formation does differ from Bahariya in its apparent lack of flora. Vegetation was so uncommon in the Kem Kem beds that wood and pollen are extraordinarily rare (Cau 2012c; Läng et al. 2013; Läng 2014). Whether this is a genuine environmental signal or the result of a taphonomic or preservational bias is debatable.

Although most of this formation is barren, plants are found in some locations. Engel and colleagues (2012) discovered a variety of plants from areas near the coastline, and Ibrahim and colleagues (2020a) document a gallery forest that grew around a large lake or pond. Murray and colleagues (2013) also record a grove of plants from the Agoult region that comprises "horsetails, ferns, Dicotyledons and monocotyledons."

Interestingly, the plants discovered by Engel and colleagues (2012) were not preserved in situ. Engel and his team suggest that this material was washed inland during a tidal surge from one of the hurricanes that battered the North African coastline, or possibly during a tsunami.

Off the northwestern coastline were the Idrissides islands. While most researchers acknowledge their existence, no one has made any attempt to quantify their exact size and position. Canerot and colleagues (2003) refer to another large island near what is now Tarfaya. Given that this is the most proximal proposed location for the Idrissides islands, they might be the same.

The region initially had a wet and humid climate but became progressively more arid and seasonal. Although some areas—the Idrissides islands and Agadir Basin—still had localized humid climates because the prevailing trade winds brought in rain, these were effectively isolated microhabitats within the broader ecosystem (Gertsch et al. 2010b).

Ultimately, the Kem Kem region would have been a "relatively bleak environment" (Cau 2012c), a mix of tropical and flooded savanna—an arid and open scrubland, with isolated groves, which were flooded with fresh water for most of the year, despite the dry climate.

Libya

Libya has proved to be a remarkably diverse country regarding geology, possessing rock from the Precambrian all the way to the Quaternary. Because Libya was an Italian colony, most of its early researchers were naturally Italians (Desio 1928, 1935, 1936a, 1936b, 1939, 1943). From 1944 onward, British and French scientists conducted most of its studies (Lelubre 1948, 1952; Muller-Feuga 1954; Christie 1955). Many of these later studies were focused on the mineral and oil potential (International Geological Congress 1952; Brichant 1952; Goudarzi 1970; Cepek 1979; Salaj 1979).

The Mesozoic rocks of Libya we're interested in are the Nefusa and the Mizdah Formations, part of the Al Hamra Group (Burollet 1960; Banerjee 1980). The Nefusa Formation is divided into three members: the Ain Tobi, Yafran, and Garian members (Christie 1955). The Ain Tobi member is composed of mainly thick layers of limestones. The Yafran member, which crosses the Turonian boundary, is made up of limestones, shales, and gypsum-rich marls, 80 meters (262 feet) thick in some areas (Goudarzi 1970). The Garian member is of Turonian age and will not be discussed further apart from a mention that the boundary between the Yafran and Garian members is difficult to identify.

The Mizdah Formation, however, is where the fun begins. The Mizdah Formation is divided into three units: the Thala, Mazuzah, and Tigrinna members (Jordi and Lonfat 1963; Goudarzi 1970). But the estimated age of each member varies greatly. The Thala member was considered to be of Campanian age by Barr and Weegar (1972), while the Mazuzah member was considered Santonian in age by Megerisi and Mamgain (1980a, 1980b). Goudarzi (1970) later estimated a Turonian to Campanian age for this formation, while Nessov and colleagues (1998) returned to the Santonian to Campanian estimates of Barr and Weegar!

Some studies were criticized for focusing only on lithostratigraphy without looking at the fossils found in that region. Rage and Cappetta (2002) identified a Cenomanian age for at least part of this formation on the basis of fossils that are conclusively of Cenomanian age, but that still leaves the question of overall stratigraphic age unanswered.

It must be noted that the Cenomanian layers of Rage and Cappetta (2002), found in an area called Draa Ubari, are in between the Thala and Mazuzah members. This means that the overlying Mazuzah member must also be of Cenomanian age or younger. It also suggests that the younger Qasr Tigrinnah and Nalut Formations are of Cenomanian age too. Rage and Cappetta also theorize that there is a sizable stratigraphic gap in the geological record between the Mazuzah and Thala members; such a gap would go a long way toward explaining the difficulty in identifying the age of this formation, although Rage and Cappetta also suggest regional variations in the stratigraphic series.

Now that the saga of the Mizdah Formation has been settled as best we can, what environment was recorded in what is now Libya during the Cenomanian? The presence of turtles and crocodiles suggests an extensive river and delta ecosystem. In the south, there was a series of lagoons that were gradually drowned by rising sea levels over time (Hallett and Clark-Lowes 2016). The increasing amounts of gypsum suggest that the salinity levels increased over time (Hallett and Clark-Lowes 2016).

While no plant fossils have been named or described from this region, the presence of wood fragments shows that trees and ferns were present along this coastline (Tawadros 2001). These beds are also rich in terrestrial animal fossils (Nessov et al. 1998), suggesting that Libyan deltas had a vibrant ecosystem.

Algeria

Surprisingly little has been published on the Cenomanian of Algeria, despite its abundant strata. However, recent work has begun to uncover this country's untapped potential, and I'm confident that Algeria could rival the fossil wealth of Morocco and Egypt.

Most of Algeria's Cretaceous strata are located in the Guir Basin in Algeria's Béchar province. Some of the work began as early as 1931, and it was determined from the start that Algerian formations form a continuous unit with the Cretaceous beds of Morocco and Tunisia, named the Continental Intercalaire (Kilian 1931; Menchikoff 1936; Deleau 1951, 1952).

The Béchar region can be divided into three formations (see fig. 1.9): the Gres rouges Formation, Marnes à gypse inférieures Formation, and the tongue-twisting Calcaires de Sidi Mohamed Ben Bouziane Formation (Benyoucef et al. 2016).

Generally, the Gres rouges Formation is composed mostly of siliciclastic sediments that grade upward into a mix of evaporates and carbonates (the base of the Marnes à gypse inférieures Formation). The evaporates and carbonates then give way to layers formed almost entirely of carbonates, which mark the base of the Calcaires de Sidi Mohamed Ben Bouziane Formation (Benyoucef et al. 2012). The difficulties that arise in separating these formations are due to the large area this compound assemblage covers and the varying thickness of the individual strata (Benyoucef et al. 2012).

The Gres rouges Formation was once the best known of the three formations (Benyoucef 2012; Benyoucef et al. 2014a) and can be correlated with the Ifezouane Formation of the Kem Kem Formation (Dubar 1948) and the Zebbag Formation of Tunisia. The best-studied sections of this formation are the outcrops at Kenadsa and Menaguir.

The Marnes à gypse inférieures Formation is divided into the marno-gypso-limestone and calcareo-marly units. The marno-gypso-limestone

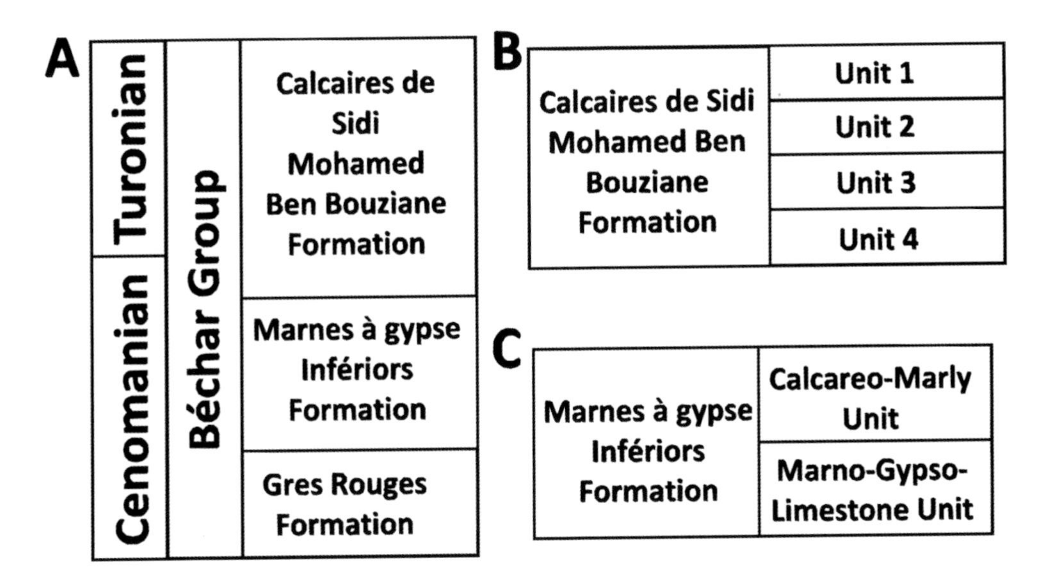

Fig. 1.9. *A*, stratigraphic chart for the Béchar Basin; *B*, the Calcaires de Sidi Mohamed Ben Bouziane Formation; and *C*, the Marnes à gypse inférieures Formation. Based on Benyoucef et al. (2012, 2014b).

is composed of gray limestones and red and green marls, with periodic layers of wackestone, sandstone, packstone, and mudstone. Toward the top of this unit, gypsum, iron, and silt deposits become prevalent. The unit is capped by a thick layer of green marl. This unit is rich in crab burrows and fish, bivalve, and gastropod remains, and it was formed in an intertidal salt marsh that was periodically flooded by the sea and then dried out again. The southern end of this marsh was repeatedly hit with storms (Benyoucef et al. 2014b).

The calcareo-marly unit is formed of green marls periodically punctuated by thick slabs of brown limestone. In one section exposed in Berridel, there are three clay layers containing gypsum nodules. The depositional environment was a shoreline that preserved the intertidal to supratidal zones, with phases of sea level rise and fall. This unit is coeval with the Aoufou Formation of Morocco (Benyoucef et al. 2014b).

A great deal of work has recently been done on the Calcaires de Sidi Mohamed Ben Bouziane Formation, giving a much clearer picture of the faunal and environmental change in this region. Unlike the two formations that underlie it, this formation can easily be divided into four units (Benyoucef et al. 2012).

Unit 1 consists of various limestones, rich in fossil material. All of the fauna is oceanic, suggesting that this unit was deposited in a near-shore environment, with mudflats along the coastline (Benyoucef et al. 2012).

Unit 2 is also composed of limestones. This unit shows signs of heavy bioturbation, with the burrows of *Thalassinoides* being especially prominent. Stromatolites (rocky mounds formed from cyanobacteria) are also known. This unit represents a marginal-littoral environment (Benyoucef et al. 2012).

Unit 3 is composed of thick layers of limestone, riddled with fossil corals. This unit records an open-water environment formed by the flooding

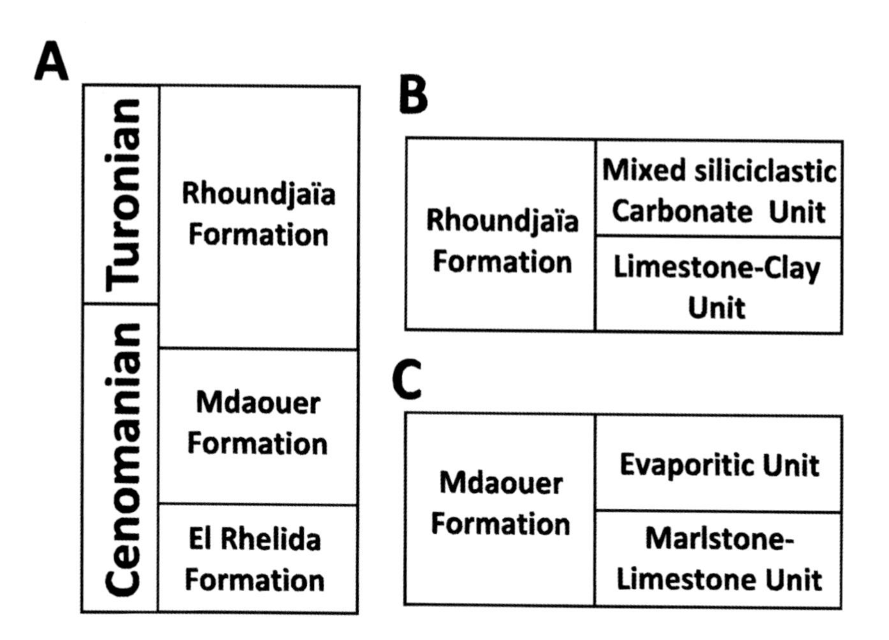

Fig. 1.10. *A*, stratigraphic chart for the Ksour Mountains; *B*, the El Rhelida Formation; *C*, the Mdaouer Formations. Based on Benyoucef (2017).

of the region by the sea, migration of the coastline farther to the south, and the development of offshore coral reefs (Benyoucef et al. 2012).

Unit 4, composed of thin layers of laminated mudstones, crosses the Cenomanian/Turonian boundary. This unit recorded continued sea level rise, which drowned the coral reefs and formed a deep-water environment (Benyoucef et al. 2012).

Three other Cenomanian formations from Algeria's Ksour Mountains are also known: the El Rhelida, Mdaouer, and Rhoundjaïa Formations (see fig. 1.10). The El Rhelida Formation is coeval with the Gres rouges Formation and has two units: the mixed siliciclastic-carbonate unit and the limestone-clay unit. The mixed siliciclastic-carbonate unit is composed of alternating sandstone and clay, periodically interbedded with microconglomerates and limestone. This unit was formed along a shallow-water shoreline (Benyoucef 2017).

The limestone-claystone unit is composed mainly of green and red clays punctuated by occasional limestones. This unit was deposited along a large series of mudflats with supratidal deposits, suggesting that this coastline was battered by periodic hurricanes (Benyoucef 2017).

The Mdaouer Formation also has two units: the evaporitic unit and the marlstone-limestone unit. The evaporitic unit is composed mainly of marlstones, dolostones, mudstones, sandstones, and gypsum deposits. The middle and upper layers are interspersed with layers of limestone. The depositional environment was a flat coastline dominated by sabkhas. A sabkha is a lagoon that has dried out and is now only periodically flooded by seawater, which creates salt pans and other evaporite deposits in a shallow basin (Benyoucef 2017).

The marlstone-limestone unit is composed of two limestone layers separated by a layer of marlstone. This unit was formed during rising sea

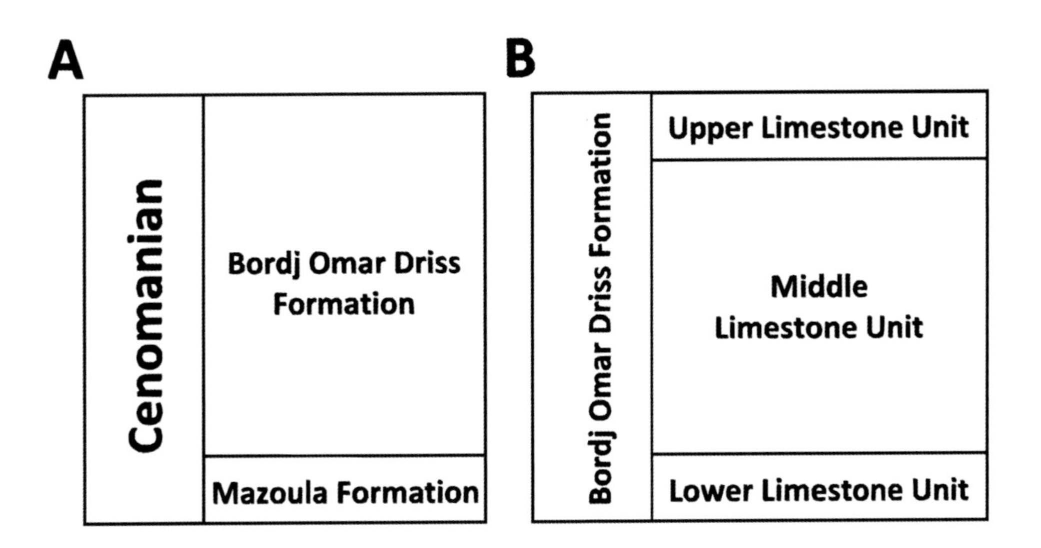

Fig. 1.11. *A*, stratigraphic chart for the Tinrhert Plateau; *B*, the Bordj Omar Driss Formation. Based on Benyoucef et al. (2019).

levels, when the sabkhas gave way to shallow lagoons and bivalve reefs, although mud crack suggests that parts of the coastline were still periodically exposed to the air (Benyoucef 2017).

The Rhoundjaïa Formation has three units. The lower limestone unit is made of homogeneous brown limestone, although the section exposed at Djebel Rhoundjaïa and Djebel Mdaouer is subdivided into two layers: the lower blue limestone and the upper gray limestone, with the blue and gray limestones representing a patch reef (Benyoucef 2017).

The middle mixed unit is composed of alternating layers of marlstone and limestones. This unit recorded continuing sea level rise, which drowned the reefs and created an offshore marine environment (Benyoucef 2017).

The upper limestone unit is subdivided into three layers, but as this unit crosses the Cenomanian/Turonian boundary, only the lower layer will be discussed here. The lower layer is composed of mostly homogeneous limestone, although petrographic analysis reveals thin layers of interspaced mudstone and wackestone. This unit was formed when sea levels reached their maximum height and created an open marine basin (Benyoucef 2017).

Tinrhert Plateau in the west of Algeria also has two Cenomanian sites: the Mazoula and Bordj Omar Driss Formations (see fig. 1.11). The Mazoula Formation is middle Cenomanian in age and is equivalent to the El Rhelida or Mdaouer Formations. The formation is mostly composed of red and green clays, rich in gypsum with three to five wackestone-packstone layers. The top of the formation is capped by a thin layer of limestone interbedded with green marls and mudstones (Benyoucef et al. 2019). It too was deposited in a coastal environment dominated by sabkhas.

The Bordj Omar Driss Formation has three units. The lower limestone unit is composed of limestone rich in fossils. The depositional environment appears to have been a shallow reef, hit with occasional storms (Benyoucef et al. 2019).

The middle limestone unit is missing in some areas because of erosion, and it is composed of fossiliferous limestones. The top layer is made of thin nodular limestone. This unit was deposited in an oyster reef that was forming on recently flooded coastline, with normal levels of oxygen and salinity. It too shows signs of frequent storm activity (Benyoucef et al. 2019).

The upper limestone unit is also composed of gray/white limestones, but the fauna has changed, with ammonites and siliceous sponges being especially abundant. This unit records an open sea environment, deep enough that the sediments were deposited below the ocean's photic zone (the layers of the ocean that sunlight can penetrate). The explosion of planktonic foraminifera known as heterohelicidids suggests a sudden influx of low-oxygen water at the Cenomanian/Turonian border, marking the start of the Cenomanian mass extinction we'll discuss in chapter eight (Benyoucef et al. 2019).

In general, Algeria was an arid coastal floodplain (Benyoucef et al. 2014a), with vast river networks in the Kenadsa region flowing into a series of coastal swamps, mudflats, lagoons, and reefs (Benyoucef et al. 2012; Benyoucef 2017). The region also possessed extensive forest cover in some areas. Huge, petrified trees are known from a site called Wargla, which is presumably in the Ouargla province (de Lapparent 1960). It's even been theorized that this wetland was also connected to the Paleo-Niger delta (Delfaud and Zellouf 1995). The rich fossil content shows that this region, before repeated marine transgressions, was teeming with life. As the Cenomanian progressed, however, the sea flooded more and more of this habitat (Benyoucef et al. 2014a; Benyoucef et al. 2019).

Tunisia

Tunisia has a long history of paleontological discoveries dating back to the 1950s (de Lapparent 1951), with some of its vertebrate fauna known from reasonably complete skeletons (Fanti et al. 2013). However, nearly all of these discoveries are from periods predating the Cenomanian, as the Continental Intercalaire records a vast period (Fanti et al. 2013).

The Cenomanian portion of the Continental Intercalaire in Southern Tunisia is confined to the Zebbag Formation, which is currently subdivided into the Kerker and Gattar members, which are of Cenomanian age. While the age of the Zebbag's formerly recognized basal Rhadouane member has been debated (Busson 1967; Benyoucef et al. 1985; Ouaja et al. 2004), it's now believed to be of Albian age and to actually belong to the Aïn El Guettar Formation, and it lies at the Albian/Cenomanian boundary (Fanti et al. 2013).

Both the Kerker and Gattar members are predominantly carbonatic. The Kerker member varies from mudstones to ocean evaporites and other marine carbonates (Fanti et al. 2013), while the later Gattar member consists primarily of limestones riddled with small chert nodules (Contessi

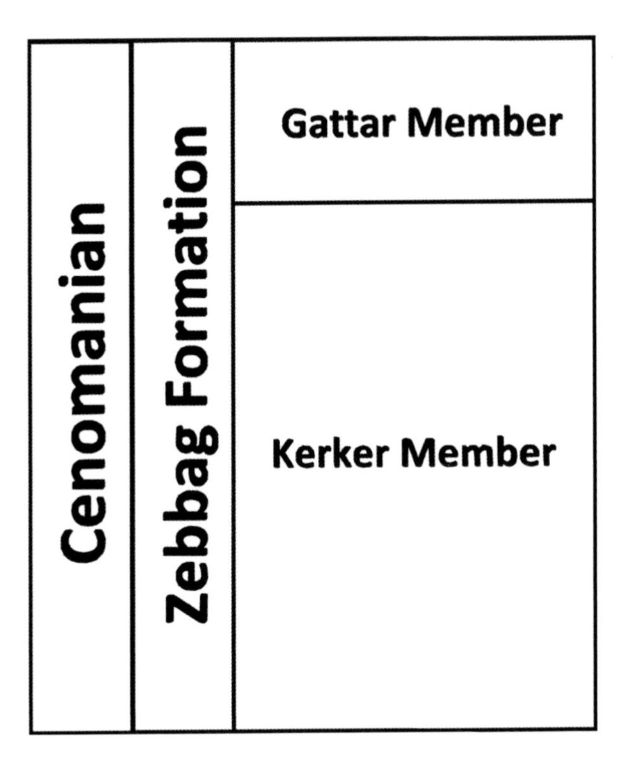

Fig. 1.12. Stratigraphic chart for the Zebbag Formation. Based on Fanti et al. (2014) and Contessi and Fanti (2012b).

and Fanti 2012b). The Zebbag Formation (see fig. 1.12) also outcrops at the Gafsa region, where a third section, the Bahloul member, is recognized. Here, the sediments are primarily marls and limestones and represent a shallow-water patch reef (Chikhi-Aouimeur et al. 2008).

These depositional environments were created during a marine transgression as the Tethys and the Trans-Saharan Seaway drowned North Africa. As a result, no dinosaur skeletal remains have ever been found in this region (Contessi and Fanti 2012a). So, if the only known terrestrial vertebrate fossils are from the Albian, what does this have to do with the topic of this book? Everything, since the creatures living during the Albian would have been the ancestors of the animals inhabiting subsequent ecosystems. It's also become apparent that many of the unusual ecological patterns we see began in the Albian (Fanti et al. 2014), making it essential to our understanding of how these ecosystems functioned.

And while we have yet to find their fossils, it's clear that terrestrial vertebrates still inhabited this region. Two sites, Ksar Ayaat and Boulouha (both part of the Kerker member), have been a bountiful source of footprints (Contessi and Fanti 2012a, 2012b; Contessi 2013a). Some of these prove the existence of groups not previously documented in the fossil record and help fill in the gaps in our knowledge of the North African ecosystems noted above.

The environment recorded in this region was an arid coastline known as the Ougarta and Hoggar Peninsulas, along the western shore of the Trans-Saharan Seaway. Limited plant material suggests that this part of the North African coastline was desertified, although there was a more diverse flora farther inland, with both gymnosperms and angiosperms known (de Lapparent 1960; Benton et al. 2000; Ouaja et al. 2004).

A vast series of shallow lagoons, patch reefs, sandbanks, and mudflats made this part of the coastline unique. This formed a feeding station for shorebirds (Bodin et al. 2010; Petti et al. 2011; Contessi and Fanti 2012a; Contessi 2013b). These mudflats were presumably created by the same process as the Broome mudflats of Western Australia; the shape of the coastline acted as a sediment trap.

To the northeast was the huge lagoon and its ring of islands we discussed earlier (Herkat 2003). Even farther to the east was the island of Gargal (Lüning et al. 2004). To the southeast was the Chotts Hanging Reef, created through a combination of sea level change and active fault lines (Tlïg et al. 2008).

This region is also the first to record an extensive network of carbonate platforms in the north, stretching from Africa to Europe (Contessi 2013b). These subaerial platforms existed from the Albian through to Cenomanian times and may have even formed land bridges between mainland Gondwana and Laurasia (Fanti et al. 2013). The implications of this connection between Africa and the Mediterranean region will be discussed later. For now, it's enough to say that the harsh Tunisian margin was still home to a rich fauna.

Fig. 2.1. A speculative reconstruction of *Weichselia reticulata*. Artwork by Joschua Knüppe.

The Flora of North Africa

2

While studying in Liverpool, I once tried to explain the complexities of paleoecology to a flatmate who believed that paleontology was merely a matter of discovering a fossil, naming a new species, and then moving on to the next. In reality, no species lives in a vacuum; the environment it lives in and the creatures it lives alongside have profound impacts on its anatomy and behavior.

I've attempted to account for all the animals and plants recorded in this region. But because many originated from undated beds or were briefly noted once and never looked at again, I freely admit that this list, while extensive, may be missing species that have slipped my notice. It is also subject to change as new species are named and old taxa redescribed.

Plants

Plants are the foundation of almost all known terrestrial ecosystems. While they have a complex evolutionary tree, for simplicity, we shall stick to the three basic groups: the polypodiopsids, which reproduce via spores; the gymnosperms, which reproduce via seeds; and the angiosperms, which produce flowers and fruit as well as seeds.

The Cenomanian of North Africa has proved to be a veritable Garden of Eden. Some of the best fossils come from Bahariya, where more than a hundred different genera are known (Darwish and Attia 2007). Most of this material has yet to be properly described, with many being little more than names on a list. This makes it impossible to give a proper description or offer a reconstruction for many of them.

Some of this Bahariyan material has been assigned to the orders Cornaceae, Proteaceae, Vitaceae, Sapindales, Piperaceae, Poaceae, Lauraceae, Saxifragales, and Platanaceae (Lejal-Nicol and Dominik 1990; Lyon et al. 2001; Darwish and Attia 2007). Stromer (1936) also mentions evidence of Magnoliopsida and Nymphaeaceae. Dicotyledons and monocotyledons are also known from the Agoult beds of Morocco (Murray et al. 2013).

Polypodiopsida (Ferns)

Weichselia

Of all the plants found in Egypt, *Weichselia reticulata* (fig. 2.1) is undoubtedly the most iconic. *Weichselia* is a widespread taxon, yet only recently have we gotten a clear understanding of its anatomy, although reconstructions by Alvin (1971) and Bommer (1911) were close. It even took a while to realize that all the isolated specimens belong to the same species (Kräusel 1939). The isolated stem and root fragments were once

Fig. 2.2. Fragmentary specimen of *Weichselia reticulata*; showing why it's so difficult to gain a clear understanding of how this species looked in life. *1*, Leaf; *2*, the trunk and the top of the buttress roots; *3*, branch?; and *4*, the pinna (parts of the leaf). Credit Ghedoghedo / CC-BY-SA-2.0. See Wikimedia Commons.

given the name *Paradoxopteris stromeri* (see fig. 2.2), although (Blanco-Moreno et al. 2020) think *Paradoxopteris* should be reinstated as a valid taxon. So, while kept as *W. reticulata* for now, the African specimens may be transferred to a new genus in the future.

Weichselia reticulata was larger than many often suppose, growing to heights of 6 to 8 meters (20 to 26 feet), with leaves several meters in length (El-Khayal 1985; Nothdurft et al. 2002; Blanco-Moreno et al. 2020). It was also supported by a series of false trunks and buttress roots that rose out of the sediment like stilts, much like modern mangrove plants.

Weichselia reticulata also had both vegetative and fertile fronds; the vegetative fronds conducted the business of photosynthesis and other functions necessary for the plant's survival, while the fertile fronds were just for reproduction. The larger vegetative fronds would have grown at the top of the trunk, while the fertile fronds would have grown lower down (Sendera et al. 2015).

Indeed, *W. reticulata* is so unusual it was once given its own family, the Weichseliaceae (Zimmermann 1959; Alvin 1971). However, it's now seems that it's a highly derived member of either the Matoniaceae (van Konijnenburg-van Cittert 1993; Diez et al. 2005; Sendera et al. 2015) or Marattiaceae (Blanco-Moreno et al. 2020).

Weichselia reticulata was a xerophyte, a plant evolved to live in environments that lack water—cactus being the best-known living example.

This led to some confusion over its environmental preferences. The first hints of the truth were discovered by Daber (1968), who suggested that *W. reticulata* dwelled in marine environments. El-Khayal (1985) also came to this view, suggesting that the dunes preserved alongside *Weichselia* were a coastal dune system.

These adaptations for a waterless environment were adaptations for a lack of *fresh water*. A mangrove plant faces the same problem as a desert plant: how to find water. The seawater they grow in is osmotically stressful because of its salt content.

Weichselia has specialized cells for storing water and bears thick layers of waxy cuticle on the leaves to limit the rate of transpiration (the loss of moisture through the pores on the leaf). It's been suggested that the thick cuticles were an adaptation to cold environments (Blanco-Moreno et al. 2020), which I don't find credible given that Africa was not cold at this point. The roots also have an incredible microfiltration system for removing salt (Nothdurft et al. 2002).

Weichselia was a keystone species in Bahariya and other places in Africa. It formed extensive gallery forests along the rivers and coastline (Nothdurft et al. 2002). Its prop roots captured sediments and stored nutrients in this habitat.

Cladophlebis

Another of the ferns native to North Africa is *Cladophlebis* (fig. 2.3). At least twenty-three species of *Cladophlebis* are known, although the Bahariyan species is currently unnamed and has yet to be described.

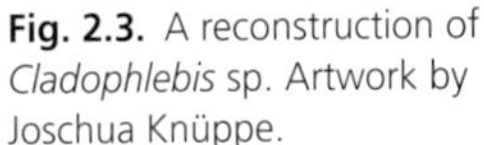

Fig. 2.3. A reconstruction of *Cladophlebis* sp. Artwork by Joschua Knüppe.

Fig. 2.4. A specimen of *Cladophlebis nebbensis*. While almost certainly not the same species of *Cladophlebis* as the one found in Egypt, it would have been similar in general form. Image provided by Daderot Universal Public Domain. See Wikimedia Commons.

Usually, we can learn a lot about a species by studying related species found elsewhere, but this is one time when such comparisons fail us. Barbacka and Bodor (2008) have conducted a review of many of the *Cladophlebis* species found in Europe and have found that the ability to define them is limited, with many specimens showing a great deal of individual variation (see fig. 2.4).

This led to a straight-up "lumper versus splitter" fight over whether a specimen is distinct enough to warrant the erection of a new species (Kiritchkova 1962; Barbacka and Bodor 2008), leaving us uncertain about how many species there were or what features can be used to identify them (Pole 2014).

Even the family *Cladophlebis* should be assigned to is disputed, although features of the stem and foliage indicate it could be the Osmundaceae, Dicksoniaceae, or Schizaeaceae (Pole 2014). Complicating matters, some specimens have been labeled as *Todites*. Isolated stems were also given their own name, *Osmundites*. In that respect, *Cladophlebis* is used less like a genus and more like a descriptive term for a specific type of fern morphology (Pole 2014).

This use of *form taxa* for plant parts is commonplace. Only more complete specimens would allow us to connect the leaves, pollen, and woody parts and deduce which belong to a single species and which don't. Regardless, *Cladophlebis* was an ecological generalist. While not a true mangrove specialist, *Cladophlebis* only grows near water, from rivers to brackish coastal waters (Barbacka and Bodor 2008).

Fig. 2.5. A reconstruction of *Dicksonia* sp. Artwork by Joschua Knüppe.

cf. *Dicksonia*

Dicksonia (fig. 2.5) is a taxon with modern-day representatives. Although it's now restricted in the wild to the Southern Hemisphere, some of the more easily cultivated species, such as *D. antarctica*, are now used as decorative plants around the world (Large and Braggins 2004).

Moroccan specimens are identifiable as *Dicksonia* by the presence of a bivalved membranous covering on the outer sori (the structure on the leaf that produces spores) and the structure of the individual segments of the leaf blade (pinnules). Other morphological features, such as the shape of the leaf and the asymmetrical position of the pinnules, are unique.

The anatomical characteristics of the Moroccan *Dicksonia* are more similar to *Cibotium*, *Culcita*, or *Davallia* (Krassilov and Bacchia 2013) and may warrant the creation of a new genus when the Moroccan material is fully described. Fragments possibly belonging to this genus are mentioned by Martill and colleagues (2011). More specimens are known from Egypt, but it is not known whether these are the same species as the one found in Morocco (Kräusel 1939).

Marsilea

Marsilea (fig. 2.6) is another extant species, superficially similar to a four-leaf clover. There are at least sixty-five modern species, with an almost worldwide distribution. The extinct North African form has never been named or adequately described (Lyon et al. 2001).

Marsilea is an aquatic fern, although some species can easily grow on land, as long as the soil is moist. In water, the plants stand upright with the leaves floating on the surface, although periods of total submersion seem to do them little harm (Mabberley 1997; Chaffey 2002).

Marsilea would have been eminently adapted for life in extreme climates of Cenomanian North Africa, as it not only flourishes in wet conditions but can also survive prolonged drought. Some specimens have still germinated after being desiccated for one hundred years (Mabberley 1997).

Fig. 2.6. A reconstruction of *Marsilea* sp. Artwork by Joschua Knüppe.

Fig. 2.7. A reconstruction of *Agathis* sp. Artwork by Joschua Knüppe.

Agathis

Gymnosperm

The genus *Agathis* (fig. 2.7) is also still around today. *Agathis* belongs to the Araucariaceae, a family of conifers that achieved a global distribution during the Jurassic period. But those glory days are gone, and *Agathis* is now almost wholly restricted to the Southern Hemisphere, where it forms part of a global belt of relic Gondwana forest (see fig. 2.8).

A currently unnamed species of *Agathis* is known from North Africa, with much of the material coming from Bahariya (Lyon et al. 2001; Darwish and Attia 2007). This is rather strange, as living *Agathis* can only survive short periods of waterlogging and mild salt spray (Tomlinson 1986; Thomson 2006)—hardly the tree you would expect in a mangrove swamp.

Fig. 2.8. Living specimens of *Agathis* from New Zealand. While certainly not the same species as the one found in North Africa, it does provide an example of this taxon's general appearance. Credit Pseudopanax Universal Public Domain. See Wikimedia Commons.

Agathis forests were found primarily in the mountains and drier regions of Egypt during the earlier Aptian age (Mahmoud and Moawad 2002; Moustafa and Lashin 2012), and this may very well have been the case in the Cenomanian. For this reason, we can assume that *Agathis* was probably rare in the coastal swamps and far more common inland.

The leaves are noted for their dense, leathery texture. Unlike other conifers, *Agathis* leaves are flat, broad, and oval (Earle 2012). Interestingly the juvenile leaves are larger than those of the adult trees and are sometimes coppery red in color (de Laubenfels 1988).

Agathis is a long-lived genus, with some specimens in New Zealand being an estimated two thousand years old. Some *Agathis* can achieve heights of 50 meters (164 feet), so *Agathis* sp. would have formed the primary canopy and emergent layers of the North Africa forests.

Dadoxylon

Dadoxylon aegyptiacum (fig. 2.9) is a poorly known species from Egypt and Tunisia. Whether to include this taxon was a bit of a debate as not everyone agrees the specimens are of Cenomanian age (Stewart 1898). However, I've decided to follow Kräusel (1939), McCabe and Parrish (1992), and Darwish and Attia (2007) and include this species.

Fig. 2.9. A reconstruction of *Dadoxylon aegyptiacum*. Artwork by Joschua Knüppe.

We must take notice of the suffix *xylon* at the end of the name. Botany has its own nomenclature—the International Code of Nomenclature for Algae, Fungi, and Plants (ICN)—distinct from the nomenclature we use for animals (the International Code of Zoological Nomenclature—ICZN). And under the ICN rules, *xylon* is added to a genus name when the exact identity of a fossil is unknown. Thus *xylon* is used to signify a general similarity to a known genus of plant (Hinsley 2010).

Stromer (1936) identified coniferous wood he thought was from *Cupressinoxylon*, but Kräusel (1939) believes it may also belong to *Dadoxylon*, although he also points out that the poor preservation prevents any positive identification. Ironically, Philippe and colleagues (2003) conclude that *D. aegyptiacum* is itself a likely synonym of *Metapodocarpoxylon*, but the specimens in Paris were too poor to say for sure. If not a synonym, it is then clearly related to either *Metapodocarpoxylon* or *Araucarioxylon*.

Fig. 2.10. A reconstruction of *Pseudotorellia ensiformis*. Artwork by Joschua Knüppe.

Pseudotorellia

Pseudotorellia ensiformis (fig. 2.10) from Morocco is known from leaves and shoots. While the leaf shape varies significantly among *Pseudotorellia* species (Watson and Harrison 1998), *Pseudotorellia ensiformis* leaves are bilaterally flattened and needlelike (Krassilov and Bacchia 2013).

The taxonomy for *Pseudotorellia* above the genus level is unclear. *Pseudotorellia* was once treated as a Ginkgo (Krassilov 1969; Hughes 2010), but studies now suggest that it's a type of conifer (Watson and Harrison 1998). Yet Miroviaceous conifers are probably polyphyletic, with specimens coming from various families. Even the distinction of

Pseudotorellia from the leaf taxon Miroviaceae is unsupported (Krassilov and Bacchia 2013).

Interestingly, the axis of the shoots is similar to *Sulcatocladus robustus* from Great Britain. Presumably, this represents a closely related species, or possibly even the same species (Krassilov and Bacchia 2013).

Regardless, *Pseudotorellia* was a wetland plant, growing in swamps and along freshwater river basins (Gomez et al. 2012). That may seem unusual, but coniferous swamps are well known (Prince 1997; Kost 2002).

Frenelopsis

Known only from the Cretaceous, *Frenelopsis* (fig. 2.11) currently has fifteen species (Gomez et al. 2002). As no stems are known for this genus, we don't know whether *Frenelopsis* actually grew a trunk or was merely a large shrub, although large logs have been found in association with the leaves of *F. ramosissma* from the US (Axsmith and Jacobs 2005), suggesting the former.

You may be wondering why the North African species *Frenelopsis* cf. *teixeirae* has the letters *cf.* before its name. The expressions *aff.* and *cf.* are used when a specimen, which may or may not be a new genus or species,

Fig. 2.11. A reconstruction of *Frenelopsis* cf. *teixeirae*. Artwork by Joschua Knüppe.

is similar to a known genus or species. *Aff.* is an abbreviation of *Species Affinis*. *Cf.* is an abbreviation of *conferatur* and is used when the identity of a specimen is suspected but cannot be deduced with certainty. In this case, the species is uncertain but believed to be *teixeirae*.

What makes *Frenelopsis* easy to identify is the way the branches grow in whorls, forming a series of connected, cylindrical structures (Watson 1977). The whorl morphology could be a defining characteristic on a species level, but some researchers now believe this trait to be too variable for that (Mendes et al. 2010). The North African species has whorls growing in opposite pairs.

It's even been suggested that the whorls and nodal branching seen in this taxon are an ecological adaptation connected to changes in the seasons (Hluštík 1987), although no correlation between morphology and environment has ever been documented in any species that have similar branch morphology (Daviero et al. 2001).

Frenelopsis was an ecological generalist, found in a wide range of habitats, although all seem to be water rich (Gomez et al. 2001, 2002). This taxon has even been documented growing in shallow marine environments (Mendes et al. 2010).

Podozamites

Podozamites is a genus of evergreen conifer. Darwish and Attia (2007) identified a specimen of *Podozamites* sp. from Bahariya. While they provided a photo of the specimen, no description was given. With so little published on these specimens, it's difficult to discuss the African taxa further.

Nehvizdya

Nehvizdya andegavense is a member of the Ginkgophyta, a once-diverse group that now only has one surviving member, *Ginkgo biloba*. Darwish and Attia (2007) mention a specimen of *N. andegavense* from Bahariya but give no further details.

Fortunately, better specimens of *N. andegavense* are known from Spain (Gomez et al. 2000). *Nehvizdya andegavense* was a shrubby tree that can be differentiated from other species of *Nehvizdya* by the number of stomata (pores that the plant breathes through) on the leaf, as well as their distribution and structure (Gomez et al. 2000).

Nehvizdya grew in a wide range of habitats but was also salt tolerant and grew around coastal lagoons and salt marshes, often in association with *Frenelopsis*, *Weichselia*, and *Pseudotorellia* (Gomez et al. 2000).

Ginkgo

As stated above, *Ginkgo* is the only extant member of the Ginkgophyta. Once common throughout the Mesozoic, the sole surviving species is currently endangered, growing wild only in a few places in China (Shen et al. 2005).

Darwish and Attia (2007) identified specimens assigned to *Ginkgo* sp. While they offered no description, the figures provided show the fan shape typical of *Ginkgo*. But they did note that these specimens suffered insect damage, which left small rings on the leaf.

Angiosperms

Liriodendrites

Liriodendrites (fig. 2.12) is a genus of tree distantly related to modern Magnolia, noteworthy for its bisexual flowers that grow in whorls (Zomlefer 1994). The Egyptian species is currently unnamed (Lyon et al. 2001). *Liriodendrites* is identifiable by its horseshoe-shaped leaves (see fig. 2.13). Such morphology is rare among modern angiosperms. However, these leaves are similar to the leaves of *Liriophyllum*, which can make identification difficult (Alekseev 2009).

Fig. 2.12. A reconstruction of *Liriodendrites* sp. Artwork by Joschua Knüppe.

Fig. 2.13. A specimen of *Liriodendrites* sp., showing the classic horseshoe morphology. The undescribed Egyptian specimens were probably similar in general appearance. Image provided by Michael Donovan and the Denver Museum of Nature and Science.

Alekseev (2009) has suggested that *Liriodendrites* was an extremophile, capable of surviving in extremely high temperatures, on the basis of the features of its leaves. This would explain its presence in Egypt, where the daytime temperature could reach a maximum of 34°C (Amiot et al. 2010b; Murray et al. 2013). The specimens from Egypt are also the oldest known representatives of this genus and may suggest an African origin for this taxon.

Terminalioxylon

This genus was once known as *Evodioxylon*, on the basis of its supposedly close relationship with *Evodia*. Later work would show that this taxon is related to the genus *Terminalia*, hence the change in name (Mädel-Angeliewa and Müller-Stoll 1973).

Terminalioxylon intermedium (fig. 2.14) is known from wood samples (Kräusel 1939; McCabe and Parrish 1992). There is some dispute as to whether the specimens from Egypt are indeed Cenomanian in age

Fig. 2.14. A reconstruction of *Terminalioxylon intermedium*. Artwork by Joschua Knüppe.

(El-Saadawi et al. 2014), but Ehrendorfer (2013) and Darwish and Attia (2007) consider them to be of Cenomanian age, and that view is followed herein.

Terminalioxylon intermedium would have been right at home in Bahariya, as some members of this genus can survive in salt water. This ability may be related to the fact that some species can store water, which comes pouring out when the trunk is penetrated (Shyamaprassad 2004, Gaire 2020).

Terminalioxylon intermedium presumably would have been similar in basic form to living species of this genus, although, given the significant number of species and variations seen in the genus, it's impossible to reconstruct its appearance in detail on the basis of extant examples (El-Saadawi et al. 2014).

Fig. 2.15. The extant species of *Terminalia argentea*. This specimen gives a clear example of the taxon's unusual branching morphology. Credit João Medeiros / CC-BY-SA-2.0. See Wikimedia Commons.

Rogersia

Rogersia longifolia is an obscure taxon that was previously documented from the US (Ward et al. 1905). Darwish and Attia (2007) assigned a leaf from Bahariya to this species but offered no further comment.

Salix

Salix is another extant genus, known colloquially as willow trees. Darwish and Attia (2007) refer a leaf to this genus but, while providing an illustration, offer no description or reason for this referral. Personally, I doubt that the specimen truly belongs to this extant genus. However, the Saxifragales are an ancient group, having evolved sometime between the Turonian and Campanian periods (Hermsen et al. 2003), so it's perfectly possible that it does indeed belong to this order. So for the moment I will keep this specimen as *Salix* herein, but I suspect that it will warrant its own genus when properly described.

Asiatifolium

Asiatifolium elegans is a species known from the Aptian/Albian of China and the Russian far east (Sun and Dilcher 2002; Golovneva et al. 2018). Darwish and Attia (2007) refer what appears to be a partial specimen from Bahariya to this genus, but while providing a photograph, they offered no further description.

Magnolia

I'm sure everyone is familiar with this tree genus. *Magnolia* contains 210 species and is also a remarkably old taxon, dating back at least ninety-five million years. This genus is a bit of a paradox, as it's a flowering plant whose evolution predates the evolution of the first bees. So exactly what was pollinating this plant is a mystery, although beetles are presumably the vector (Crane 1988).

One species, native to the freshwater river networks of Bahariya, is *Magnolia barthouxi* (Barthoux and Fritel 1925; McCabe and Parrish 1992). Sometimes this species is spelled *barthoux*, but I've decided to follow the spelling given by Barthoux and Fritel (1925). Given such poor representation in the literature, it's impossible to say much about this species, beyond acknowledging its existence.

Magnoliaephyllum

A specimen cataloged as *Magnoliaephyllum* sp. is documented from Bahariya by Darwish and Attia (2007). While the specimen is never described, the illustration provided shows a broad and rounded leaf. I suspect it may be referable to *M. palaeocretacicus* from the Turonian of Jordan and Israel (Dobruskina 1997), although a re-description will be needed before we can say for certain.

Macclintockia

Whether to include this taxon was a bit of a debate, as I initially doubted that the specimen assigned to this taxon was of Cenomanian age. However, Darwish and Attia (2007) consider the beds in question to be of Cenomanian age, and I've decided to follow them herein.

Macclintockia cretacea is a magnoliopsid from the Cretaceous of Greenland, but Kräusel (1939) also reported a leaf from Bahariya. Regardless, the sole specimen is missing the base, tip, and most of the edge. The illustration of the specimen does superficially resemble *M. barykovensis* (Moiseeva 2011). The primary veins are straight, with a slight curve, and the outer edge is smooth (Kräusel 1939).

However, the taxonomy is confusing, with Krassilov (1979) suspecting that *M. cretacea* may be a synonym of *M. kanei*. The specimen is now lost, so no re-description will ever be possible.

Hibiscoxylon

One of the more beautiful plants found in this region would have been *Hibiscoxylon niloticum* of Egypt (fig. 2.16). Kräusel (1939) assumes that *Hibiscoxylon* was similar to the modern genus *Hibiscus*. However, Hinsley (2010) suggests that *H. niloticum* is a species of dombeyoid. As both belong to the same family—the Malvaceae—we can make educated inferences about *H. niloticum* regardless of its affinities.

Much as with *Frenelopsis*, it's currently unknown whether *H. niloticum* formed a trunk or was a large bush; both growth habits are known among the Malvaceae (Bornhorst 1991), with some species reaching heights of 10 meters (33 feet), while other species are pushed to reach 3 meters (9 feet).

Members of this group stand out because of their flowers; the Malvaceae are known the world over for their ornate appearance and vivid

Fig. 2.16. A reconstruction of *Hibiscoxylon niloticum*. Artwork by Joschua Knüppe.

colors. Even more incredible is the fact that some extant taxa are pollinated by lizards (Bègue et al. 2014). Regardless, *H. niloticum* would then have produced either nuts or fruits.

Laurophyllum

Laurophyllum (fig. 2.17) is yet another magnolia tree. Its family is part of the order Magnoliales, best known from the Cenozoic era but also present in the Cretaceous of North and South America (Berry 1914). Mcabe and Parrish (1992) note the presence of *Laurophyllum* leaves, from the

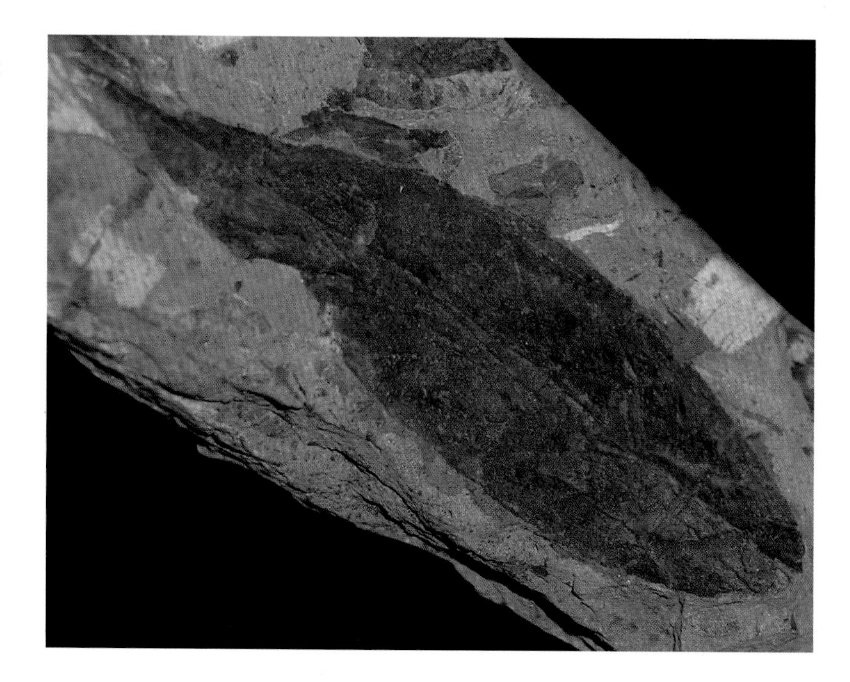
Fig. 2.17. A specimen of *Laurophyllum* sp. While almost certainly not the same species as the one from North Africa, this specimen demonstrates the taxon's basic morphology. Credit GyikToma / CC-BY-SA-2.0. See Wikimedia Commons.

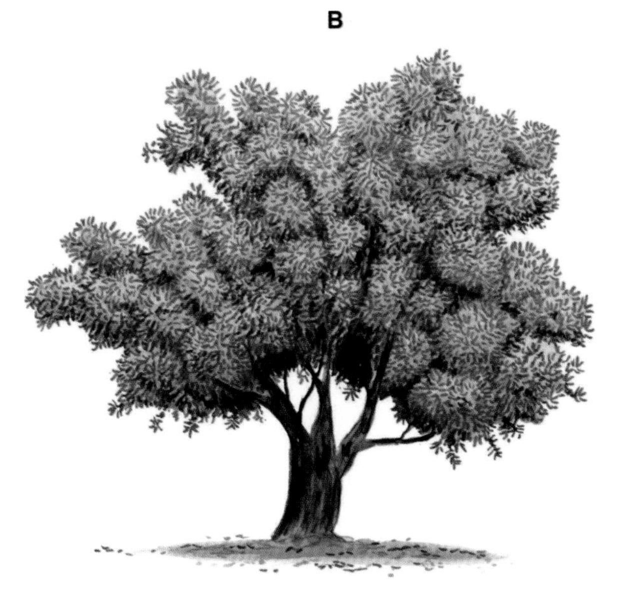

Plate 1. The trees of North Africa. *A*, *Magnolia barthouxi*; *B*, *Laurophyllum* sp.; *C*, *Dryophyllum* cf. *Subcretaceum*; and *D*, *Barykovia* cf. *Tschukotica*. Images not to scale.

Nubian shield, near Bahariya. Darwish and Attia (2007) also discovered *Laurophyllum* leaves from Bahariya itself and identified them as *Laurophyllum* aff. *lanceolatum*.

With more than three thousand species among fifty genera, this family is so diverse that its members can even form their own ecosystems, which are known as Laurel Forests (Little et al. 2009). Of course, with so little published on these specimens, it's difficult to discuss the African taxa in any detail.

Dipterocarpophyllum

Dipterocarpophyllum is a genus of tree known from the Oligocene of Egypt (Darwish et al. 2000), but it was also found in the Cenomanian Kisciba Formation (Darwish and Attia 2007). The specimens were never described, and *Dipterocarpophyllum* itself has been used by previous authors as a wastebasket taxon for much of the Dipterocarpaceae (Darwish et al. 2000).

Dryophyllum

Dryophyllum is poorly known in North Africa, although it is well documented elsewhere in the world (Jones et al. 1988). Known from their small but elongated leaves, the Moroccan specimens have never been given a full description and are left as *D.* cf. *subcretaceum* (Krassilov and Bacchia 2013).

While *Dryophyllum* was long considered a member of the Fagaceae, it is now believed to be a member of the Juglandaceae, a type of walnut tree (Jones et al. 1988). These had only just evolved in the Cenomanian, but they would later become one of the dominant groups in the subtropical forests at the end of the Mesozoic and early Cenozoic eras (Krassilov and Bacchia 2013).

Populus

Populus is a genus of tree from the Northern Hemisphere that includes the aspen and cottonwood trees. Four *Populus* leaves have been documented from Bahariya, one of which is identified as *Populus wilmattae*, with the rest left as *Populus* sp. (Darwish and Attia 2007). These specimens have also never been described.

Barykovia

One of the smaller species in this region was *Barykovia*. Known from Russia, it was also present in the Cenomanian of North Africa (Krassilov and Bacchia 2013). *Barykovia* may belong to the Fagaceae; indeed, many of these were once classified as species of oak. Moiseeva (2012), however, notes similarities in leaf morphology with the Ulmaceae (elm trees).

Barykovia is easily identifiable because of its broad leaves, which are small and crenulated (possessing rounded projections along the leaf margins). The exact species present in Morocco is uncertain; Krassilov and

Bacchia (2013) identify this species as *B.* cf. *tschukotica*, but it may warrant its own species in the future. One of the reasons for this uncertainty is the considerable variation in leaf shape within this genus (Moiseeva 2012).

Ficus

Ficus is a genus of tree cultivated for its fruits, which are sold as figs. Darwish and Attia (2007) tentatively identified some specimens from Bahariya as cf. *Ficus pandurifolia*. But because no description of this specimen was given, just a photograph, it's difficult to discuss this taxon further, especially as *F. pandurifolia* may be a synonym of another *Ficus* species.

Nelumbites

Given the extent of these wetlands, the presence of water lilies is to be expected. One of these is *Nelumbites schweinfurthiii* (fig. 2.18). It was first recorded in Egypt by Stromer (Kräusel 1939), but more specimens have been found since (Lyon et al. 2001; Darwish and Attia 2007).

Fig. 2.18. A reconstruction of *Nelumbites schweinfurthiii*. Artwork by Joschua Knüppe.

Fig. 2.19. A modern grove of *Nelumbites* sp. Credit Lago Mar / CC-BY-SA-3.0. See Wikimedia Commons.

Nelumbites is characterized by its broad, circular leaves and extensive aerenchyma system (aerenchyma being the spongy air channels in the stem and roots). *Nelumbites* also anchored itself using roots that grew from the stem (Kräusel 1939; Dobruskina 1997; Kvacek and Herman 2004; Wang and Dilcher 2006; Moiseeva 2012).

While the pad usually floats on the surface, *Nelumbites* can survive periods of submersion (see fig. 2.19). *Nelumbites* is also capable of surviving in both fresh water and salt water; its only requirement appears to be stagnant or slow-moving water (Wang and Dilcher 2006).

Aquatifolia

A close relative of *Nelumbites* is *Aquatifolia*, a taxon previously recorded from the Albian of the US (Wang and Dilcher 2006). Darwish and Attia (2007) apparently identified a specimen of *A. fluitans* from Bahariya. But apart from a photograph, no description of these specimens was ever given.

Nymphaea

Darwish and Attia (2007) also note the presence of the water lily *Nymphaea*. While listing its presence in the Bahariya Formation, they give no description. This makes it difficult to offer a reconstruction, beyond generalized comments based on extant members of this genus.

Brasenites

Brasenites kansense is another aquatic angiosperm from the US; it is found in association with *Aquatifolia* and is presumably related to the extant Cabombaceae (Wang and Dilcher 2006). Darwish and Attia (2007) assigned a leaf from Bahariya to this taxon, but with so little published on this specimen, it's difficult to discuss the African taxa in any detail.

Garasbahia

One of the more recent additions to the local flora is *Garasbahia fexuosa* (fig. 2.20). With a long thin stem, *G. fexuosa* has both rhizomes and adventitious roots. The branches grow distichously: one branch on either side in an alternating arrangement, one above the other (Krassilov and Bacchia 2013).

The leaves are small and peltate in shape (the stem attaching to the center of the leaf, not the base). A cursory glance at the fossil suggests that it had flowers, but these floral cups are an artifact of preservation, a result of the leaves being crushed together during burial and fossilization (Krassilov and Bacchia 2013).

Krassilov and Bacchia (2013) suggest that *G. fexuosa* is related to the genus *Cabomba*, although it has similarities to *Brasipelta*, *Limnobium*, and *Limnochari*. The Cabombaceae are closely related to the water lilies and are considered to be one of the most basal of all the flowering plants (Iles 2014). The presence of *G. fexuosa* in the Cenomanian of North Africa adds to our impressive Mesozoic record for this family (Mohr et al. 2008; Friis et al. 2011) and may contribute to our understanding of Cabombaceae global distribution.

Fig. 2.20. A reconstruction of *Garasbahia fexuosa*. Artwork by Joschua Knüppe.

How *G. fexuosa* reproduced is unknown, as some species of Cabombaceae are wind pollinated while others rely on insects (Taylor and Williams 2009); regardless of the method used, the presence of rhizomes suggests that *G. fexuosa* could also produce new plants asexually. On the basis of living cabombacid taxa, we can assume that *G. fexuosa* would have grown in slow or stagnant water (Friis et al. 2011).

Typhaephyllum

Another water plant from Bahariya is *Typhaephyllum* sp., a type of water reed or bulrush (Darwish and Attia 2007). Although previous authors (Lyon et al. 2001) have referenced them, no thorough description has been given of these specimens.

Cocculophyllum

One of the harder to classify plants is *Cocculophyllum* cf. *furcinerve* (fig. 2.21), which could belong to either the Lauraceae or Menispermaceae (Kvacek 1983), even though those two groups are not closely related. While Krassilov and Bacchia (2013) note similarities to the extant *Cocculus*, a type of Menispermaceae, Friis et al. (2011) suggest that the matter requires more research before a definitive identification can be made.

The North African specimens have been tentatively assigned to C. cf. *furcinerve*. These specimens have acrodromous laterals (leaf veins) along the leaf margin, connecting to irregular transverse laterals (Krassilov and Bacchia 2013). More importantly, this leaf morphology suggests that

Fig. 2.21. A reconstruction of *Cocculophyllum* cf. *furcinerve*, shown here as a climbing vine. Artwork by Joschua Knüppe.

Cocculophyllum may have been a climbing vine (Krassilov and Bacchia 2013).

Cocculophyllum is characterized by oblong or ovate leaves (Kvacek 1983). The leaves also have glands used to secrete substances. Kvacek (1983) theorized that these produced noxious substances used in defense against herbivores, but it's also possible that these were used to secrete excess salt (Barhoumi et al. 2008).

Vitiphyllum

Vitiphyllum is a genus from the Albian of the US (Fontaine 1889; Friis et al. 2003), although specimens possibly belonging to this taxon have also been found in Kazakhstan (Vakhrameev and Krassilov 1979). The leaves of this taxa have a reticuled, ternate pattern, with several veins on each lobe (Fontaine 1889; Friis et al. 2003). Darwish and Attia (2007) assigned a specimen from Bahariya to V. aff. *multifidum*. This specimen was not described, and no reason for this referral was given.

Plantae
Incertae Sedis

Avicennia

Avicennia is an extant mangrove genus. Darwish and Attia (2007) refer three leaf specimens from Bahariya to *Avicennia* sp. While the illustration provided looks superficially like *Avicennia*, the lack of description and reason for this referral makes me hesitant about it, especially as the oldest *Avicennia* specimens hail from the Eocene, with the first diversification event for this genus occurring only 6 mya (Renema et al. 2008; Li et al. 2016). So I'd rather leave these specimens here until they are properly described.

Bambusa

As the name suggests, *Bambusa* is a genus of Bamboo. Darwish and Attia (2007) refer three leaves from Bahariya to this genus. However, the oldest known *Bambusa* are only 5.7 million years old (Teodoridis et al. 2015), while the Bahariya specimen is roughly 100.5 million years old. So I'm dubious about this identification, as it would require a large ghost lineage and the specimen was also never described in detail.

However, the Bambusoideae does have a Cretaceous origin (Prasad et al. 2005). So it's possible that these specimens actually are bambusoids. I favor leaving this specimen as Bambusoideae *incertae sedis* until it's given a description.

Celtis

Celtis is a genus of deciduous tree with a near pan-global distribution. Darwish and Attia (2007) tentatively assigned a single leaf to *Celtis mccoshii*. If that's accurate, it would represent another ghost lineage as *C. mccoshii* is from the Eocene period and the oldest confirmed specimens of this genus only go as far back as the Paleocene (Manchester et al. 2002). Given the cautious nature of the referral, the lack of description for this specimen, and the large gap in age, I prefer to leave this specimen as Cannabaceae *incertae sedis*.

Cyperites

Kräusel (1939) made a brief mention of reedlike stems and leaf fragments from Bahariya that he identified as *Cyperites* sp. However, Kräusel seems to have been using *Cyperites* as a form taxa, as *Cyperites* is a lycophyte, but he considered these specimens to belong to either the dicotyledons or the monocotyledons. These specimens were destroyed, so a re-description is impossible.

Nelumbo

Darwish and Attia (2007) have identified a specimen of water lily from Bahariya as *Nelumbo lutea*, a close relative of *Nelumbites*. Personally, I find this identification unlikely, given that *N. lutea* is an extant species native to North America. But, until this specimen is properly described, there is little else we can say on this subject.

Omphalopus

Darwish and Attia (2007) record a leaf from Bahariya they identify as *Omphalopus* sp. *Omphalopus* is a dicotyledont belonging to the family Melastomataceae. It's also a scrambler, a plant with long, adhesive roots that allow it to grow on rock walls or over other plants (Clausing and Renner 2001).

However, *Omphalopus* has since been found to be a junior synonym of *Dissochaeta*, an extant taxon native to Indochina, Malesia, and New Guinea (Kartonegoro et al. 2018). This leaves the identity of the Bahariyan specimen an open question. We don't have the data to justify transferring it over to *Dissochaeta*, since no reason for the initial identification was ever given. So I prefer to leave this specimen as Melastomataceae *incertae sedis*.

Parataxodium

Parataxodium is a conifer known from the US and Europe, and it was one of the most common taxa in the polar regions (Arnold and Lowther 1955). Darwish and Attia (2007) identified a branch with several leaves from Bahariya as *Parataxodium* sp. The specimen remains undescribed. The photograph Darwish and Attia (2007) provide does look similar to the holotype from Arnold and Lowther (1955), but the image quality is low, and with no accompanying description, it's difficult to say for sure.

This is problematic because *Parataxodium* is apparently chimeric; its sole species, *P. wigginsii*, containing at least four different conifer taxa (Rothwell and Stockey 2018). This research is currently pending publication, but until then we have little idea what constitutes *Parataxodium*; only the holotype specimen may remain valid (Rothwell 2020). Since *Parataxodium* is about to be split and it's impossible to transfer the

Egyptian specimen to any of the subsequent new taxa given the lack of data, I favor leaving these specimens as Taxodiaceae *incertae sedis*.

Parnassia

Parnassia is an extant genus of herb, known colloquially as bog-stars given it's generally a wetland plant, although it's also found in dunes and arctic climates. Darwish and Attia (2007) hesitantly assigned a leaf to *Parnassia glauca*. Given the tentative identification and lack of description, I support leaving this specimen as Celastraceae *incertae sedis*.

Platanus

Platanus is an extant genus of tall deciduous trees, found throughout the Northern Hemisphere. Darwish and Attia (2007) assign a leaf from Bahariya to *Platanus marginata*, with another left as *Platanus* sp. However, these specimens have yet to be described—a situation made more complicated by the fact that *P. marginata* has been declared a junior synonym of *Tasymia*, along with five other species of *Platanus* (Golovneva 2008). Given the lack of description and the uncertain status of *P. marginata*, I believe these specimens are best left as Platanaceae *incertae sedis*.

Tristemma

Tristemma is an obscure taxon belonging to the family Melastomataceae, an edible plant currently studied for its potential antibiotic properties (Nguenang et al. 2018). Darwish and Attia (2007) assign a leaf from Bahariya to the extant species *Tristemma incompletum*. I find this identification doubtful, especially since it's unclear whether this is even a valid species (Plant List 2013). So I prefer to leave this specimen as Melastomataceae *incertae sedis* until it's properly described.

Zizyphoides

Darwish and Attia (2007) tentatively referred a single specimen to the Eocene taxa *Zizyphoides*. But no description or illustration of the specimen was ever given. So, as with the others in this section, I prefer to leave this specimen as Magnoliopsida *incertae sedis* until it's given a full description.

Specimen 87087 and 87088

Given the tropical climate, it's no surprise Darwish and Attia (2007) identified palm leaves from Bahariya. These may be referable to *Palmoxylon*, which is also known from Bahariya (Said 1962). However, where those *Palmoxylon* specimens came from was never stated, and we can't be sure that they are of Cenomanian age, as *Palmoxylon* was found throughout the Cretaceous and all the way through to the Miocene period (Dutta et al. 2011). So until these specimens are described, there is little else we can say.

Seeds, Fruits, and Miospores

Khalloufi and colleagues (2010) noted the presence of seeds, less than 10 mm long, from Jbel Tselfat in Morocco. While no description was given, they were briefly noted as being "tear-shaped" and similar to those produced by some types of cedar tree. Darwish and Attia (2007) also mention preserved fruit from Bahariya, tentatively identified as belonging to a monocot.

Miospores, small spore or pollen grains, are also known from this region. This branch of earth science is known as palynology. Palynology not only gives us further information about what plant life grew in this region but also allows us to track changes in plant diversity over time. The Bahariyan biome underwent four significant changes during the Cenomanian, correlating to a warming climate and rising sea levels (Tahoun et al. 2013).

Palynology also allows us to identify specific ecosystems within the overall biome. On the basis of the miospore data, we have currently identified three ecosystems in Bahariya and its surrounding area: mangrove swamps (Tahoun et al. 2013), upland forests (Moustafa and Lashin 2012), and riverine gallery forests (Baioumi et al. 2012). A full compendium of palynomorphs found in North Africa is given in the appendix.

The Fauna of North Africa: Invertebrates

3

Invertebrate refers to an animal that lacks a spine. But the word lacks any real scientific meaning because many of the groups considered to be invertebrates aren't closely related to one another. For that reason, I'll use the term here in its general sense: the animal groups that lack a bony skeleton.

Microfauna and Meiofauna

The most diverse groups in extant wetlands belong to the microfauna and meiofauna (Alongi 1987; Kathiresan and Bingham 2001). These are microscopic invertebrates small enough to slip through a 1 mm mesh. The microfauna live in soil, while the slightly larger meiofauna are aquatic.

But we have a poor fossil record of these creatures for parts of North Africa, and some areas, such as the Aoufous beds of Morocco, have low levels of diversity (Cavin et al. 2010), although foraminiferans are known from the Upper Cenomanian units of the Souss Massa Drâa and Meknes Tafilalet provinces (Lézin et al. 2012).

Foraminiferans are best known from Egypt's Sinai Peninsula (Gertsch et al. 2010a; Shahin and El Baz 2014). Foraminifera also help in ecological reconstructions, as the most common foraminiferan taxa in the Wadi El Ghaib section of the Galala/Raha Formation are all adapted for shallow-water, low-oxygen conditions and low levels of salinity. That indicates that this site, north of Sharm El Sheik, was near the ancient coastline and that there were large influxes of fresh water. Most of these foraminiferans also appear to have lived near or even in the seabed (Gertsch et al. 2010a).

Toward the Cenomanian/Turonian boundary, foraminifera become rarer. Those that are present, such as *Ammobaculites plummerae* and *Gaudryina* sp., indicate a harsher environment in which the oxygen levels in the water had decreased even further (Gertsch et al. 2010a). A full list of the known North African microfauna and meiofauna taxa will be given in the appendix.

Porifera

Sponges are one of the most primitive extant multicellular organisms. They have no nervous, digestive, or circulatory systems. Instead, they are full of channels that seawater flows through and that absorb food and oxygen from the passing water (Vacelet and Duport 2004). Most North African sponges belong to a group called the sclerosponges or coralline sponges, which can produce calcareous skeletons like coral (Reitner

1992). Unfortunately, the sclerosponges do not form a natural group, as many are not related to one another (Vacelet 1985).

Actinostromarianina

One type of sclerosponge found in this region is *Actinostromarianina*. Pandey and colleagues (2011) first mentioned the presence of *Actinostromarianina* in Egypt but gave no formal description. Ayoub-Hannaa and Fürsich (2012b) later identified two different species of *Actinostromarianina* (both of which are still unnamed), on the basis of the basic morphology and internal microstructure.

The first is identifiable as a nodular type and is 4 inches high, while the second is dendroid in shape with a polygonal meshwork, standing at 2 inches tall (Ayoub-Hannaa and Fürsich 2012b). Besides that rather terse description, little else has been written about either species.

Anthozoa

The Anthozoa includes the corals, sea pens, sea anemones, and sea fans. Today we would expect the tropical waters of Northern Gondwana to be filled with coral reefs; instead, the seas were dominated by rudists, a type of bivalve mollusc. While they technically didn't form *true* reefs—they only accumulated sediments around their shell instead of producing carbonate structures like coral (Gili et al. 1995)—the distinction is academic, as the rudists were practically the sole reef builders in the tropics by the Cretaceous (Hartshorne 1989; Johnson 2002; Benyoucef et al. 2012).

This rise to dominance was due to the hot, saline waters of the Tethys, which favored rudists over corals (Johnson 2002). Yet the corals would have the evolutionary last laugh; the rudists ran afoul of the mass extinction that wiped out the dinosaurs, leaving the corals as the primary reef builders in the present day.

Aspidiscus

One of the survivors in a rudist world was *Aspidiscus cristatus*, which was the most prolific coral in Egypt, Tunisia, and Algeria (Pandey et al. 2011). *Aspidiscus cristatus* is a colonial coral (a structure formed out of many individuals), roughly bilaterally symmetrical in form, with a concave base. It has a series of distinct ridges (monticules) running from the top to its base. These ridges also have an opening at the top, forming a "crater-like" structure (Pandey et al. 2011).

Unlike many corals, adult A. *cristatus* colonies don't have an attachment area and would have just sat on the seabed (Pandey et al. 2011). The concave base would have acted as a paperweight, holding it in place.

Some specimens of *Aspidiscus* show evidence of parasitism (Ayoub-Hannaa and Fürsich 2012a; Wilson et al. 2014), which occurred when bivalve larvae began to bore into the coral's skeleton. The coral would subsequently heal, and new tissue would grow over the hole, with the bivalve safe inside.

Aspidiscus cristatus appears to have very precise habitat requirements and has only been found in shallow-water environments with low energy and soft substrates (Pandey et al. 2011). The dome shape itself was an adaptation to such an environment: sand and other sediments would have slid down its surface, preventing the colony from being buried alive (Pandey et al. 2011).

Polytremacis

Africa was also home to the octocoral *Polytremacis*. Despite the confusion surrounding its taxonomy (Gregory 1899), *Polytremacis* is important because it provides the "missing link" that connects the modern Helioporato and the extinct Heliolitidse. However, its exact placement between the two groups is a matter of contention, and some species of *Polytremacis* end up being classified as members of the heliolitids and others as basal helioporids (Gregory 1899).

The African species usually goes unnamed in most papers, although Said (1962) mentions *P. chalmasi* in passing. The calices (bowl-like structures on the coral's surface) are also undescribed for the North African species, which is unfortunate because this trait seems to vary among species and may help with identification (Gregory 1899).

Polytremacis lacks symmetry, with its branches growing seemingly at random. The septae (plates that form within the walls of a coral's skeleton) have a toothlike structure that gives them a rough surface. *Polytremacis* also has a series of ridges on its outer surface. This is another trait that varies between species, although *Polytremacis* tends to lose parts of the septa as it ages (Gregory 1899).

Tortoflabellum

Tortoflabellum is a rare taxon in North Africa, currently recorded only in Egypt. It's found elsewhere in the Mediterranean region, however, with some species found as far west as Jamaica and as far east as New Zealand (Squires 1958; Baron-Szabo 2006).

The species found in Egypt is currently unnamed, but it inhabited this region from the late Albian to the end of the Cenomanian (El Qot et al. 2009; Ayoub-Hannaa 2011; Pandey et al. 2011). Little has been published on this taxon apart from the fact that it was rare and inhabited shallow regions with a soft, sandy substrate.

Cladocora

The last of the North African corals was *Cladocora*. One species, *C. caespitosa* (fig. 3.1), still survives in this region, where it forms the only true reefs found in the Mediterranean (Kružić and Požar-Domac 2003). The African species is still unnamed but reasonably well known (Ayoub-Hannaa and Fürsich 2011).

Cladocora can grow either as individual colonies, known for their dome shape, or as interconnected formations called a *bank* (Morri et al.

Fig. 3.1. The extant *Cladocora*. This gives an idea of what the extinct North African species may have looked like in life. Credit Esculapio / CC-BY-SA-3.0. See Wikimedia Commons.

2001). The extant species dwell in a variety of habitats, as well as both shallow and deep water. The Cenomanian species are found only in shallow reefs.

Anthozoa Trace Fossils

Some might remember seeing sea anemones, a group of Anthozoans, in rock pools at the beach. While some sea anemones are capable of swimming, most species attach themselves to the seafloor via a pad called a basal plate. Because of their lack of skeleton or other hard parts, anemones are rarely fossilized, but the imprints left by their basal plates can be preserved.

This is what's known as a trace fossil. A trace fossil, or ichnofossil, to use the scientific term, is the evidence left behind of an animal's or plant's presence: things like footprints, leaf imprints, burrows, root traces, droppings, and tooth marks on a bone. Much like a police officer identifying a perpetrator by studying the evidence left at the scene, a paleontologist can learn a lot about an extinct species by studying the traces it left behind during its day-to-day activities.

Trace fossils from anemones are preserved in the Kem Kem beds, where circular imprints are interpreted as being the resting traces left behind when a sea anemone attached itself to the sediment (Ibrahim et al. 2014a). These were later identified as *Conichnus*, but no reason for this identification was ever given (Ibrahim et al. 2014a).

Some of these trace fossils are so large that some suggest that they represent lungfish burrows. However, Ibrahim and colleagues (2014a) point out that they lack the characteristic features of lungfish burrows. While such fossils prove the presence of large anemones in North Africa at this time, they tell us little else about the species that produced them other than that they were large and inhabited a wide range of aquatic habitats (Ibrahim et al. 2014a).

Annelida

The annelids include segmented worms, ragworms, and leeches. Ayoub-Hannaa (2011) mentions the presence of marine tube worms belonging to the family Serpulidae from Egypt but gives no further details. Benyoucef and colleagues (2019) document the tube worms *Filograna socialis*, *Neovermilia ampullaceal*, and *Glomerula serpentine* from Algeria but gives no reasons for these referrals or descriptions of these specimens.

Annelida Trace Fossils

Worm burrows, assigned to the ichnotaxon *Beaconites antarcticus*, are known from Morocco (Ibrahim et al. 2014a; Ibrahim et al. 2020a).

Echinodermata

The Echinodermata include the sea stars, sea urchins, sea cucumbers, brittle stars, and crinoids. The Echinodermata underwent an evolutionary radiation in the Mesozoic known as the Mesozoic Marine Revolution. During this period, a large number of durophagous (shellfish-eating) marine animals evolved. This led to the extinction of many shelled animals, with the survivors developing means of defending themselves (Vermeij 1977; Stanley 1974, 2008).

Echinoidea

The sea urchins were one of the most successful groups in Africa, with many taxa having multiple species. They ranged from herbivores to carnivores and deposit feeders. In Egypt, they serve as environmental indicators because their morphology is closely related to their habitat. For example, the presence of *Heterodiadema* and *Tetragramma* (see fig. 3.2) are indicative of shallow-water environments (Ayoub-Hannaa and Fürsich 2012b).

To describe all the species found in this region would warrant a book in its own right. For that reason, the named taxa will only be identified in the appendix, without detailed descriptions of every species.

Fig. 3.2. A specimen of the echinoid *Tetragramma* sp. from Morocco. In *A*, upper; *B*, lower; and *C*, lateral views. Image provided by Manuel.

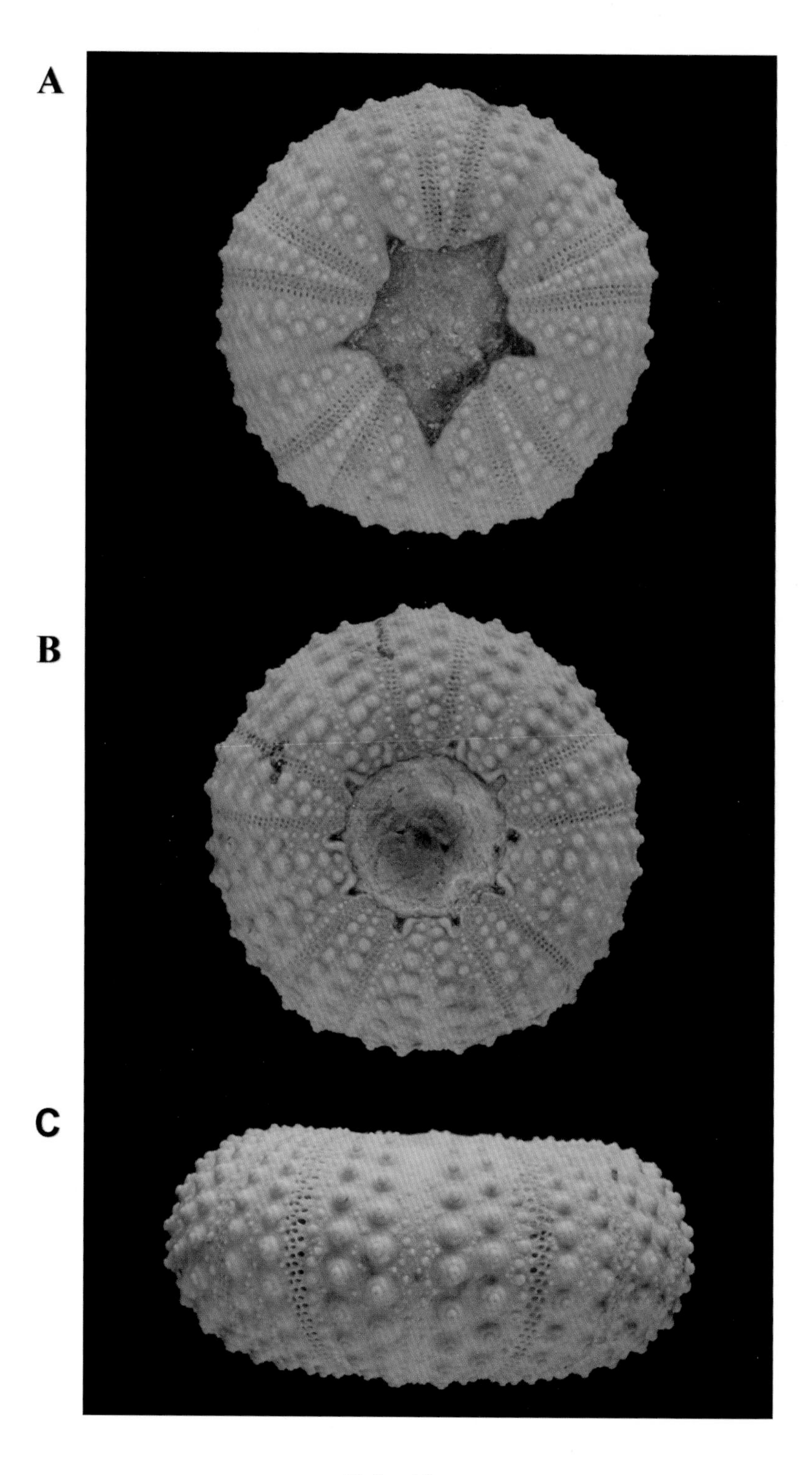

Crinoidea

The crinoids seem more like plants than animals. Crinoids are divided into two groups: the sea lilies, which anchor themselves to the seabed with a stalklike structure, and the comatulids, which lack stalks. They are a remarkably ancient group; the Ordovician period (485.4–443.8 mya) was their evolutionary heyday (Foote 1999). They are still remarkably successful, with six hundred extant species.

Roveacrinus

The best-known Cenomanian crinoids from North Africa are found in Algeria, with *Roveacrinus* being one of the most common. Unfortunately, these specimens are known only from thin sections, as they were discovered during a petrographical analysis (Ferré et al. 2017). Thin sections are created when a rock is cut into thin slices and each slice then ground flat so it can be viewed under a microscope.

This means that all the specimens are preserved in cross sections. However, they can still be positively identified. *Roveacrinus* and its kin are all informally known as *microcrinoids*, which means they are only a few millimeters in size and difficult to see with the naked eye. Three species of *Roveacrinus* are known from this region: *R. alatus*, *R.* cf. *alatus*, and *Roveacrinus* sp. (Ferré et al. 2017).

Orthogonocrinus

Another of the microcrinoids is an unnamed species of *Orthogonocrinus*. Known from three specimens, this species seems to be a transitional form between the slim-bodied *O. apertus* and the heavyset *O. janeti* of Europe and North America (Ferré et al. 2017).

Applinocrinus

Applinocrinus is the last of the named microcrinoids from this region. While only one specimen was found in Algeria's Tinrhert area, this unnamed species of *Applinocrinus* has proved more common in the Guir basin and Tunisia's western desert. This is presumably due to the taxon's environmental preferences, as *Applinocrinus* preferred muddy lagoons instead of the open waters of the continental shelf preferred by *Roveacrinus* (Ferré et al. 2017).

Echinodermata *incertae sedis*

The reefs of North Africa were home to at least one species of indeterminate comatulid, discovered by Garassino and colleagues (2008b). Their branchlike arms, covered in bristlelike structures, have earned the comatulids the name *feather stars*. While Garassino and his team never described the specimen, the only published photograph suggests a poor state of preservation (Murray et al. 2013).

Gastropoda

Of all the molluscan clan, snails are probably the group people are most familiar with. The Cenomanian aquatic snails are remarkably diverse in this region, with both herbivorous and carnivorous species known. Their habitat preferences ranged from shallow reefs to ocean depths (Gertsch et al. 2010a).

Fig. 3.3. A series of presumed gastropod tracks. Scale bar is 50 mm (5 cm). Image from Ibrahim et al. (2014a, fig. 6A).

By the time the Cenomanian ended, they were so abundant that some fossil beds were composed almost entirely of gastropods; one taxon, *Pchelinsevia coquandiana*, became especially prevalent in some areas (Ayoub-Hannaa and Fürsich 2011; Contessi and Fanti 2012a). As with echinoids, the full list of North African gastropods will be given in the appendix.

Gastropod Trace Fossils

Gastropod burrows and crawling traces are known from Morocco. They can be divided into two types. While both are rare, the second type appears more often. Several were found preserved, with the large traces believed to be the resting traces of sea anemones (Ibrahim et al. 2014a). Some of these have been assigned to the ichnofossil *Scolicia* (Ibrahim et al. 2020a).

Type one is made up of two rows of asymmetric bumps, spaced 3 mm apart. The second type, while similar, has a deep groove between the bumps (Ibrahim et al. 2014a). This trace fossil is reminiscent of the ichnogenus *Protichnites* (see fig. 3.3), which is widely believed to be made by arthropods instead of gastropods (Ortega-Hernandez et al. 2010), which makes me wonder whether this track has been misidentified.

Ammonoidea

The Ammonoidea surely needs no introduction; every museum and fossil collector will undoubtedly have a specimen somewhere. Famed for its ornate shells, the ammonite itself only inhabited the front of its structure.

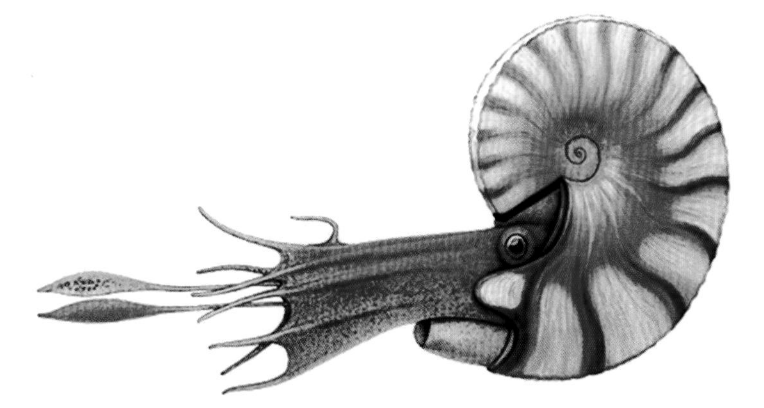

Fig. 3.4. A reconstruction of *Neolobites*. Artwork by Andrey Atuchin.

The rest was filled with air and liquid, which allowed the creature to float suspended in the water. Unlike the nautiloids, the ammonites completely died out alongside the dinosaurs at the end of the Mesozoic.

Neolobites

Neolobites (fig. 3.4) was a successful and widespread taxon (d'Orbigny 1840–42; Kennedy and Juignet 1981). Many species of *Neolobites* have been named, some of dubious validity, but three were native to North Africa: *N. fourteaui*, *N. vibrayeanus*, and *N. peroni*. Work by Wiese and Schulze (2005) shows that all three are valid species. Although *N. vibrayeanus* populations, with a shell diameter of 92.9 mm, show a great deal of physical variability, there is little overlap between them (Nagm and Wilmsen 2012).

Neolobites fourtaui predates *N. vibrayeanus* (Kennedy and Juignet 1981; Wiese and Schulze 2005). *Neolobites vibrayeanus* (fig. 3.5) has wide, elongated bullae (protruding sections of the shell) even as juveniles, along

Fig. 3.5. A specimen of *Neolobites vibrayeanus*. Image provided by Ricardo. See image at http://www.thefossilforum.com/index.php?/gallery/image/36995-neolobites-vibrayeanus-dorbigny-1841/.

with a series of ribbing that possesses nodes. The whorl section also seems more rectangular than in the other species (Wiese and Schulze 2005).

Interestingly there appears to be more than one distinct growth pattern for N. *vibrayeanus* (Wiese and Schulze 2005). Once above 50 mm in size, N. *vibrayeanus* develop either distinct ribbing or a smooth shell; the slender morphotypes lack ribbing. As both morphotypes are found together in the same beds, it's likely that this variation represents sexual dimorphism (Wiese and Schulze 2005).

Neolobites peroni can be separated from the other two species by dint of its exceptionally large, "funnel-like" umbilicus (the depressed area in the center of the shell) (Wiese and Schulze 2005). However, Wiese and Schulze also caution that, as so few N. *peroni* specimens have ever been described, we have no idea about its intraspecific variability.

Neolobites is often used as an index fossil. However, Wiese and Schulze (2005) caution that it may not be the best genus to use in some areas because it is capable of inhabiting regions where no other ammonites could survive. Therefore, the presence *Neolobites* and the absence of other "stratigraphically significant ammonites" is not necessarily proof positive that an assemblage is of a certain age.

Neolobites was a specialized taxon evolved to live in shallow, subtidal to intertidal waters hit by frequent monsoons (Powell 1989; Schulze et al. 2003; Wiese and Schulze 2005). The flat, streamlined shell of *Neolobites* suggests that it was a fast swimmer. It also seems to form a distinct fossil assemblage with the nautiloid *Angulithes*, as the two are often found together (Luger and Gröschke 1989; Meister and Rhalmi 2002; Wiese and Schulze 2005). Presumably for the same reason, they could survive in an environment where no others could (Wiese and Schulze 2005).

Baculites

Of all the ammonites, *Baculites* (fig. 3.6) is my personal favorite. Unlike other ammonites, *Baculites* has a wonderful elongated shell, reminiscent of the orthocerids of the Paleozoic era—although the two are not related.

Baculites was a cosmopolitan genus, ranging from the Middle East to North America. Because an individual *Baculites* could cover a large amount of territory during its life, Penman (2007) found evidence of gene flow between the various North American *Baculites* species. This means that many of the features used to identify each species may need reviewing.

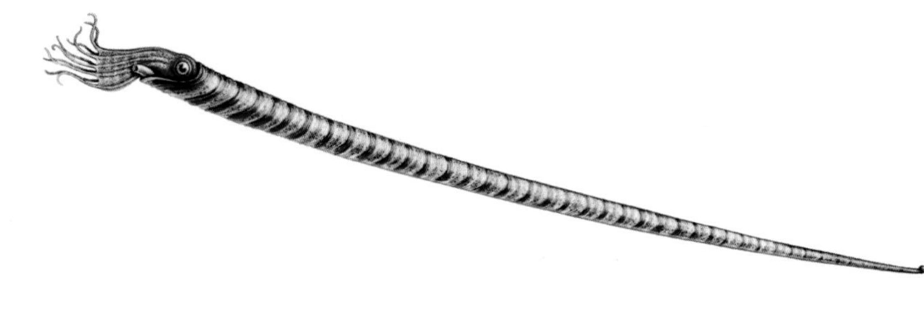

Fig. 3.6. A reconstruction of *Baculites*. Artwork by Andrey Atuchin.

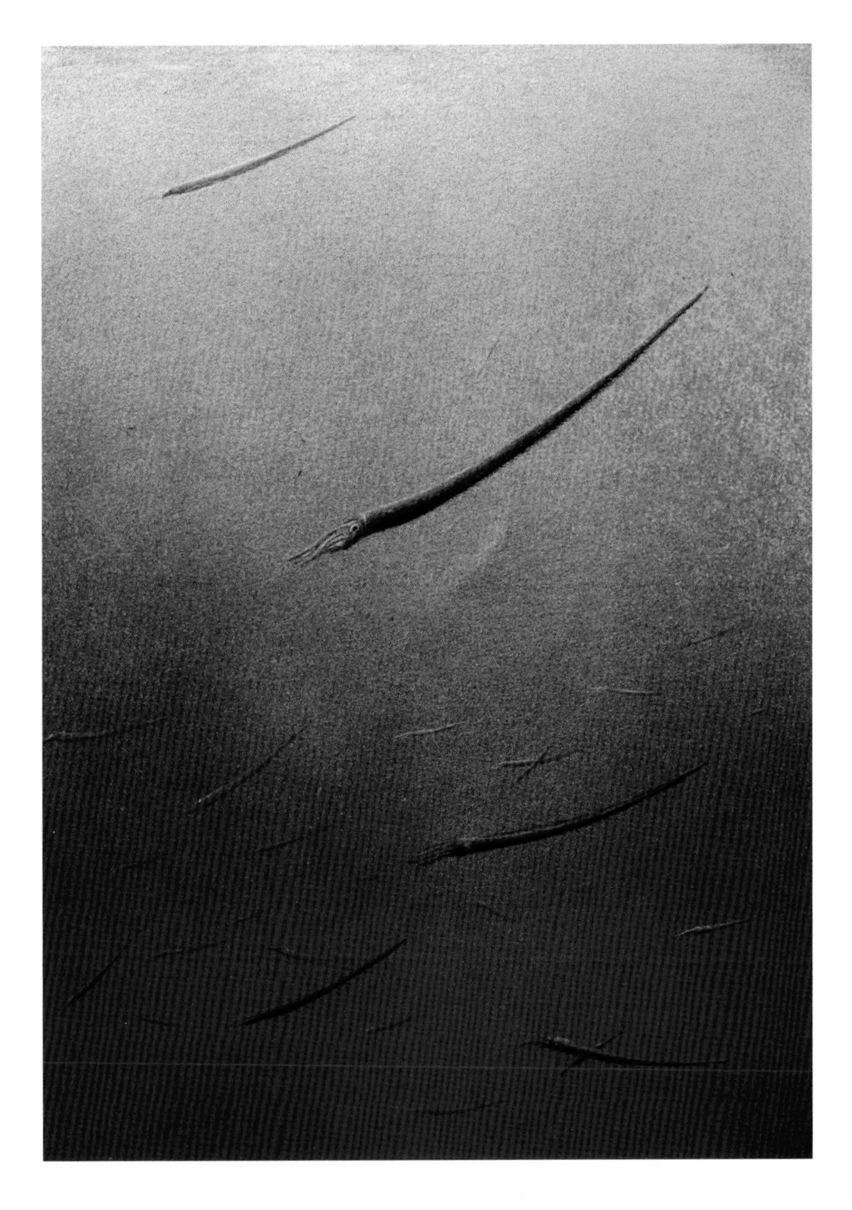

Plate 2. A shoal of *Baculites*, rising up the water column to feed in the open waters off the Bahariyan coast. Artwork by Joschua Knüppe.

Exactly what species was found in North Africa is unknown (Stromer 1936). It's likely that the Egyptian specimens can be assigned to *B. syriacus* given the proximity of Egypt to Syria. However, this cannot be verified until more complete specimens are found. There was mention of another species, called *B. aegyptiaca* (Tristram 1865), but I can find little corroboration anywhere, so it's probably a *nomen oblitum*. This means that the scientific name is no longer valid as it has become disused and forgotten (obliterated) over time. Whatever the species, *Baculites* was remarkably common (Dawson 1885).

Most reconstructions show *Baculites* floating horizontally, but some scientists have suggested that they floated vertically with the head pointed down. However, Westermann (1996, 2013) has proved that some species of *Baculites* were capable of swimming horizontally when fully grown,

Fig. 3.7. A reconstruction of *Choffaticeras*. Artwork by Andrey Atuchin.

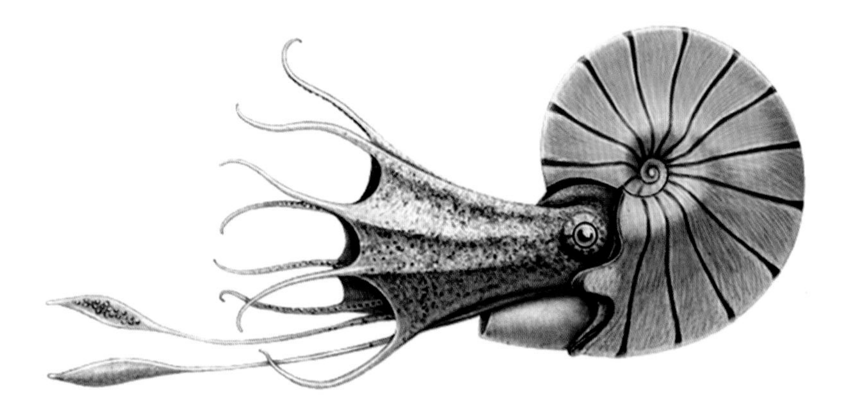

as they could achieve neutral buoyancy by flooding the shell chambers via a complex hydrostatic system (plate 2).

Arguably, the most remarkable thing about *Baculites* is that its lower jaw is twice as large as the upper jaw and has a blunt end instead of a sharp tip. The radula (a tonguelike structure) also has thin teeth on the surface (Kruta et al. 2011). This anatomy is similar to extant plankton-eating cephalopods, suggesting that *Baculites* was a filter feeder, eating both plankton and macroplankton. This is further supported by the discovery of isopods and snail larvae preserved in the jaws of an American *Baculites* (Kruta et al. 2011).

Choffaticeras

Choffaticeras (fig. 3.7) is a reasonably common taxon from Egypt, Tunisia, and Algeria, although it's also found in Europe, North America, sub-Saharan Africa, and the Middle East. Currently, three species are recognized from North Africa: *C. securiforme*, *C. sinaiticum*, and *C. segne* (Ayoub-Hannaa 2011; Benyoucef et al. 2019).

Choffaticeras securiforme is identifiable by its large size, with a shell 114.1 mm in diameter. It also has a sharply pointed venter (the outer edge of the shell) and spiral-shaped keel. *Choffaticeras segne* is characterized by its broad umbilical shoulder (which constitutes the widest part of the shell) and convex flanks (Ayoub-Hannaa 2011). The *C. sinaiticum* specimens from Algeria have not yet been described (Benyoucef et al. 2019).

Calycoceras

Calycoceras (fig. 3.8) is a difficult genus to pin down as it's historically been confused with *Mantelliceras* (Barroso-Barcenilla 2004). This confusion stems from the poor description of the holotype by Mantell (1822); when directly compared, *Calycoceras* and *Mantelliceras* have a quite distinct ornamentation (Wright and Kennedy 1984).

A small genus, *Calycoceras* has very pronounced ribbing on its shell with subspinose tubercles (Kennedy and Klinger 2010). The North African species, *C. guerangeri*, can be identified by the size and shape of the shell's ribbing, as this trait varies between *Calycoceras* species

Fig. 3.8. A reconstruction of *Calycoceras*. Artwork by Andrey Atuchin.

Fig. 3.9. A reconstruction of *Euomphaloceras*. Artwork by Andrey Atuchin.

Fig. 3.10. A specimen of *Euomphaloceras costatum*. Image from Gale et al. (2005, fig. 7E).

(Kennedy et al. 2013). Another species, *C. naviculare*, was identified from Algeria, but no description was ever given (Benyoucef et al. 2019)

Euomphaloceras

Euomphaloceras (fig. 3.9) is the type genus for the subfamily Euomphaloceratinae. The North African species, *E. septemseriatum*, has very fine ribbing on the shell. *Euomphaloceras septemseriatum* also has three rows of tuberculation (Wilmsen and Nagm 2013). Nagm and colleagues (2010) also note that *E. septemseriatum* has a depressed whorl and a weak keel, ornamented with tubercles (Kennedy and Juignet 1981).

A second species, *E. costatum* (fig. 3.10), is also known from Egypt. This species has a shell 64.8 mm in diameter and is characterized by the quadratic whorl and rounder, larger umbilicus. The shell is also flattened, with large tubercles on the main ribbing (Nagm and Wilmsen 2012).

Burroceras

Burroceras (fig. 3.11) is another obscure North African taxon whose existing specimens are poorly preserved and described. The native species, *B.*

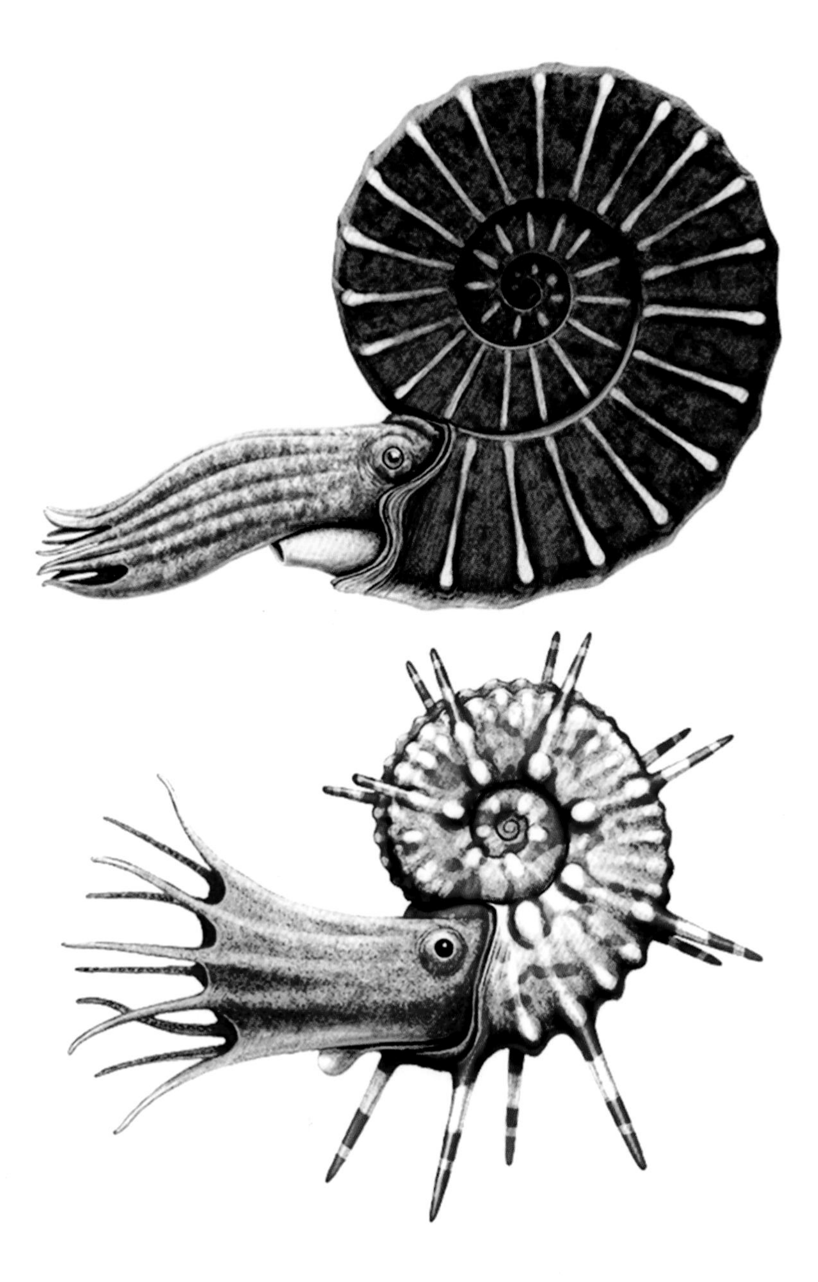

Fig. 3.11. A reconstruction of *Burroceras*. Artwork by Andrey Atuchin.

Fig. 3.12. A reconstruction of *Pseudaspidoceras*. Artwork by Andrey Atuchin.

irregulare, can be identified by the large tubercles and random striations on the shell that fade toward its living chamber (Meister and Abdallah 2005). Regardless of its taxonomic affinities, this species would have been one of the last of the Cenomanian ammonites, as it's found only toward the end of this age (Meister and Abdallah 2005).

Pseudaspidoceras

Pseudaspidoceras (fig. 3.12) was another widespread genus with species found in Asia, Africa, Europe, and the Americas (Kirkland 1996). *Pseudaspidoceras* was also a successful genus, with three or four species found in this region. These species were long-lived as well as widespread, and they are found in both the Cenomanian and Turonian ages.

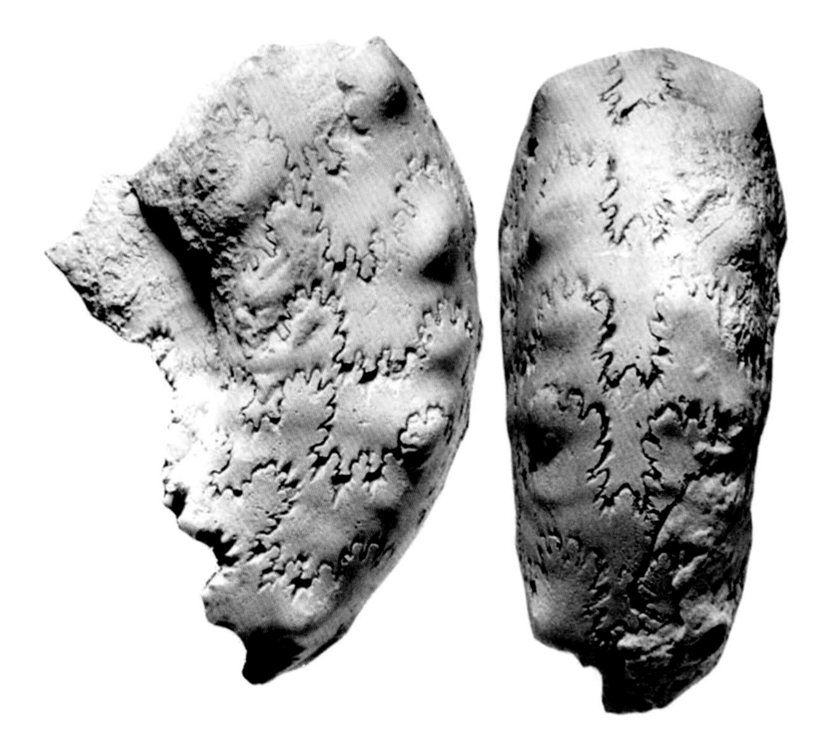

Fig. 3.13. A partial specimen of *Pseudaspidoceras flexuosum*. The spines have broken off, but their bases can still be seen. Image from Gale et al. (2005, fig. 7G).

Pseudaspidoceras barberi is the first species of the genus to appear in North Africa, and it remained rare (Meister and Abdallah 2005). *Pseudaspidoceras pseudonodosoides* only appeared at the end of the Cenomanian. Another species, *P. paganum*, is known from poorly preserved specimens and may well be a synonym of one of the other species, although Meister and Abdallah (2005) consider it valid on the basis of the current evidence.

The spines covering the shell make this genus stand out. The larger spines are usually found on the body chamber, with fewer spines on the hind chamber (Ifrim 2013). These would have rendered *Pseudaspidoceras* less streamlined and would have hindered its swimming (Teichert 1967; Jacobs and Chamberlain 1996). Such spines must have provided a real advantage to make this trait worth evolving, given the cost.

The most obvious explanation is that they were for defense. Indeed, any predator capable of biting through the shell would have found it difficult to do so without getting stabbed (Ifrim 2013). Another possibility is that the spines may have acted as sensors (Checa and Martin-Ramos 1989). Ifrim (2013) suggests that the last three pairs of spines seem the best suited for this. At the moment it's impossible to deduce which hypothesis is the more likely, although the spines could have served multiple purposes.

It was once thought that the spine pattern indicated sexual dimorphism in this genus (Kennedy and Cobban 1976). However, it's now known that there is no real difference between the ornamentation of male and female *Pseudaspidoceras* (Ifrim 2013), at least in *P. flexuosum* (see fig. 3.13).

Fig. 3.14. A reconstruction of *Nigericeras*. Artwork by Andrey Atuchin.

This genus preferred the pelagic zone of the open ocean, although it would have only dwelled in the more oxygenated surface waters (Ifrim 2013). *Pseudaspidoceras* also appears to have adapted to the expansion of low-oxygen waters toward the end of the Cenomanian by becoming smaller over time. Paradoxically, its population only increased under such harsh conditions, presumably because of lack of competition (Ifrim 2013).

Nigericeras

Nigericeras (fig. 3.14) is another obscure yet species-rich genus. While other species of *Nigericeras* are found in West Africa, only *N. gadeni*, *N. jaqueti*, and *N. jaqueti involutus* are found in North Africa (Meister and Abdallah 2012; Benyoucef et al. 2019).

Some North African specimens have been tentatively identified as *N.* cf. *scotti*, which is found in the US (Meister and Abdallah 2012). While it may seem odd that an American ammonite would be found in Egypt, many ammonite taxa have huge distribution ranges, and *N. scotti* is also known from Europe (Kennedy et al. 2003). The identification is tentatively accepted here.

Many of these species also show a great deal of morphological variability (Schöbel 1975; Meister et al. 1992). The anatomy also varies with age; *Nigericeras* loses the ornamentation on its whorls as it ages, but *N. gadeni*, for instance, loses its ornamentation long before *N. scotti* (Kennedy et al. 2003).

Fig. 3.15. A reconstruction of *Metoicoceras*. Artwork by Andrey Atuchin.

Metoicoceras

Metoicoceras geslinianum (fig. 3.15) is another ammonite species with a remarkable amount of physical variation, so much so that at one time many specimens were assigned to different taxa (Wilmsen and Nagm 2013). Kennedy and Juignet (1981) give the most thorough description of the various morphologies and the "transitional forms" between them. Adults grew to sizes of between 195 and 245 mm, and the species also shows sexual dimorphism (Wright and Kennedy 1981; Cobban and Kennedy 1990).

The *M. geslinianum* specimens from Egypt are almost identical to the *M. geslinianum* lectotype (a specimen chosen to serve as the type specimen when no holotype has been designated) of Kennedy and Juignet (1981). The only real difference is that the members of the Egyptian population have a larger whorl section and poorer ornamentation (Aly and Abdel-Gawad 2001). *Metoicoceras geslinianum* shells are involutedly coiled and characterized by their limited ornamentation, although the straight ribbing is reasonably well developed (Nagm et al. 2010; Wilmsen and Nagm 2013).

Despite its initial evolutionary success, *M. geslinianum* was a short-lived species. It became extinct before the end of the Cenomanian (Cobban and Kennedy 1990; Nagm 2009; Nagm et al. 2010).

Vascoceras

Of the numerous species of *Vascoceras* (fig. 3.16), *V. cauvini* is perhaps the best known. With a shell 98.0 mm in diameter, this species is characterized by its tall, compressed whorl and small ribs (Nagm et al. 2010; Nagm and Wilmsen 2012). As with many other ammonites, we continue to see a great deal of variability in *Vascoceras*. Two different morphotypes are known for *V. cauvini*: smooth shelled and ornamented shelled (Chudeau

Fig. 3.16. A reconstruction of *Vascoceras*. Artwork by Andrey Atuchin.

1909). Transitional forms between these two morphs are known from Egypt (Nagm and Wilmsen 2012).

The *V. cauvini* population from Egypt also differs from those of Niger and Nigeria in both the ornamentation and whorl shape (Schöbel 1975; Meister et al. 1992; Zaborski 1996). Nagm and colleagues (2010) point out that the Egyptian population, characterized by their larger size and smaller ornamentation, seem more similar to those from Israel than those from other parts of Africa. Two more species, *V. rumeaui* from Egypt and *V. gamai* from Algeria, have also been identified (Benyoucef et al. 2019).

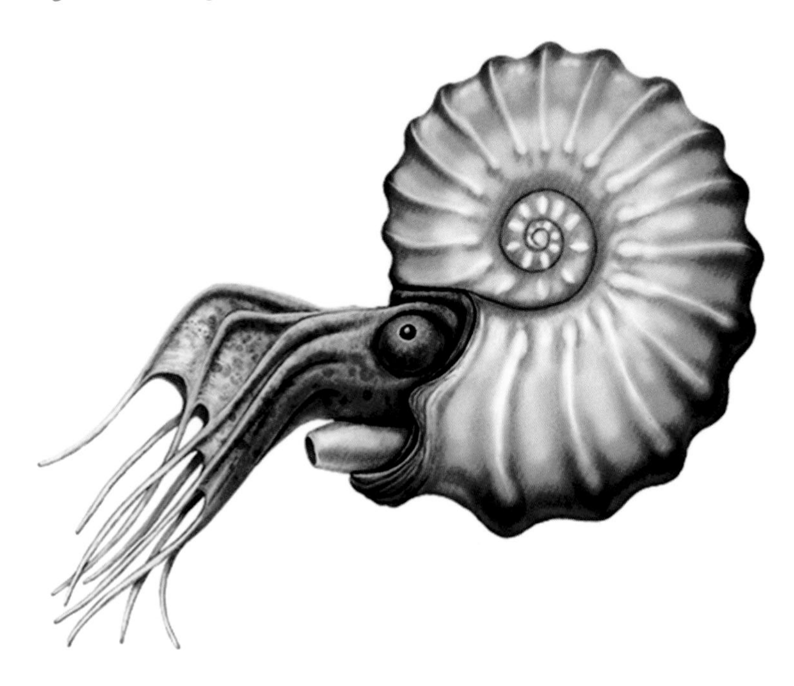

Fig. 3.17. A reconstruction of *Thomelites*. Artwork by Andrey Atuchin.

Thomelites

Thomelites (fig. 3.17) was an ammonite that achieved a nearly global distribution by the Cenomanian. *Thomelites sornayi* has a compressed whorl, and the shell's ribs form either in pairs or singularly (Wilmsen and Nagm 2013). The ribbing covers an estimated 15% of the whorl (Kennedy et al. 2003). The Tunisia specimens of *T. sornayi* are slightly different from those from Egypt, bearing a coarser ornamentation (Meister and Abdallah 2005). However, this coarser ornamentation is not considered enough to erect a new species.

A potential second species, *T. numidicus*, is also known from Tunisia. This species has an exceptionally derived morphology, characterized by compressed and involute whorls with limited ornamentation (Meister and Abdallah 2005). It's possible that *numidicus* warrants its own genus, but Meister and Abdallah (2005) consider it unwarranted.

Fig. 3.18. A reconstruction of *Metengonoceras*. Artwork by Andrey Atuchin.

Metengonoceras

Metengonoceras (fig. 3.18) is a genus best known from North America, although remains have been found in Tunisia, Morocco, Niger, Egypt, and the Middle East. Possible *Metengonoceras* specimens have also been found in Russia. While *Metengonoceras* is usually considered a small-to medium-sized taxon, the North Africa specimens had shells reaching at least 40 mm in diameter (Aly and Abdel-Gawad 2001; Aly et al. 2008).

Assigning what limited material we have to a specific species is difficult because of the fragmentary nature of the remains. However, *M. dumbli* seems a plausible candidate (Meister and Abdallah 2005; Aly et al. 2008; Pandey et al. 2011). For ornamentation, *M. dumbli* has only a series of small striations, although a few specimens show faint ribbing. The sides of the whorls are round but very narrow, giving a flattened appearance (Cobban 1987; Meister and Abdallah 2005).

Forbesiceras

Forbesiceras, known for its smooth shell, is a reasonably large ammonite that lived during the late Cenomanian. Some ammonites from the middle Cenomanian of Algeria have been tentatively identified as *Forbesiceras* cf. *largilliertianum*. Unfortunately, these specimens still await description (Benyoucef et al. 2019).

Placenticeras

Placenticeras is a widespread genus found throughout Asia, Europe, and the Americas. Benyoucef and colleagues (2019) assign some specimens from Algeria to *Placenticeras* cf. *kaffrarium*, a taxon known from India, South Africa, and the Indian subcontinent. These specimens also await description.

Cunningtoniceras

Benyoucef and colleagues (2019) have discovered specimens they identify as *Cunningtoniceras tinrhertense*, from the middle Cenomanian of Algeria. But apart from a photograph of a specimen being excavated, they offer no description of these specimens. This is problematic because *Cunningtoniceras* is often confused with *Euomphaloceras*; the two taxa were viewed as synonymous until the 1980s, the primary distinction between them being the morphology of suture lines (Cobban et al. 1989).

Eucalycoceras

Eucalycoceras is a taxon with a near global distribution (Kennedy and Juignet 1994) and with four species named, two of which, *Eucalycoceras pentagonum* and *Eucalycoceras* sp., are known from the early to middle Cenomanian of Algeria and Tunisia (Benyoucef et al. 2019). Benyoucef and colleagues gave no reasons for this referral or description of the specimens.

Rubroceras

Rubroceras is a genus known from North America, Europe, and Arabia (Callapez et al. 2018). Benyoucef and colleagues (2019) have identified specimens of *R. burroense* from Algeria, although they gave no reasons for this referral or description of the specimens. *Rubroceras burroense* and another species, *R. alatum*, are also known from Egypt (Abdel-Gawad et al. 2004).

Rubroceras burroense has a rounded whorl and thirty ribs on each whorl. The adults also have narrow ribbing on the body chamber. *Rubroceras alatum* has more distinctive ornamentation, consisting of tubercles and broad ribbing, interspaced with secondary ribs that often form pairs (Callapez et al. 2018).

Rubroceras is found in the late Cenomanian, and it survived through to the early Turonian. It inhabited shallow water reefs, often in association with *Vascoceras*, *Fikaites*, *Pseudaspidocera*, and *Nigericeras* (Benyoucef et al. 2019).

Fikaites

Fikaites is a rare ammonite from Nigeria, with two species known, *F. laffitei* and *F. subtuberculatus* (Benyoucef et al. 2019). But Benyoucef and colleagues gave no reasons for this referral or description of the specimens. *Fikaites* is found right at the very end of the Cenomanian and inhabited shallow reefs hit by frequent storms (Benyoucef et al. 2019).

Nautiloidea

Nautiloids are sometimes confused with the ammonites because of their superficial similarities. However, the two are not closely related. Only two extant nautiloid genera survive today.

Angulithes

The North African waters were home to *Angulithes* (fig. 3.19), a taxon also known from Europe and North America (Wilmsen 2000). Most are usually left as *Angulithes* sp.; however, Tawadros (2011) specifically names A. *mermei* (fig. 3.20). Two features that can identify *Angulithes* are a tightly coiled shell and a venter that is either round or angular (Wilmsen 2000).

Angulithes is often found along with the ammonite *Neolobites* (Luger and Gröschke 1989; Meister and Rhalmi 2002; Wiese and Schulze 2005). This is odd since the European species, A. cf. *triangularis*, becomes rare when ammonites dominate the marine biota (Wilmsen 2000). I suspect that the environmental conditions—storm-perturbed waters and a low-diversity biota—worked to *Angulithes*' advantage, creating a niche with little competition except for *Neolobites* (Wiese and Schulze 2005).

Angulithes would have been carnivorous, primarily scavenging although perfectly capable of hunting live prey. Modern nautiloids are deep-sea dwellers (Wells et al. 2009; Dunstan et al. 2011). However, nautiloids were more diverse in the past, as *Angulithes* has only been found

Fig. 3.19. A reconstruction of *Angulithes*. Artwork by Andrey Atuchin.

Fig. 3.20. A partial specimen of *Angulithes mermei*. Image provided by Ricardo. See image at http://www.thefossilforum.com/index.php?/gallery/image/37784-lessoniceras-angulithes-mermeti-coquand-1862/.

in shallow coastal waters (Luger and Gröschke 1989; Meister and Rhalmi 2002; Wiese and Schulze 2005).

Eutrephoceras

Benyoucef and colleagues (2019) discovered another nautiloid from Algeria, identified as *Eutrephoceras* sp., a long-lived genus found from the Jurassic all the way through to the Miocene period (Cichowolski et al. 2005). But they gave no reasons for this referral or description of the specimens. *Eutrephoceras* inhabited shallow, subtidal bivalve reefs, often in association with *Neolobites* and *Angulithes* (Benyoucef et al. 2019).

Bivalvia

Bivalvia is the group that includes the clams, mussels, and oysters. The bivalves first appeared 540–520 mya (Elicki and Gürsu 2009) and have flourished ever since. One estimate suggests that there are 9,200 species (Huber 2010), and that's not including those that have become extinct (Bouchet et al. 2010).

For much of the Cenomanian, the bivalves were the dominant marine invertebrates, composing at least half of the total benthic macrofauna in some areas of Egypt (Ayoub-Hannaa and Fürsich 2012b). The subfamilies Exogyrinae (see fig. 3.21) and Liostreinae were the most common (Ayoub-Hannaa 2011).

Of special mention is the subgroup of bivalves known as the rudists. Instead of having flat or rounded shells like most bivalves, rudists are shaped like ice cream cones. The top valve forms a lid, with the underside forming the conelike structure, which was buried in the sediment.

As reef builders, rudists were an integral part of most aquatic ecosystems (Hartshorne 1989; Johnson 2002; Benyoucef et al. 2012; Ayoub-Hannaa and Fürsich 2012b). Bivalves, in general, also act as environmental

Fig. 3.21. A specimen of *Exogyra* cf. *columba*, one of the most common bivalve species in Cenomanian North Africa. Grid squares equal 1 cm. Image provided by Thomas Magiera. See image at https://www.steinkern.de/steinkern-de-galerie/essen/exogyra-columba-lamarck-11550.html.

indicators. The changes in species diversity and abundance show that Egypt went through at least two main ecological turnovers during the Cenomanian as the reef developed (Ayoub-Hannaa 2011).

As with the other highly diverse invertebrate groups, the full species list will be given in the appendix. Otherwise, they'd take over the entire book.

Chelicerata

Despite their colloquial name, horseshoe crabs actually belong to the order Xiphosura and are more closely related to arachnids than crustaceans (Ballesteros and Sharma 2019). Horseshoe crabs are also known as living fossils: a group whose members have existed, superficially unchanged, since ancient times (Kin and Błażejowski 2014).

Mesolimulus

While two specimens of *Mesolimulus* are known from North Africa, both are poorly preserved. The lack of detail prevented their initial identification beyond their assignment to the Xiphosura (Garassino et al. 2008a). However, Lamsdell and colleagues (2019) later named these specimens *M. tafraoutensis*.

Mesolimulus tafraoutensis can be distinguished from other species of *Mesolimulus* because they have only two tubercles on the thoracetron (the large plate behind the "head"). Given the small size, less than 40 mm long, these specimens from North Africa were probably juveniles (Lamsdell et al. 2019).

Crustaceans

Crabs are common in extant mangroves, and the North Africa mangroves were no exception (Stromer 1936; Garassino et al. 2008b). The subphylum Crustacea has become a tangled mass of classes and subclasses. For

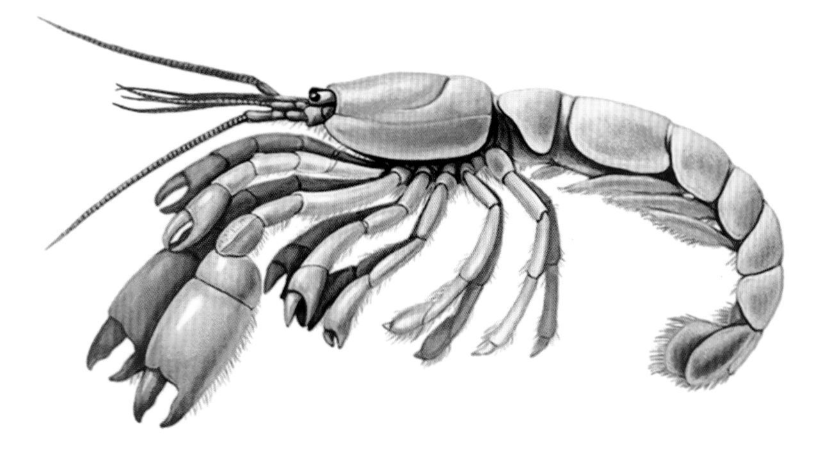

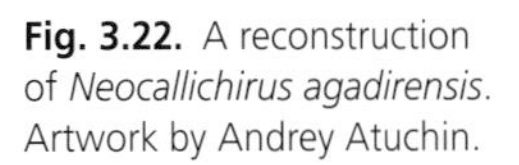

Fig. 3.22. A reconstruction of *Neocallichirus agadirensis*. Artwork by Andrey Atuchin.

the sake of clarity, we shall avoid issues of taxonomy beyond the basic groups, such as crabs, lobsters, and ostracods.

Neocallichirus

A type of ghost shrimp, *Neocallichirus agadirensis* (fig. 3.22) is known from incomplete specimens from Morocco. Although we have only the pincers to go on, the North African specimens shows numerous defining characteristics for the genus *Neocallichirus* (Garassino et al. 2011).

Neocallichirus agadirensis dwelled in the tidal zone, although related species are also found in lagoons and kelp and seagrass meadows. Shallow water appears to be the only habitat requirement for this genus. This taxon was also fossorial (burrowing) in nature and would have required sandy or muddy substrate to live in (Garassino et al. 2011).

Fig. 3.23. A reconstruction of *Cretapenaeus berberus*. Artwork by Andrey Atuchin.

Fig. 3.24. A possible specimen of *Cretapenaeus berberus*. Preservation is too poor for a precise identification, although I'm confident that it belongs to that genus, if not that species. Image provided by Thomas Bastelberge. See image at http://www.thefossilforum.com/index.php?/gallery/image/28758-cretapenaeus-berberus-garassino-pasini-dutheil-2006/.

Cretapenaeus

Another of the Moroccan crustaceans is *Cretapenaeus berberus* (fig. 3.23), a type of prawn (Garassino et al. 2006; Garassino et al. 2008b). The current specimens are not preserved in enough detail to allow a detailed description (see fig. 3.24), but we can tell that it had a subrectangular carapace. Its large eyes were held atop surprisingly short eyestalks (Garassino et al. 2006).

Unlike other prawns, *C. berberus* appears to have had a broad ecological range, inhabiting freshwater rivers and lakes as well as coastal coral reefs. The large eyes also suggest that it probably was nocturnal (Garassino et al. 2008b).

Glyphea

Morocco was also home to *Glyphea garasbaaensis* (fig. 3.25), a type of crustacean closely related to the lobsters and crayfish (Schram and Ahyong 2002, Garassino et al. 2007; Garassino et al. 2008b). The Glypheoidea was considered extinct at one time until, much as with the coelacanth,

Fig. 3.25. A reconstruction of *Glyphea garasbaaensis*. Artwork by Andrey Atuchin.

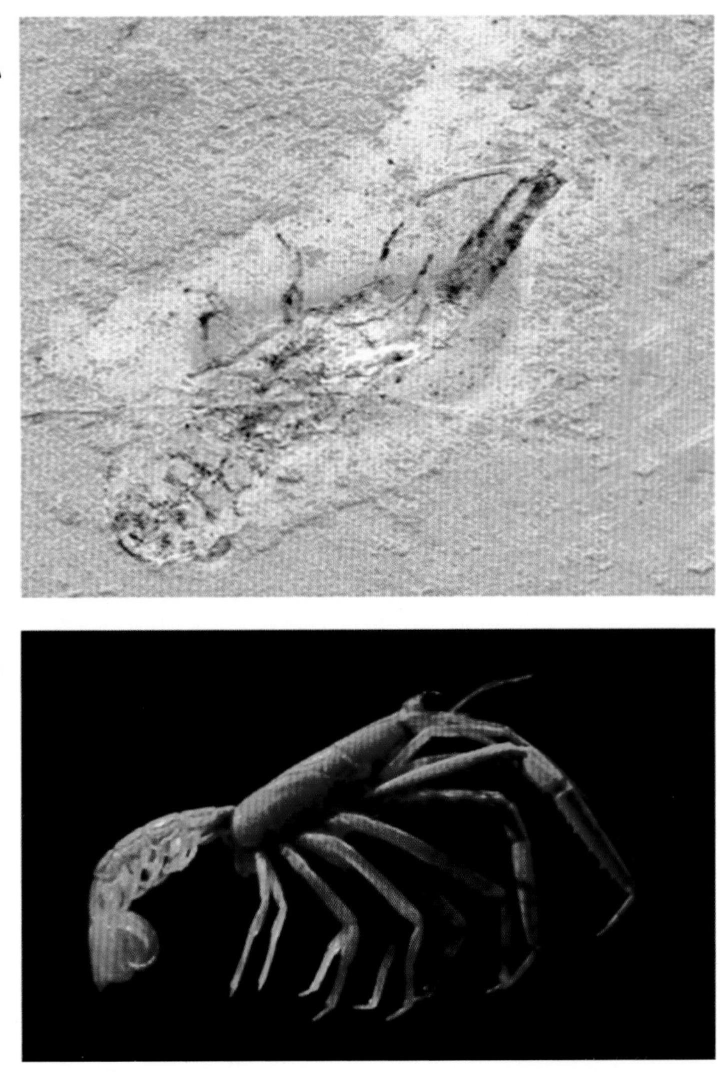

Fig. 3.26. *A*, a specimen of *Glyphea garasbaaensis*; and *B*, an extant species of *Glypheoidea* from the Timor Sea. Image *A* provided by Thomas Bastelberge. See image at http://www.thefossilforum.com/index.php?/gallery/image/27557-glyphea-garasbaaensis-garassino-de-angeli-pasini-2008/. Image B provided by Citron / CC-BY-SA-3.0. See Wikimedia Commons.

a modern species was discovered in the Philippines (Forest and de Saint Laurent 1975). Another species was discovered in New Caledonia in 2005 (Richer de Forges 2006). For a dead superfamily, the Glypheoidea seems to be doing all right for itself.

If we use the modern glypheoid species as a comparison (see fig. 3.26), *G. garasbaaensis*, an estimated 10–50 mm long, was probably active during the day (Forest and de Saint Laurent 1989). The extant species shows marked sexual dimorphism, both physically and behaviorally (Forest and de Saint Laurent 1989). The same may have been true of *G. garasbaaensis*, although we will need more specimens to prove this.

Muelleristhes

One of the crabs with a complicated history is *Muelleristhes africana* (fig. 3.27). This species was initially assigned to the genus *Paragalathea* in 2008 (Garassino et al. 2008a). Later that same year, Garassino and colleagues

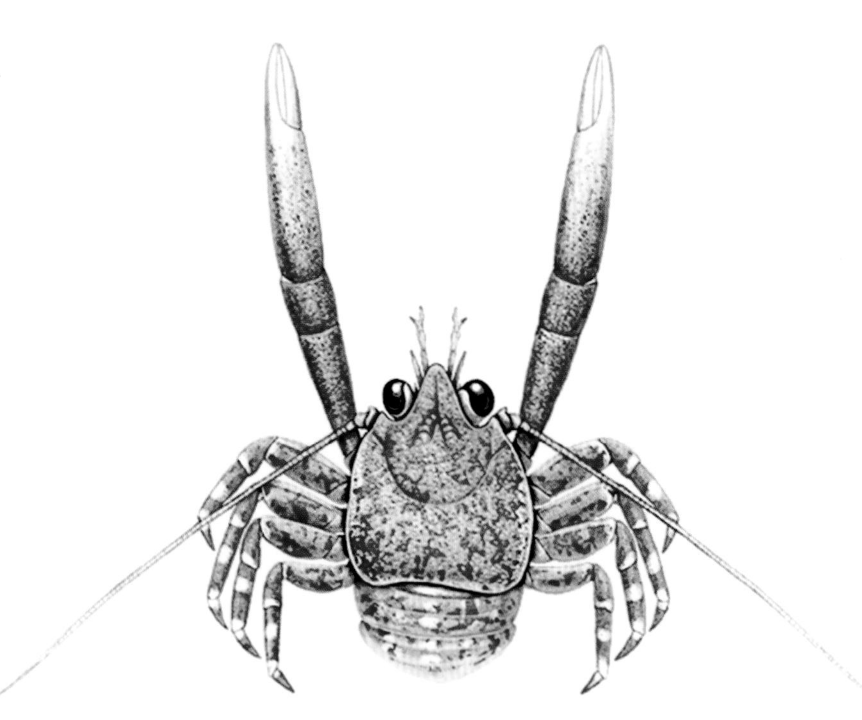

Fig. 3.27. A reconstruction of *Muelleristhes africana*. Artwork by Andrey Atuchin.

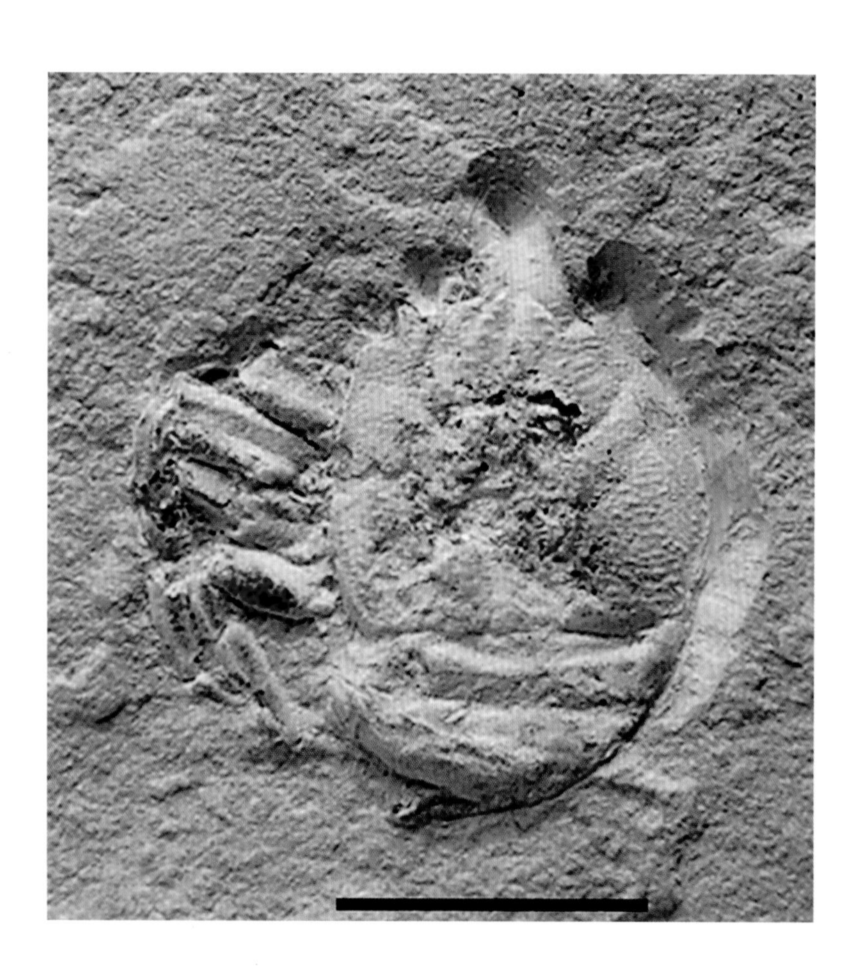

Fig. 3.28. A specimen of *Muelleristhes africana*. The scale bar is 5 mm. Image by ScriptaGeologica / CC-BY-SA-3.0. Image matches the holotype specimen published in *Scripta Geologica* in 2014.

(2008b) then decided that *africana* was distinct enough to be given its own genus, *Muelleristhes*.

A type of squat lobster (Ahyong et al. 2010), *M. africana* itself is characterized by its small size, about 7 mm long (see fig. 3.28). The carapace is rectangular and is wider than it is long. The legs are large and strong, but the pincers are not preserved (Garassino et al. 2008b).

Extant species are remarkably territorial (Denny and Gaines 2007). Whether or not the same was true for *M. africana* is impossible to say, as the claw-bearing limbs are not preserved. Like other members of the Porcellanidae, *M. africana* would have been a de facto filter feeder, straining organic particles from the water.

Galathea

Another of the squat lobsters is *Galathea sahariana* (fig. 3.29). Another genus with living examples, *Galathea* has at least forty-six extant species (Benedict 1902; Garassino et al. 2007; Garassino et al. 2008b).

Diagnostic characteristics for *G. sahariana* include the short but wide carapace, 3.8 mm long (see fig. 3.30). The lateral and median spines are not preserved on most specimens, but their presence is confirmed by one specimen, the holotype, which has partial spines on its left margin (Garassino et al. 2008b).

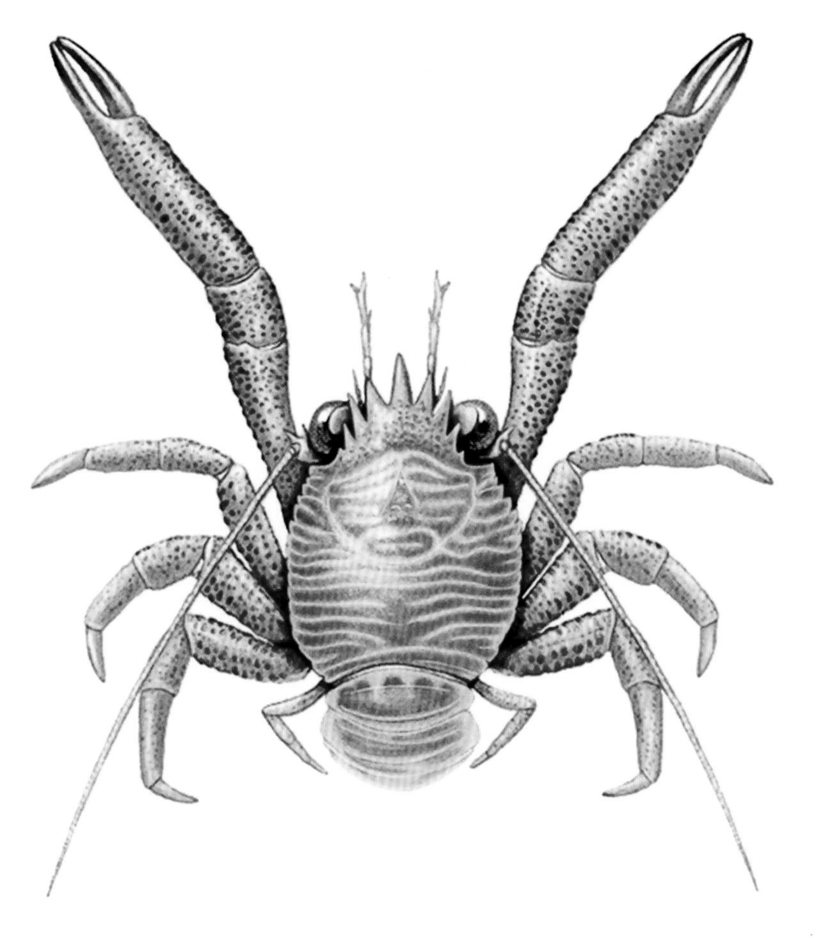

Fig. 3.29. A reconstruction of *Galathea sahariana*. Artwork by Andrey Atuchin.

Fig. 3.30. A specimen of *Galathea sahariana*. Image provided by Thomas Bastelberge. See image at http://www.thefossilforum.com/index.php?/gallery/image/34653-galathea-sahariana-garassino-de-angeli-pasini-2007/.

While the pincers are average in size, G. *sahariana* has powerful and elongated first limbs. In this regard, it resembles the extant G. *intermedia*. This suggests that G. *sahariana* might have been an opportunistic detritivore, feeding on any material it came across (Nicol 1932; Hoyoux et al. 2012).

Cretagalathea

Another member of the Galatheoidea was *Cretagalathea exigua* (fig. 3.31). Beyond that, the taxonomy is uncertain. Garassino and colleagues (2008b) originally cataloged this genus as a galatheid, but Poore and colleagues (2011) consider C. *exigua* to be a member of the Munididae. From an ecological perspective, the Galatheidae and Munididae, despite both being part of the superfamily Galatheoidea, occupy very different

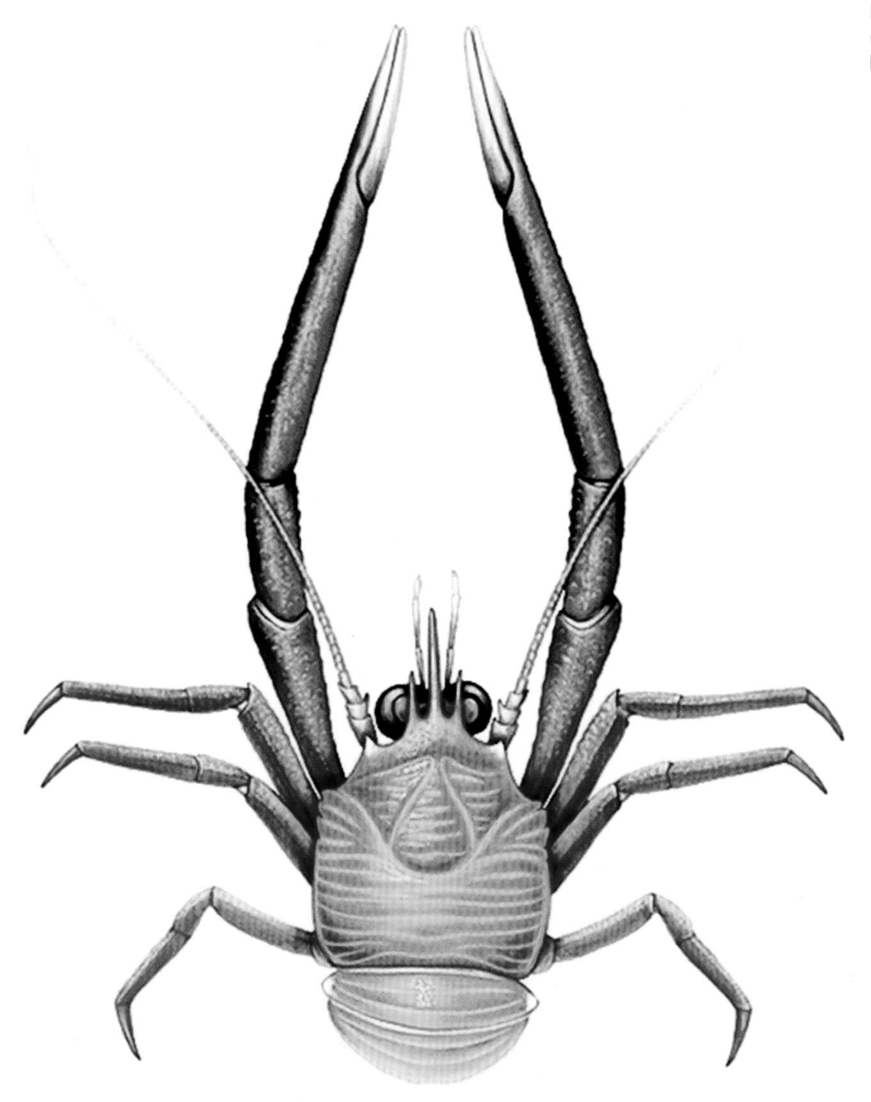

Fig. 3.31. A reconstruction of *Cretagalathea exigua*. Artwork by Andrey Atuchin.

niches. The majority of munidids inhabit deep water, while the galatheids prefer shallow water.

The morphology of the carapace and legs, as described by Garassino and colleagues (2008b), do match those of the Munididae. *Cretagalathea exigua* is 4.7 mm long and has small frontal spines above the eyes (Garassino et al. 2008b), which are a defining characteristic for the Munididae (Ahyong et al. 2010). However, it will require a re-description of the current specimens to be sure.

Corazzatocarcinus

Morocco was also home to *Corazzatocarcinus* cf. *hadjoulae* (fig. 3.32). Known from two incomplete specimens, the Moroccan form shows remarkable similarities to *C. hadjoulae* from Lebanon (Roger 1946; Larghi

Fig. 3.32. A reconstruction of *Corazzatocarcinus*. Artwork by Andrey Atuchin.

2004). The specimens are therefore referred to that species with some reservation (Garassino et al. 2008b).

Poorly known, the Moroccan specimens are noted for their spines along the delicate limbs. Other features, such as the partially complete sternum and carapace, were not described in detail (Garassino et al. 2008b). Beyond this, nothing much has been published on the Moroccan variant of *Corazzatocarcinus*.

Telamonocarcinus

Another of those widespread taxa is *Telamonocarcinus* (fig. 3.33), first recorded from the Cenomanian of Lebanon (Larghi 2004). For the moment, Garassino and colleagues (2008b) have cautiously assigned the Moroccan taxon to *T. gambalatus*, the same species found in the Middle East.

While a detailed description is still pending, *T.* cf. *gambalatus* is characterized by a small carapace with a pronounced dorsal region and large orbits. The most obvious feature is the two pairs of limbs that are longer

Fig. 3.33. A reconstruction of *Telamonocarcinus*. Artwork by Andrey Atuchin.

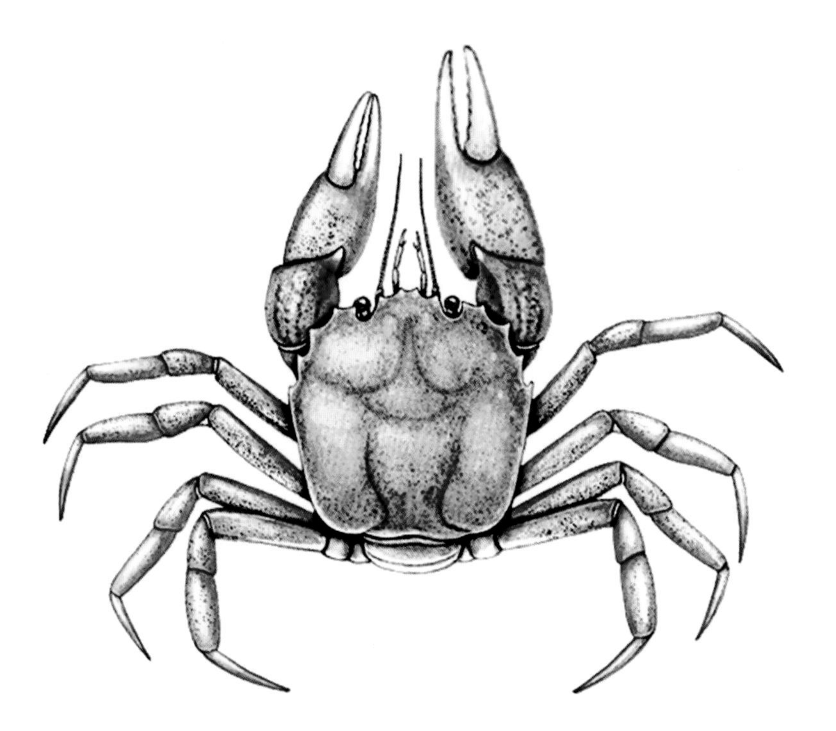

Fig. 3.34. A reconstruction of *Marocarcinus pasinii*. Artwork by Andrey Atuchin.

than its own body. The remaining pair are shrunken and were probably useless for walking. The reason for this adaptation is not readily apparent (Garassino et al. 2008b).

Garassino and colleagues also suggest that the unnamed crab from the Bahariya Formation of Egypt could belong to this genus. While I have nothing against this identification, it will take a detailed analysis to prove. Until then, I think it best to keep the Moroccan and Egyptian taxa separate, as the only shared characteristics mentioned are that they are both small and have long legs (Garassino et al. 2008b).

Marocarcinus

As the name suggests, *Marocarcinus pasinii* (fig. 3.34) is native to Morocco. It deserves a special mention because its discovery resulted in the erection of a new family, the Marocarcinidae (Guinot et al. 2008).

Marocarcinus pasinii is characterized by a 16.4-mm carapace that's wider than it is long. This species also demonstrates heterochely (in which one pincer is larger than the other) in both males and females—a rare trait in crabs, as usually the male alone has an enlarged pincer. The reason for the enlarged pincer is a mystery, although Guinot and colleagues (2008) suggest it may have been used to acquire food or defend against predators.

Amazighopsis

Known from three complete specimens, *Amazighopsis cretacica* is a unique crustacean that also warranted the creation of a new family, the Amazighopsidae. Looking superficially like a lobster, *A. cretacica* indeed belongs to the infraorder Astacidea, a group that includes lobsters and crayfish (Garassino and Pasini 2018).

Amazighopsis cretacica can be identified by its subcylindrical carapace. But the limbs are poorly preserved. While the arms seem to be similar in size, the left chela (pincer) might be slightly larger and more robust than the right one. Such heterochely is well documented in crustaceans, but the preservation of the pincers is too poor to tell if this is a genuine feature. The pincers also have pronounced daggerlike projections along their edge for gripping (Garassino and Pasini 2018).

Unusuropode

Isopods are the group of crustaceans that includes woodlice. This group first appeared during the Carboniferous era and now has a global distribution and at least ten thousand species (Schram 1970). At least one species of isopod is native to Morocco (see fig. 3.35). Although there are many specimens, often preserved in three dimensions, there was no formal description until 2018.

Fig. 3.35. The North African isopod. Image provided by Thomas Bastelberge.

While leaving the specimens unnamed, Garassino and colleagues (2008b) did refer them to the family Sphaeromonidae. Corbacho and colleagues (2018) agreed with this identification and assigned the specimens to *Unusuropode castroi*, a taxon previously known from Brazil.

However, the line diagrams given in Corbacho and colleagues (2018) do show some differences between the *U. castroi* from Brazil and the *Unusuropode* from Morocco. But it's unclear whether these differences are taxonomically important—a situation made worse by the apparent existence of some degree of sexual dimorphism in this taxon (Corbacho et al. 2018). Also, Corbacho and his team gave only a very brief description. Hopefully, a full description will be forthcoming.

Crustacea *Incertae Sedis*

Tanaidacea

Garassino and colleagues (2008b) also note the presence of tanaids in this region. Tanaids are shrimplike crustaceans noted for their small size, with most species being a mere 2–5 mm in length. No formal description of these specimens has been given.

Ostracoda

The ostracods are arguably the most diverse of all the crustaceans, with a total of seventy thousand known species (Brusca and Brusca 2003). They stand out because they have a shell with two valves and superficially look like clams.

North Africa had a diverse ostracod assemblage (Ismail and Soliman 1997; Gebhardt 1999). Some of the best come from Egypt's Raha Formation. These are especially important because of the environmental data they provide for the west central Sinai.

The dominance of cytherellids, particularly the genus *Cytherella*, suggests the development of intertidal reefs and large lagoons in this region, 10 to 50 meters deep, with increasingly dysoxic waters of average salinity (Lézin et al. 2012). The extensive list of known species is given in the appendix.

Nephropidae

Stromer (1936) also mentioned lobster remains from Egypt but gave no further details. As clawed lobsters first appeared in the Valanginian age (Tshudy et al. 2005), the presence of such creatures in the Cenomanian of North Africa is entirely possible.

Decapoda

SPECIMEN: CGM 81126 THROUGH CGM 81128

The crabs from Bahariya have yet to be named. Because of poor preservation, their identification was limited to their initial placement in the Necrocarcinidae (Schweitzer et al. 2003). However, a recent review of the Necrocarcinoidea concluded that the Bahariyan crabs are closer to the family Tepexicarcinidae (Luque 2015; Schweitzer et al. 2016).

Fig. 3.36. An example of *Thalassinoides* burrows from Israel. Near identical specimens are known from Morocco and Egypt. Note the huge size and large number of burrows. Image provided by Mark Wilson. See Wikimedia Commons.

With a body averaging 7.4 mm long, these crabs have long appendages, with the second and third limbs being the longest and the fourth and fifth being short—a classic necrocarcinid trait. Indeed, the second and third appendages are so long—25.2 mm being the maximum length—it is considered a diagnostic trait for this species (Schweitzer et al. 2003). According to phylogenetic bracketing, this species may be an omnivore.

Crustacean Trace Fossils

Crab burrows are extremely common in this region. In Morocco, crab burrows, 40 mm in diameter on average, can cover 100% of the surface of some layers. That's roughly 690–760 burrows/m^2 (Ibrahim et al. 2014a). These trace fossils are called *Thalassinoides* (see fig. 3.36). While other groups can produce similarly shaped burrows, the majority are produced by crustaceans (Myrow 1995).

The burrows' shapes vary in cross section from subcircular to asymmetrical ovoids. This difference may be the result of sexual dimorphism for this species of crab (Ibrahim et al. 2014a). Ibrahim and colleagues (2014a) also mention that a species of crab that's the right size to produce these burrows is known from the same area, but they don't give any further details.

The situation is the same in Egypt, where the paleosols also show signs of extensive burrowing (Nothdurft et al. 2002; Schweitzer et al. 2003; Tanner and Khalifa 2009). *Thalassinoides* also occur, alongside other crustacean burrows identified as *Ophiomorpha* and *Diplocraterion*, in Algeria (Benyoucef et al. 2019), proving that North Africa, in general, had a high crab population and species diversity.

Insects

Insects are arguably the most successful group of animals on the planet, composing more than half of all living organisms (Erwin 1997; Chapman 2006). However, you wouldn't know it from looking at the fossil record from North Africa, which has no named species yet. However, this is no doubt due to preservational bias; there were certainly more insect species present than we currently have evidence for.

Chrysidoidea

One of the best known is a hymenopteran insect discovered by Martill and colleagues (2011). This specimen is easily recognizable as a wasp and has even been tentatively identified as belonging to the Chrysidoidea, a member of either the Bethylidae or Amiseginae (Heads, personal communication in Martill et al. 2011).

This identification would add a touch of color to North Africa, if accurate, as the Chrysidoidea are generally beautiful creatures, possessing bright, metallic coloration (as reconstructed on plate 3). Alas, looks are deceiving. The chrysidoids are mostly parasitic; they lay their eggs in the nests of other insects, and their larvae kill the unwitting host's real eggs and young (Agnoli and Rosa 2022).

I'm surprised that this specimen has not been named or described in greater detail over the subsequent years. It's preserved in enough detail to identify not only its family but possibly also its subfamily. Hopefully, a more detailed description will be forthcoming.

Homoptera

Another insect specimen was mentioned in passing by Garassino and colleagues (2008a). But they chose not to describe this specimen, perhaps because of its poor preservation. They assigned this specimen to the Homoptera but gave no explanation for the assignment.

The problem becomes worse because the Homoptera is now considered to be paraphyletic, which means that its members don't share a common ancestor and therefore don't constitute a real group (Dohlen and Moran 1995; Gullan 1999). Thus, it's impossible to say where this poorly known specimen belongs.

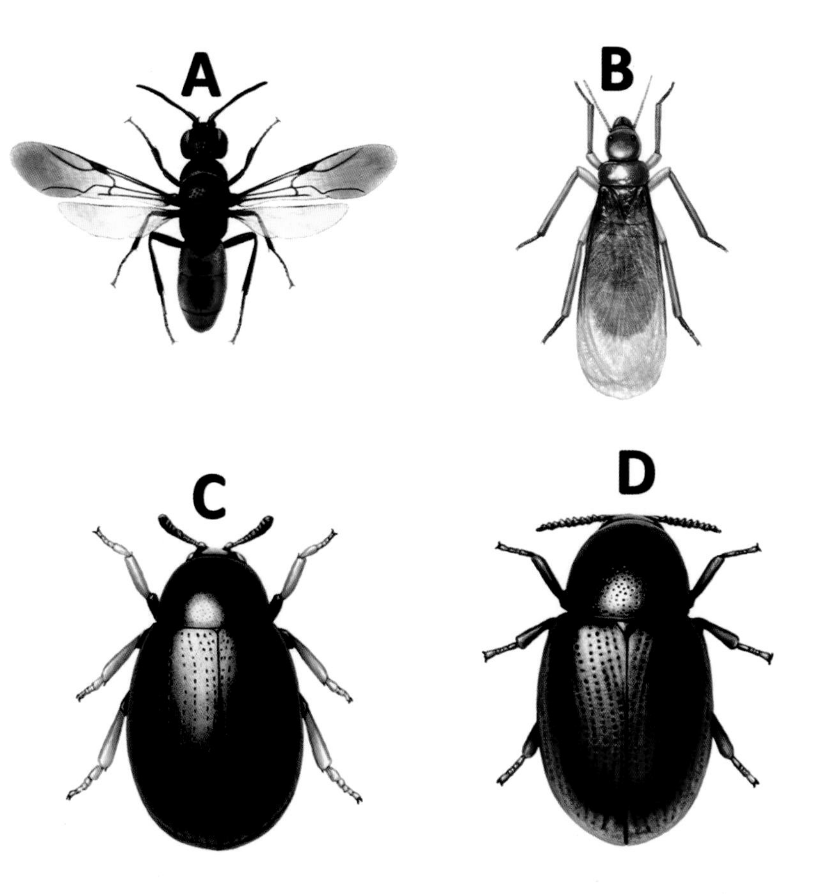

Plate 3. The insects of North Africa. *A*, the unnamed Chrysididae; *B*, the unnamed Isopteran; *C*, Specimen P1603; and *D*, Specimen P1604. Images not to scale. Artwork by Andrey Atuchin.

Coleoptera

SPECIMEN P1603

This ecosystem was also home to beetles. One specimen is preserved upended on its back. This was a hefty species, 9.1 mm long and 4.2 mm wide, with a prothorax (the segment of the thorax directly behind the head and before the wings) that's broader than its head (Engel et al. 2012). Interestingly, the pronotum (a shieldlike structure that covers the thorax) did not cover the head itself, which makes it distinct from P1604 (Engel et al. 2012).

This species has been tentatively identified as a member of the Scarabaeiformia, which includes the scarab beetles (as reconstructed on plate 3). But a lack of information about the antennae and other features makes it impossible to identify its family with certainty (Engel et al. 2012).

SPECIMEN P1604

This beetle species was 6.4 mm long, smaller than the previous species. The pronotum was broad and also covered the head like a cowl. The mouthparts on the head itself point downward. This is a trait seen in mostly herbivorous species (Engel et al. 2012).

This specimen also has distinct pits on the pronotum, which may be a defining characteristic for the species (as reconstructed on plate 3). These pits are small on the pronotum, but the elytra (the structure that covers the back and wings) has ten or eleven large pits in symmetrical rows (Engel et al. 2012).

Engel and colleagues (2012) have provisionally assigned this species to the Bostrichiformia, the group that contains the fungus, furniture, and skin beetles.

Isoptera?

The most exciting insect from North Africa is the possible termite described by Engel and colleagues (2012). At 23.1 mm in length, the specimen does not have well-preserved head, antennae, or mouthparts. The preservation of the lower legs is also poor, which makes definitive identification difficult. However, some features, such as the shape and size of the pronotum, the structure of the wings, and the poorly preserved orthopteroid-style mouthparts, suggest that this could be a termite (Engel et al. 2012), and it has been reconstructed as such on plate 3.

Termites are believed to have first evolved sometime during the Jurassic period (Xing et al. 2013; Vrsanky and Aristov 2014). Generally herbivorous, termites are essential in modern ecosystems as detritivores. Xing and colleagues (2013) have shown that termites were capable of recycling carrion by the Jurassic and wood by the Cretaceous (Poinar 2009; Colin et al. 2011). Thus, if positively identified as a termite, this species could be of great importance in understanding North African ecology.

Odonata

The Odonata is a group of carnivorous insects that includes dragonflies and damselflies (Lohmann 1996; Rehn 2003). While no adult specimens are known, odonatan larva are known from Morocco (Dutheil 2009).

Insect Trace Fossils

While insects themselves may be rare, their handiwork can be clearly seen. Engel and colleagues (2012) discovered a leaf from Morocco that shows signs of damage caused by insect feeding. These traces vary in size, ranging from 0.1 to 1.6 mm in depth, and occur along the entire leaf. The sides of the leaf are badly gnawed, but the veins are untouched (Engel et al. 2012).

Krassilov and Bacchia (2013) also report damaged leaves. These show signs of an endoparasitic insect, a species that lived within the leaf while it fed. The specimen shows long, looping "mine tracks" that lead to the leaf's surface, forming what may be an entrance or exit hole (Krassilov and Bacchia 2013). Lyon (2001) and Darwish and Attia (2007) also report plant specimens from Bahariya suffering extensive insect damage, although the details have never been published.

Fig. 3.37. A dinosaur bone from Morocco showing clear signs of insect borings. Scale bar is 50 mm (5 cm). Image from Ibrahim et al. (2014a).

Far more sinister is the presence of insect traces in dinosaur bone (see fig. 3.37). Most of these burrows are simple tubes, usually 2.5–5.0 mm in diameter, although some form U-shaped bends (Ibrahim et al. 2014a). These prove the existence of species that feed on the flesh of dead animals or directly on the bones themselves (Goff 1993). Recent work suggests that these were made by either termites or beetle larvae (Shears-Ozeki 2017).

Fig. 4.1. A reconstruction of *Onchopristis numidus*. Artwork by Joschua Knüppe.

A

Fig. 4.2. Once believed to be a true sawfish, *Onchopristis* is actually a skate; the similarities are a result of convergent evolution. When seen from above, *A*, the pectoral fins, are a giveaway as they extend almost the length of the body and join with the head, giving the impression that the skate lacks a neck. *B*, what was once believed to be the rostral tooth of *Peyeria lybica* but is now identified as an *Onchopristis* spine. Artwork by Joschua Knüppe and photo provided by Jean-François Lhomme.

B

The Fauna of North Africa: Vertebrates (Fish)

4

The most conspicuous faunal components of most ecosystems are the vertebrates. Traditionally including all animals with a backbone, this group contains the fish, reptiles, mammals, birds, and amphibians.

Fishes make up the largest group of vertebrates in the Cenomanian age of North Africa, with some species growing to enormous sizes. Because fish evolutionary history is complicated (Lecointre and Le Guyader 2007), we will stick to the basic divisions used by most researchers. The Chondrichthyes (fish with a cartilage skeleton, like sharks) and the Osteichthyes (fish with a bony skeleton, like tuna).

Chondrichthyes

Rays

ONCHOPRISTIS

Rays, or batoids, are the most diverse of all the cartilaginous fishes, typified by the well-known manta ray and stingrays. While many Cenomanian taxa are poorly known, *Onchopristis numida* (fig. 4.1) is the only fish that could challenge *Mawsonia* for the role of most iconic Cretaceous African fish species. It was originally called *Gigantichthys* (Haug 1905), but the genus name had already been used, so Stromer (1917) changed it to *Onchopristis*.

A giant, sawfish-like skate (see fig. 4.2), *O. numida* was the most common batoid taxa in North Africa. It has been placed within the Sclerorhynchoidei, but some researchers also give it its own family, the Onchopristidae (Villalobos-Segura et al. 2021). Like other sawfish, *O. numida* is known primarily from its rostrum, an elongated structure with a sharp row of rostral teeth along each side. These structures are not true teeth like the ones found in the mouth for processing food (oral teeth) but are modified scales (Welten et al. 2015). In *O. numida*, the large rostral teeth are interspaced with smaller teeth, both with a distinct arrow-shaped barb on the end (Stromer 1925; Arambourg 1940; Dutheil and Brito 2009).

However, some rostral teeth have two barbs instead of one. At one time, these were thought to represent a distinct genus (Stromer 1927), but it's now known that they result from a developmental anomaly or pathological condition (Landemaine 1991; Martill and Ibrahim 2012).

The remains from Egypt are typified by a wonderfully preserved, complete rostrum and partial skull (Stromer 1925, 1936), but the holotype no longer exists. Fortunately, more specimens have been found since then. One specimen from Morocco includes the oral teeth, Meckel's cartilage, and some of the vertebrae (Dutheil 2000; Dutheil and Brito 2009).

Like modern pristiformes, *O. numida* would have been capable of surviving in both salt and fresh water (Wueringer et al. 2012). A carnivore, it would have used the rostrum not only to capture prey but also to locate them by using specially evolved nerve clusters (electroreceptors) to detect the electrical signals generated by the targets' central nervous system (Wueringer et al. 2012).

SCHIZORHIZA

First known from Egypt (Stromer and Weiler 1930), *Schizorhiza stromeri* (fig. 4.3) is a widespread taxon found throughout North Africa, the Middle East, and the Americas (Weiler 1930; Kirkland and Aguillón-Martínez 2002; Knight et al. 2007).

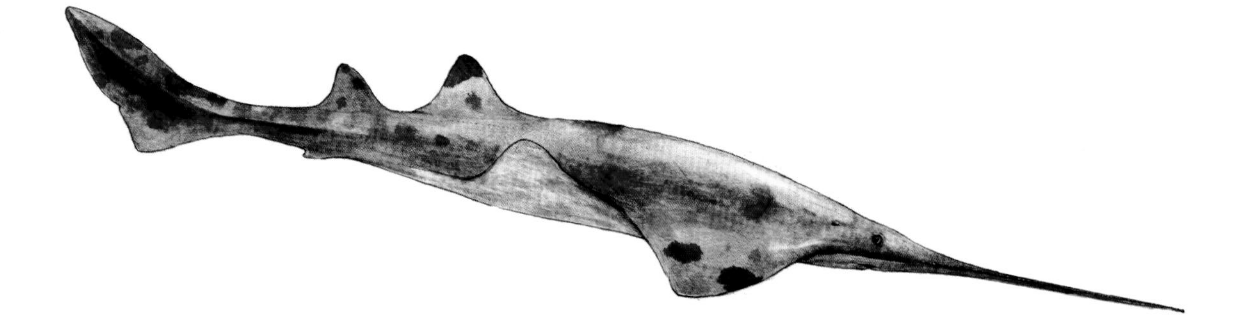

Fig. 4.3. A reconstruction of *Schizorhiza stromeri*. Artwork by Joschua Knüppe.

Schizorhiza stromeri had an estimated length of 1.5 meters (4 feet) (Kirkland and Aguillón-Martínez 2002). Its rostral teeth are shaped like equilateral triangles, with an enlarged base and a laterally compressed crown (Arambourg 1940). Its rostral teeth were not shed and replaced throughout the animal's life. Instead, the new teeth grew in underneath the older and pushed them out (Kirkland and Aguillón-Martínez 2002).

BAHARIPRISTIS

North Africa was also home to a host of other rays. Unfortunately, most of the following taxa are poorly known, and I suspect that some will prove to be junior synonyms of others should anyone review the topic. *Baharipristis basteriae*, for instance, is known only from rostral and oral teeth (Werner 1989).

MARCKGRAFIA

Marckgrafia libyca is yet another taxon named by Weiler (1935). Although it is known only from fragmentary remains, Arambourg (1940), Werner (1989, 1990), and Dutheil (2000) still consider it a valid genus.

A member of the Ganopristidae, *M. libyca* is known mainly from its rostral teeth, one measuring 20 mm long (Dutheil 2000). Two of the holotype rostral teeth from *M. libyca* survived World War II and are still kept in the Bavarian State Collection of Palaeontology (Smith, Lamanna, et al. 2006).

PEYERIA

Peyeria lybica was described on the basis of rostral (Weiler 1935) and oral teeth (Werner 1989) from Egypt. Some researchers still consider this genus to be valid (Dutheil 2000; Wueringer et al. 2009). However, the rostral teeth are now believed to be denticles from a large sclerorhynchoid, like *Onchopristis* (Sternes and Shimada 2019). So *P. lybica* is now known only from its oral teeth.

ISIDOBATUS

Isidobatus tricarinatus is a member of the Rhinobatidae (Werner 1989). Known only from oral teeth, it is considered valid by Vullo and Courville (2014). Still extant, the Rhinobatidae are commonly known as guitarfish. Unlike other rays, they are known for their flattened heads, short pectoral fins, and thick subrounded tail (Stevens and Last 1998). They also inhabit shallow tropical seas (Werner 1989).

CF. *RHINOPTERA*

cf. *Rhinoptera* is a genus known from Bahariya (Weiler 1935; Stromer 1936; Werner 1989). Twelve isolated *Rhinoptera* teeth are also among the few Bahariyan fossils to survive World War II, and they are now housed in the Bavarian State Collection of Palaeontology (Smith, Lamanna, et al. 2006).

Like many taxa in this book, it has living species for comparison (Smith 1997). Known as "cownose" rays, living species of *Rhinoptera* are migratory and oceanic, with some species known to form congregations ten thousand strong (Cervigón et al. 1992).

While capable of feeding on a range of items, *Rhinoptera* seems to favor molluscs and bivalves, although Collins and colleagues (2007) consider at least one species to be opportunistic in its diet. Their prey preference changes during ontogeny (Smith and Merriner 1985; Michael 1993).

GYMNURA

Gymnura is another taxon with living representatives. The species native to Egypt was *G. laterialata* (Werner 1989, 1990). *Gymnura* is found in shallow coastal waters, often in river mouths, estuaries, and hypersaline lagoons (Robins and Ray 1986; Cervigón et al. 1992; Smith 1997; Murch 2015). *Gymnura laterialata* is assumed to have fed mainly on bivalves and crustaceans (Murdy et al. 1997), although other members of this genus have more varied diets (Jacobsen et al. 2009).

RENPETIA

Renpetia labiicarinata is another of the poorly known rays from Egypt (Werner 1989). Known only from its oral teeth, this taxon could be a member of either the rhinobatids or the sclerorhynchids (Wueringer et al. 2009). The oral teeth once assigned to *Onchopristis* (Werner 1989) likely also belong to *Renpetia* sp. instead (Villalobos-Segura et al. 2021).

However, Slaughter and Thurmond (1974) believe *R. labiicarinata* to be a synonym of *Squatina* sp.

PTYCHOTRYGON

Ptychotrygon henkeli, named by Werner (1989), is from the family Ptychotrygonidae (Kriwer et al. 2009). It was originally known only from its bilaterally symmetrical oral teeth, characterized by triangular root lobes, with lateral foramen on each lobe (Werner 1989; McNulty and Slaughter 1972).

Sharks

SQUALICORAX

Sharks were also abundant throughout North Africa. One of these sharks was *Squalicorax* (Stromer 1927; Slaughter and Thurmond 1974; Vullo et al. 2007). Currently, eighteen species of *Squalicorax* are known, with an African species called *S. baharijensis* (reconstructed on plate 4). Like most prehistoric sharks, *Squalicorax* is known primarily from its teeth. It wasn't until relatively recently that we found articulated skeletons from three different species of *Squalicorax* to work with (Shimada and Cicimurri 2005).

Generally, *Squalicorax* teeth have a roughly rectangular root with a finely serrated tooth blade (see fig. 4.4). Usually, tooth size is used as an indicator of the overall size of a shark, but Shimada and Cicimurri (2005) point out that this calculation does not work well in the case of *Squalicorax*, as tooth size and body length vary greatly between the eighteen known species.

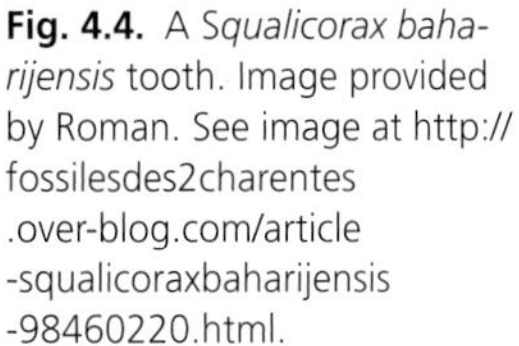

Fig. 4.4. A *Squalicorax baharijensis* tooth. Image provided by Roman. See image at http://fossilesdes2charentes.over-blog.com/article-squalicoraxbaharijensis-98460220.html.

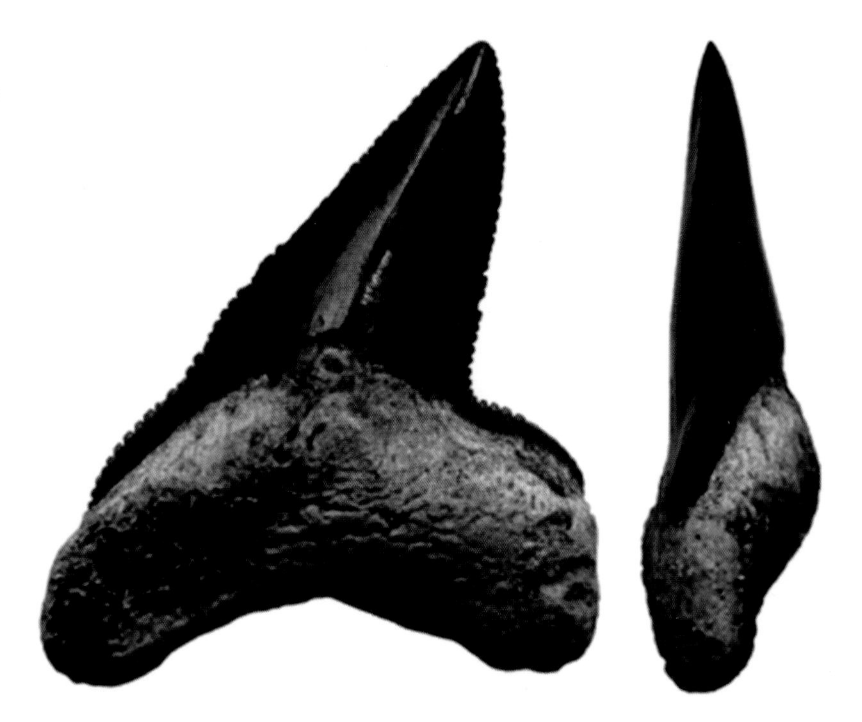

Squalicorax had poor vision compared with the other sharks it coexisted with, such as *Cretoxyrhina*. However, it did possess a powerful sense of smell (Shimada and Cicimurri, 2005). This may tie in with the extensive scavenging that *Squalicorax* is known for (Schwimmer et al. 1997), although the shark also would have hunted prey when possible (Shimada and Cicimurri 2005).

Shimada and Cicimurri (2005) note that a large attachment area for the adductor muscles in the jaw and damaged teeth are also common for this genus (Becker and Chamberlain 2012). This shows that the taxon had an aggressive feeding strategy with a bite powerful enough to damage its own teeth (because sharks are capable of constant tooth replacement, that didn't pose a problem).

CRETODUS

Cretodus is another widespread shark taxon, found thought North America, North Africa, Europe, and Asia. First discovered in Great Britain, this taxon was named *Oxyrhina* (Dixon 1850), before being renamed *Cretodus* by Sokolov (1965). A type of mackerel shark, one species, *C. longiplicatus* (Werner 1989), is known from Egypt.

The teeth are large with well-developed roots and are designed to puncture and grip struggling prey. Each tooth has at least three cusps: one large central cusp with two smaller cusps, one on either side (Shimada 2006).

Little has been published on the ecology of *C. longiplicatus*, except that this genus preferred near-shore environments (Ouroumova et al. 2016). As its fossils are almost always found with other large shark genera, suggesting niche partitioning would have been a must, although *C. longiplicatus* seems to differ from other sharks in its environmental preferences rather than in its prey preferences (Shimada et al. 2011).

CRETOLAMNA

Cretolamna one of the most successful shark genera (see fig. 4.5). The genus first evolved in the Barremian age of the Cretaceous period (130.0 mya) and only became extinct in the middle of the Miocene age of the Cenozoic (13.65 mya). *Cretolamna appendiculata* (reconstructed on plate 4), while known primarily from North America, was also found in North Africa (Stromer 1936).

The teeth of *Cretolamna* are adapted for cutting and slicing (Shimada 2007). *Cretolamna appendiculata* had a set of powerful quadratomandibularis muscles, which would have provided a powerful bite (Shimada 2007). *Cretolamna appendiculata* was a predator of large to medium-sized prey when fully grown (1 Kent 994) and was a generalist in both its diet and its habitat (Shimada 2007).

SQUATINA

Squatina is another prehistoric survivor in a modern world, but all the extant species are now critically endangered (Dumeril 1806; Lythgoe and

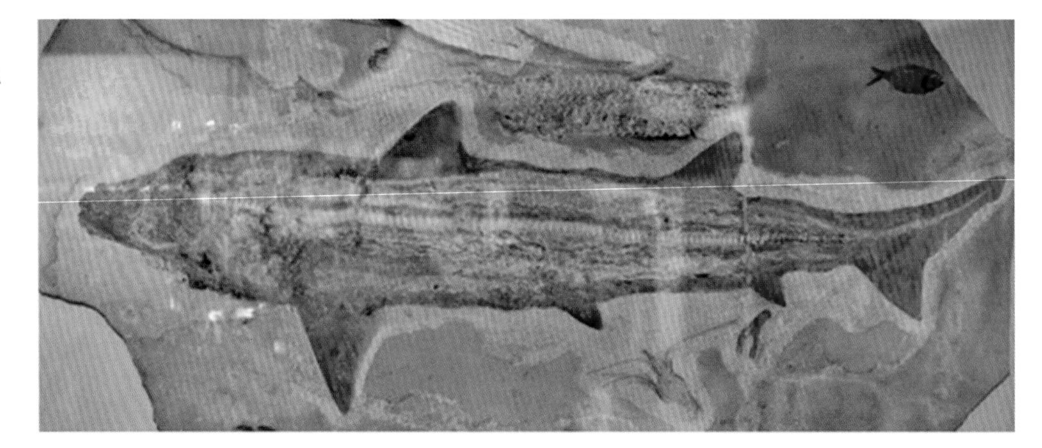

Fig. 4.5. *A*, a complete *Cretolamna appendiculata* skeleton; and *B*, a *Cretolamna* tooth. Images provided by Roman and Citron / CC-BY-SA-3.0. See *A* at http://www.wikiwand.com/en/Cretolamna; see *B* at http://fossilesdes2charentes.over-blog.com/article-cretolamnaappendiculata-98406282.

Lythgoe 1991). A species of *Squatina* is reported from the Cretaceous of Africa (Slaughter and Thurmond 1974). But the genus has become a wastebasket taxon.

Stromer (1927) named the Bahariyan species *S. aegyptiaca*, but Werner (1989) later renamed it *Sechmetia aegyptiaca*. However, these teeth likely belong to *Onchopristis* instead (Cappetta 1987; Villalobos-Segura et al. 2021), leaving it a *nomen nudum*. The name exists but has no material assigned to it. For that reason, any other specimens are left as *Squatina* sp. (Underwood and Mitchel 1999).

Squatina is noted for its broad and flattened anatomy, which is superficially similar to its distant relatives the rays. A cryptic predator, it buries itself under the seabed and remains motionless for hours before lunging out at passing prey (Compagno 1984; Murch 2015).

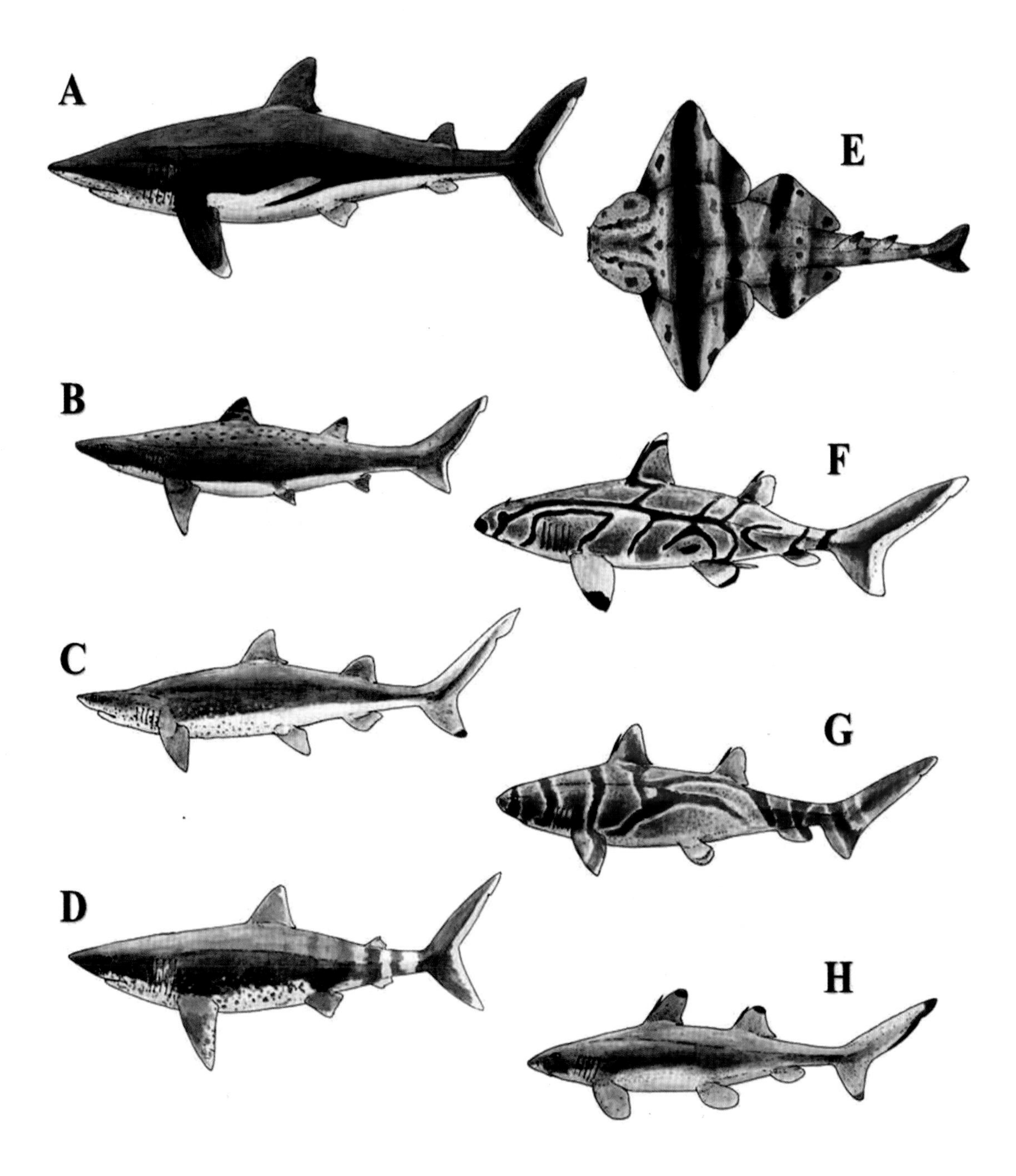

Plate 4. The sharks of North Africa. *A*, *Cretodus*; *B*, *Cretolamna*; *C*, *Haimirichia*; *D*, *Squalicorax*; *E*, *Squatina*; *F*, *Asteracanthus*; *G*, *Tribodus*; and *H*, *Steinbachodus*. Images not to scale. Artwork by Joschua Knüppe.

HAIMIRICHIA

Haimirichia amonensis (reconstructed on plate 4) has been attributed to various genera over the years: first *Odontaspis* (Cappetta and Case 1975), then *Serratolamna* (Landemaine 1991), and finally *Carcharias* (Cappetta and Case 1975). Yet it did not belong there, either, and Vullo and colleagues (2016) coined the new name *Haimirichia* after a study of a wonderfully preserved partial specimen from Morocco.

The teeth of *H. amonensis* were designed for cutting (posterior teeth) and tearing (anterior teeth). Given such tooth morphology, the shark was probably not capable of feeding on hard-shelled prey (Vullo et al. 2016).

The sediments the new specimen was found in suggest that *H. amonensis* may have tolerated both marine and brackish conditions (Vullo et al. 2016). The abundance of *Haimirichia* fossils also suggests that they may have gathered in shoals.

The new specimen also shows clear evidence of having been partially eaten by another animal, with only the front half of the body remaining. The specimen was attacked on its right flank, a blow that ripped off its right fin and may have even beheaded the animal. A lamnid shark is considered to be the most likely suspect, although it's also possible that a crocodile or dinosaur might be to blame (Vullo et al. 2016).

TRIBODUS

The hybodontids are some of the best-known Mesozoic sharks. Unfortunately, the North African taxon, *Tribodus aschersoni*, is not as well represented. Stromer (1927) and Werner (1989) both based their identification on a solitary fin spine.

Werner (1994) later realized that the teeth she once assigned to *Aegyptobatus kuehnei* belonged to a hybodont and combined them with the hybodont fin spines discovered by Stromer (1927) to create the new combination *Tribodus aschersoni*. Lane (2010) concurred with this reassignment. This means that the Egyptian taxon, *Aegyptobatus kuehnei*, is now a junior synonym of *T. aschersoni* (Werner 1994; Cappetta et al. 2006; Lane 2010).

Another species, *T. tunisiensis*, is known from the preceding Aptian/Albian of Tunisia (Cuny et al. 2004). Teeth from the Cenomanian of Tunisia match *T. tunisiensis*, but Cuny and colleagues (2007) left these specimens as *Tribodus* sp. Other *Tribodus* teeth, matching those from Bahariya, are known from Morocco (Ibrahim et al. 2020a).

Ecologically, *T. aschersoni* was a durophagous (Lane and Maisey 2012), marine species (Cuny 2012). This genus also required warm waters (Vullo and Neraudeau 2008).

ASTERACANTHUS

Another hybodont from this region was *Asteracanthus*. While some have commented on its presence, they have not offered an opinion on its validity (Werner 1990). The matter was resolved by Lane (2010), whose phylogenetic analysis confirmed the validity of this genus, as a relative of *Tribodus*. Thus *A. aegyptiacus* is accepted as valid herein.

As with *Tribodus*, the jaws were short and didn't extend past the end of the snout (Lane and Maisey 2012). Ecologically, *A. aegyptiacus* would have been similar in diet and habitat to its relative, *Tribodus*.

STEINBACHODUS

The last hybodont taxa is *Steinbachodus bartheli*. Originally placed in the genus *Lissodus* by Werner (1989), it was given its own genus by Rees and Underwood (2002). This conclusion is followed herein.

The fin spines and tooth morphology of *S. bartheli* suggest that it belongs to, or at the very least is somehow related to, either the Steinbachodontidae or the Lonchidiidae (Rees and Underwood 2002). The figures published suggest that it was similar to other species of *Steinbachodus* such as *S. estheriae* (Rees and Underwood 2002).

Osteichthyes

Actinopterygii

ADRIANAICHTHYS

The actinopterygians, or ray-finned fishes, are not only the largest group of osteichthyans but also the largest vertebrate group on the planet. The first actinopterygian we'll be looking at is *Adrianaichthys pankowskii* (fig. 4.6). The holotype specimen was purchased privately in Morocco and later donated to the British Natural History Museum by Mark Pankowski. The species was subsequently named in his honor.

Initially, Forey and colleagues (2011) placed this species in the genus *Lepidotes*. But if any taxon requires an urgent overhaul, it is *Lepidotes*, so López-Arbarello (2012) had a mountain to climb when attempting just that.

As a result of that study, many former *Lepidotes* species were reassigned to new genera, although *A. pankowskii* was not included in the analysis. The name *Adrianaichthys* was finally coined for this species by Meunier and colleagues (2016); the genus name is a tribute to López-Arbarello. This makes *A. pankowskii* rather unique in that both its generic and its species names are in honor of people.

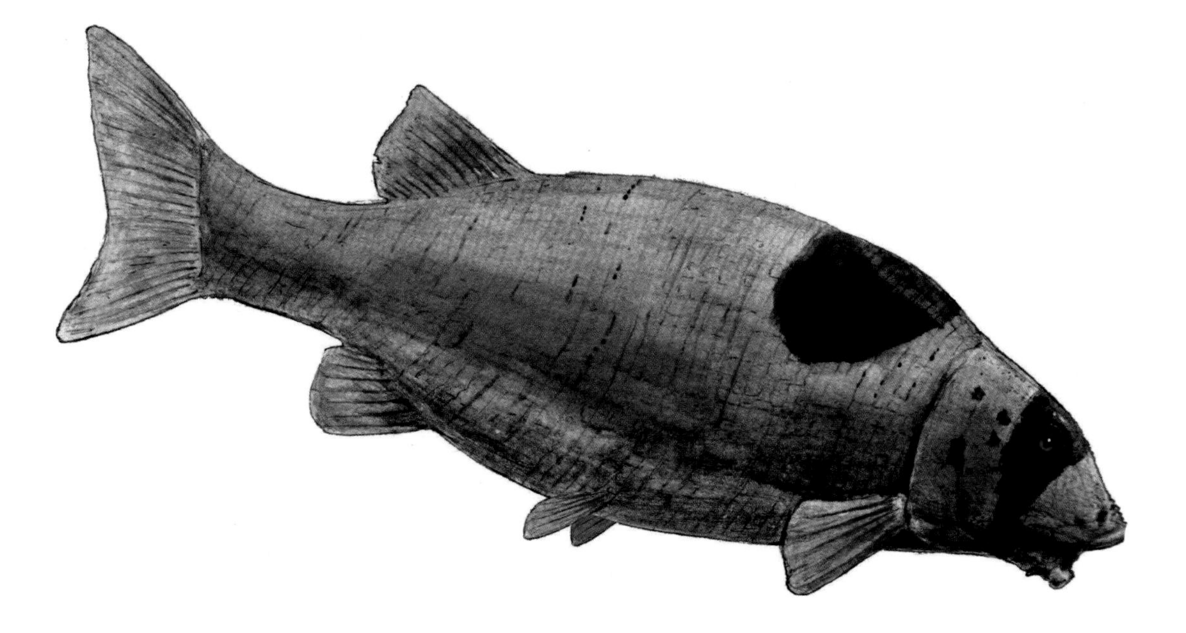

Fig. 4.6. A reconstruction of *Adrianaichthys pankowskii*. Artwork by Joschua Knüppe.

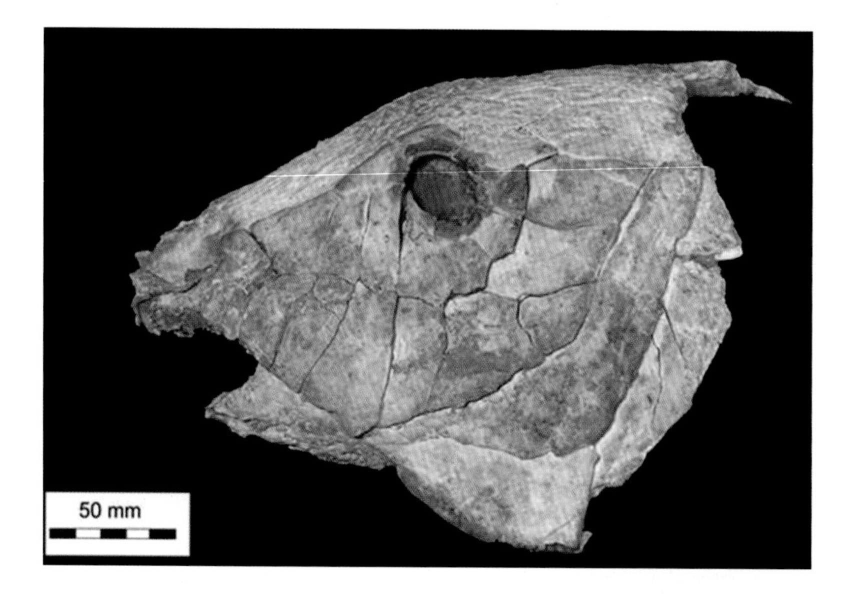

Fig. 4.7. The holotype specimen of *Adrianaichthys*. From Forey et al. (2011, fig. 4). Reproduced by permission of the Palaeontological Association.

Isolated "*Lepidotes*" scales have also been found in Bahariya and Morocco. In a review of this topic, Meunier and colleagues (2016) deduced that while most of these scales belong to *Bawitius* and *Obaichthys*, some do indeed belong to *A. pankowskii*. *Adrianaichthys pankowskii* scales can be identified by having only two tissue layers. The outer layer is composed of ganoine—a type of shiny, hypermineralized tissue (Richter and Smith 1995)—on top of a thick bony basal plate (Meunier et al. 2016).

Adrianaichthys pankowskii was effectively the ecological analogue of carp in the North African waterways, feeding on insects and hard-shelled invertebrates on the riverbed or lakebed. It was also a large species, while known only from its head (see fig. 4.7); Forey and colleagues (2011) give an estimated length of 1.6 meters (5 feet) for this species.

ATRACTOSTEUS

Atractosteus africanus (fig. 4.8) is yet another species in our list that has living relatives. In this case, they even share the same genus. As so little is known of *A. africanus* from North Africa (see fig. 4.9), much of what we know of this taxon is based on comparisons with its relatives and specimens of *A. africanus* found in Europe!

Fig. 4.8. A reconstruction of *Atractosteus africanus*. Artwork by Joschua Knüppe.

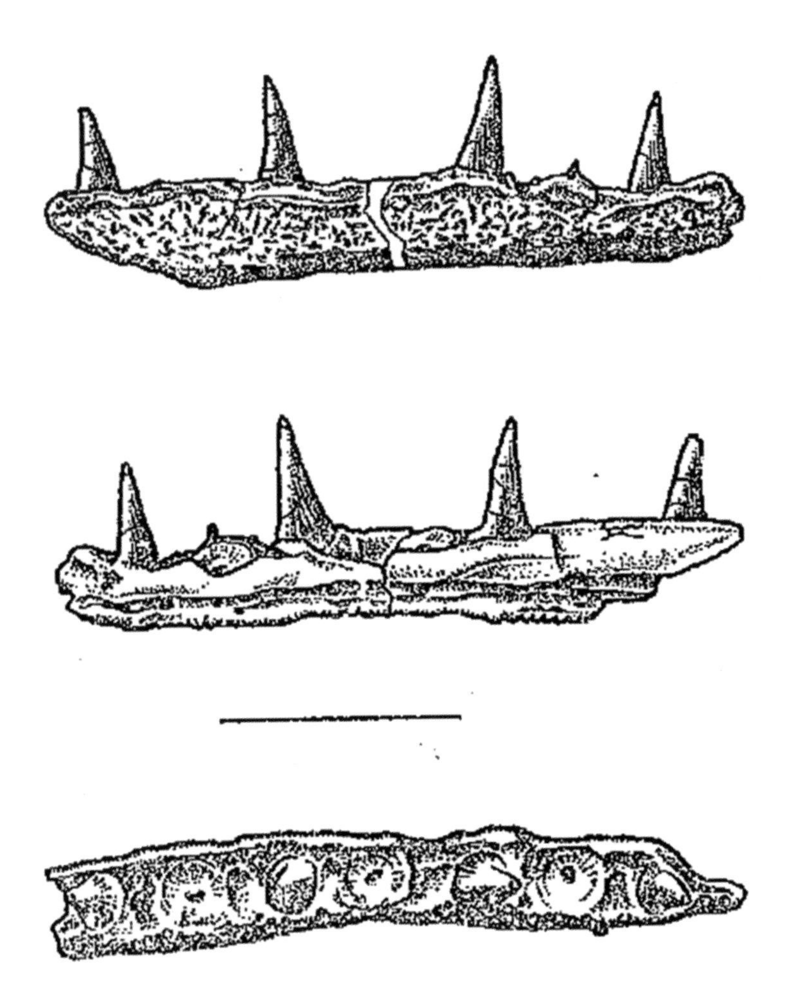

Fig. 4.9. An illustration of the *Atractosteus africanus* holotype. Image from Cavin et al. 1996 (fig. 2).

In some ways, A. *africanus* was a relict even back in the Mesozoic, as the gar family retains many primitive features, one of which is a spiral valve intestine. This means that the lower parts of the intestine develop a natural corkscrew shape, which increases nutrient absorption by increasing the surface area of the intestine (Frías-Quintana et al. 2015). *Atractosteus africanus* was also capable of breathing air, using the swim bladder as a primitive lung (Graham 1997; Buckmeier 2008). This would have been a useful trait in an arid environment hit with occasional droughts (Nothdurft et al. 2002; Krassilov and Bacchia 2013).

Atractosteus africanus was a euryhaline species, capable of surviving in a range of salinities (Goodyear 1967), as its remains are found in environments ranging from river systems to mangrove swamps. This helps explain how it can be found in both Africa and Europe, as such dispersal would have required it to cross the Tethys Sea.

All gars are ambush predators, designed for short bursts of speed, feeding on anything small enough to swallow. They also tend to hunt at night (Goodyear 1967). Modern gars are long-lived and take between ten and fourteen years to reach sexual maturity; this probably applied to A. *africanus* as well (Buckmeier 2008).

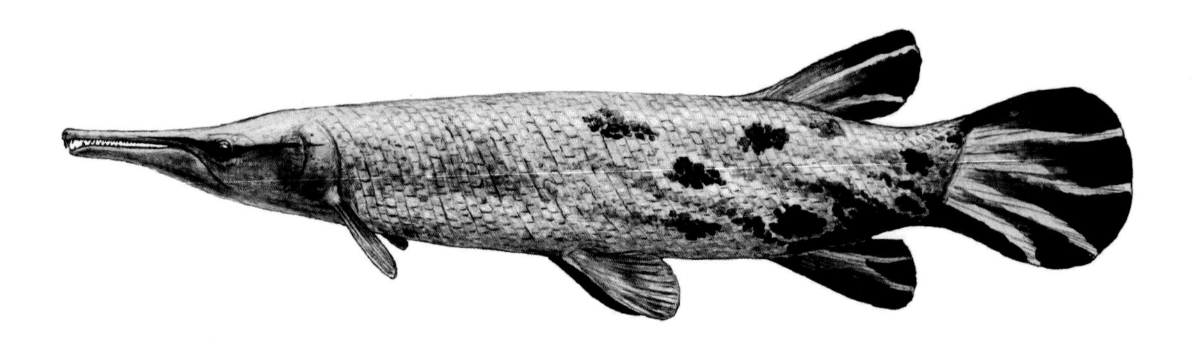

Fig. 4.10. A reconstruction of *Oniichthys falipoui*. Artwork by Joschua Knüppe.

ONIICHTHYS

Oniichthys falipoui (fig. 4.10) is another species of gar from Morocco. Known originally from two specimens, it was given its own genus because it seemed distinct from the two recent gar genera, *Lepisosteus* and *Atractosteus* (Cavin and Brito 2001). The most important of these differences was the presence of teeth on the maxillae (part of the upper jaw).

It was later decided that *Oniichthys* could not possibly be a species of *Atractosteus* or any known lepisosteid (Grande 2010). Cavin and colleagues (2015) decided that *Oniichthys* cannot be a member of the obaichthyids either because *Oniichthys* shows no other autapomorphies from this group. For that reason, they believe *Oniichthys* to be a stem lepisosteid according to the stem-based definition of lepisosteids proposed by López-Arbarello (2012).

Cavin and colleagues later found a new specimen, MDE F-13. This specimen is articulated and mostly complete. Only part of the skull, tail,

Fig. 4.11. *A*, an incomplete specimen of *Oniichthys falipoui*; and *B*, a close-up view of the head. Scale bar is 20 mm in *A* and 10 mm in *B*. Image from Cavin et al. (2015, fig. 7).

and fin are missing, and it can confidently be assigned to O. *falipoui* (see fig. 4.11).

Given such taxonomic confusion, no one has attempted to discuss behavior beyond the assumption that O. *falipoui* was probably similar to its relative, A. *africanus*, in both behavior and diet. However, its habitat might have been different, as it's only been found in fresh water (Grande 2010).

CF. *DENTILEPISOSTEUS*

Dentilepisosteus kemkemensis (fig. 4.12) is known only from isolated scales and four vertebrae (Grande 2010). Very little else is known about this taxon, apart from the fact that it is a species of gar (Grande 2010; Cavin et al. 2015).

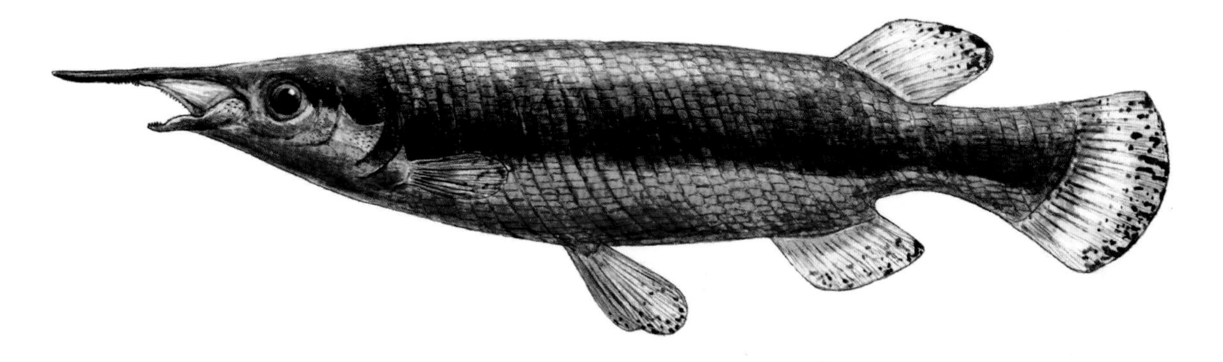

Fig. 4.12. A reconstruction of *Dentilepisosteus kemkemensis*. Artwork by Joschua Knüppe.

The scales assigned to *D. kemkemensis* possess elongated patches of ganoin, separated by a groove, with a visible bony basal plate (Grande 2010; Cavin et al. 2015). Other material has been assigned to *D. kemkemensis* over the years (Benyoucef et al. 2015; Benyoucef et al. 2016), but this has done little to illuminate the taxon better.

Similar scales have been found in Portugal and were named *Paralepidosteus cacemensis* and *Lepidotes minimus* by Jonet (1981). *Paralepidosteus cacemensis* is now considered a *nomina dubia* because not only did Jonet fail to provide any rationale for erecting a new genus but the material in question might not even belong to a single species (Cavin et al. 2015).

It should be noted that *kemkemensis*, irrespective of the genus to which it is ultimately assigned, is not a *nomina dubia*. The type material is diagnosable, and nearly identical material has been found and assigned to this genus (Cavin et al. 2015).

ENCHODUS

Enchodus (fig. 4.13) was one of those taxa that have to be seen to be believed, a saber-toothed herring! *Enchodus* was a prolific genus, with between twenty and thirty species known from across the world (Holloway 2017). Currently, there are two species known from the Cenomanian of

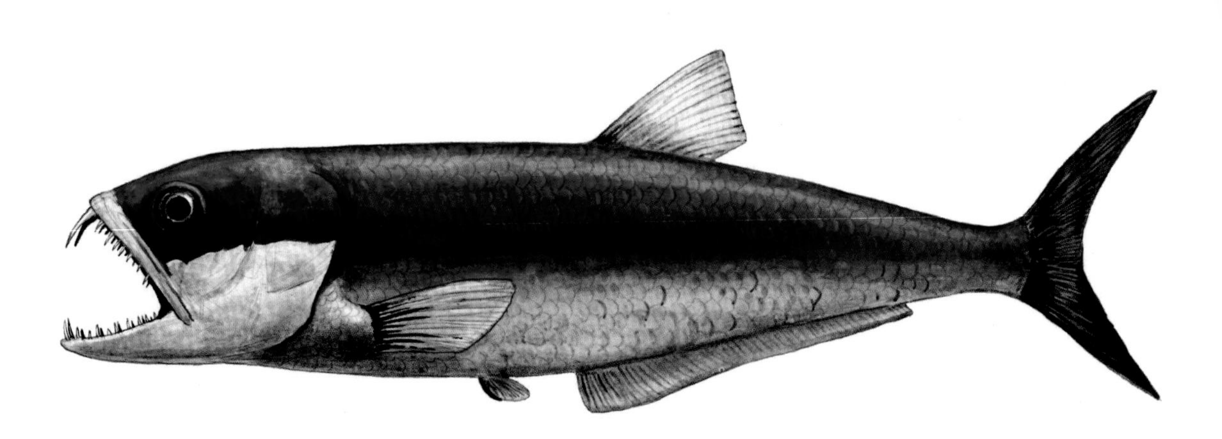

Fig. 4.13. A reconstruction of *Enchodus venator*. Artwork by Joschua Knüppe.

Africa, *E. venator* (Khalloufi et al. 2010) and a poorly known unnamed species, both from Morocco (Cavin 1999b).

The Egyptian specimens, known only from isolated teeth, represent some of the few specimens of the Stromer collection to survive World War II (Smith, Lamanna, et al. 2006). Given the paltry nature of these remains, it's impossible to deduce whether they belong to *E. venator*, the unnamed species from Morocco, or a third species.

The size of *E. venator* is difficult to determine because of the incomplete remains found at Jebel Tselfat in Morocco, but it was one of the smaller species. The skull was only 40 mm (4 cm) long on average (Cavin 1999b).

Another species is known from the Goulmima region of Morocco (Cavin 1999b). However, these beds are part of the upper Akrabou Formation and are of Turonian age. While this species will not itself be included in our discussion, it does provide general information on *Enchodus* behavior. This locality preserves mainly subadult individuals. This suggests that Goulmima was an *Enchodus* spawning ground (Cavin 1999b). Many fish travel large distances to reproduce in specific areas; clearly, *Enchodus* did the same.

While *Enchodus* is often portrayed as an analogue to the modern piranha, the viperfish is a better comparison in my mind. This fish uses its long teeth to stab and puncture prey (Seasky 2016). This suggests *Enchodus* had a diet of fish and soft-bodied creatures to avoid damaging its long teeth (Everhart 2013).

ICHTHYOTRINGA

Ichthyotringa is a genus of fish best known from North America and the Middle East. Originally named *Rhinellus* by Agassiz (1833–44), it was renamed *Ichthyotringa* by Cope (1878). Currently, there are seven species of *Ichthyotringa*, only one of which, *I. africana* (fig. 4.14), is from North Africa.

Fig. 4.14. A reconstruction of *Ichthyotringa africana*. Artwork by Joschua Knüppe.

The most detailed work on *I. africana* was conducted by Taverne (2006) on specimens held in Paris. But the holotype and paratypes have been damaged by neglect. In spite of this, Taverne was able to redescribe this species. The most obvious feature is the elongated, beaklike snout. *Ichthyotringa africana* would have been an estimated 300 mm (30 cm) long (Taverne 2006).

Ichthyotringa africana was carnivorous, with its elongated jaws filled with both large and small fangs. It is superficially similar to the modern barracuda in body form. Fielitz and Rodríguez (2008) suggest that it is an extinct analogue of gars and needlefish that specialized in preying on small fish.

CALAMOPLEURUS

Calamopleurus is a genus of bowfin fish, which are best known from South America. The bowfins are a group of freshwater fish that can breathe air like the gars (Hedrick and Jones 1999). They also show high tolerance for acidic water (Stauffer 2007).

While the African species, *C. africanus* (fig. 4.15), clearly differs from the South American species, *C. cylindricus* (Forey and Grande 1998; Cavin et al. 2015), it's questionable whether it's distinct from *C. mawsoni* from Central Africa (Martill and Brito 2000). However, Cavin and colleagues (2015) consider it valid on the basis of a study of their specimen (see fig. 4.16). Martill and colleagues (2007) also note that the two species were separated not only by large distances but also by millions of years, so *C. africanus* is retained herein as a valid species pending further revisions.

Fig. 4.15. A reconstruction of *Calamopleurus africanus*. Artwork by Joschua Knüppe.

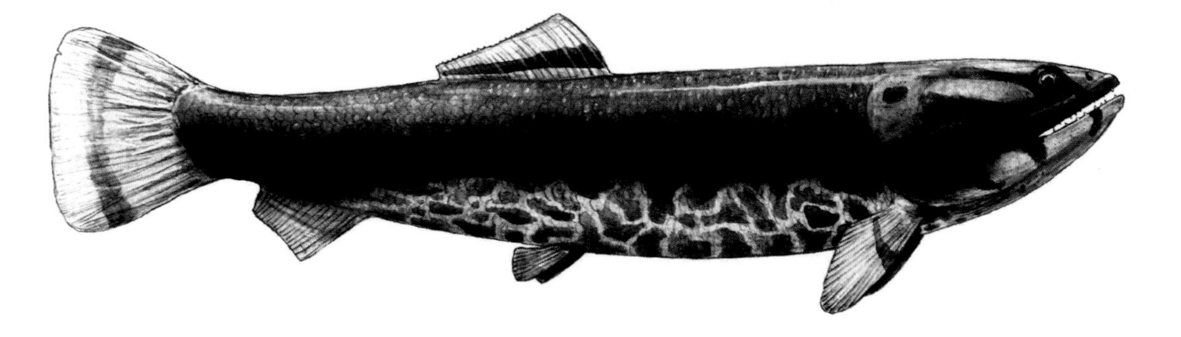

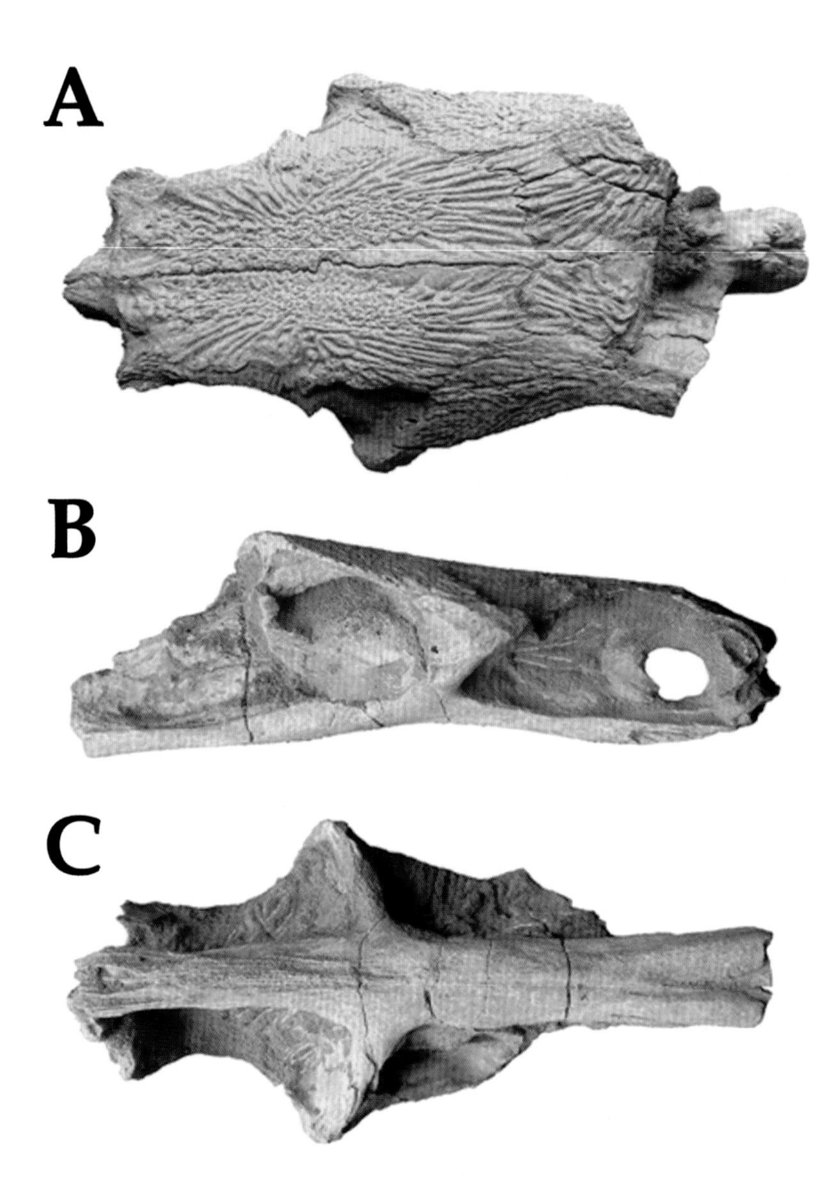

Fig. 4.16. The braincase of *Calamopleurus africanus*: *A*, dorsal view; *B*, lateral view; and *C*, ventral view. Image from Cavin et al. (2015, fig. 8).

The exact environmental preferences are uncertain, although *Calamopleurus* is often thought to be a marine or brackish-water dweller (Maisey 1991); others suggest that it lurked in fresh water (Grande and Bemis 1998). Cavin (2017) believes that *Calamopleurus* was a euryhaline taxon and that different species had different environmental ranges, with *C. africanus* living in deltaic environments. This is supported by the fact that although the Kem Kem Formation is a mostly freshwater environment, the rare remains of the marine snake found in this formation show that the coast was nearby (Rage and Dutheil 2008).

BAWITIUS

Bawitius bartheli (fig. 4.17) is a huge species of polypterid from Egypt and Morocco. Stromer (1925) was the first to mention polypterid scales from

Fig. 4.17. A reconstruction of *Bawitius bartheli*. Artwork by Joschua Knüppe.

Bahariya. Weiler (1935) disagreed with his identification, but Stromer (1936) stood his ground on the issue, correctly as it later turned out.

While that material was destroyed in World War II, save for one single scale, other material was found subsequently. Schaal (1984) would name this specimen cf. *Polypterus bartheli*. While confident that he had a new species, Schaal was not certain of the genus. There the matter rested until Smith, Grandstaff, and Abdel-Ghani (2006) and Grandstaff and colleagues (2012) conducted a new study and confirmed that this material warranted a new genus: *Bawitius*.

More *Bawitius* material, a maxilla (upper jaw) with teeth and isolated scales, is known from Morocco. While only an estimate, it is suggested that *B. bartheli* would have been roughly 2.9 meters (9.8 feet) in length. *Bawitius* would have been an active carnivore. Its huge size means that it could hunt prey of considerable size, even large vertebrates.

Bawitius scales (see fig. 4.18), the largest being 18.16 mm in length, can be differentiated from other polypterids by their shape, tiny articular processes, and distinctive ganoine layer (Smith, Grandstaff, and Abdel-Ghani 2006; Grandstaff et al. 2012). Meunier and colleagues (2016) have used some features of these scales, such as the partial ganoine layer, to suggest that *Bawitius* was an unranked member of the Polypteriformes.

Fig. 4.18. *Bawitius* scales. Image from Cavin et al. (2015, fig. 4B, fig. 4C).

Bartschichthys and *Sudania* may also be synonyms of *Bawitius*. At the moment, it's impossible to confirm or refute this without more material (Calvin et al. 2015).

Such thick armor might be used for protection against its own predators (despite its huge size, *Bawitius* was probably itself prey for a host of large theropods and crocodilians). But as growing such thick armor is metabolically expensive, Grandstaff (2018) suggested that *Bawitius* could have evolved the discontinuous ganoin layer as a type of metabolic saving.

Bawitius remains are found in coastal or nearshore marine sediments (Smith, Grandstaff, and Abdel-Ghani 2006; Grandstaff et al. 2012), suggesting that it inhabited coastal rivers, deltas, and estuaries.

SERENOICHTHYS

Serenoichthys kemkemensis (fig. 4.19) represents one of the oldest members of the Polypteriformes ever found (Dutheil 1999a). It is a small and stout species, possessing thirty scales along its mid-lateral line and ten rows running from dorsal to the abdominal margin. It also has an estimated nine or ten predorsal scales (Dutheil 1999a).

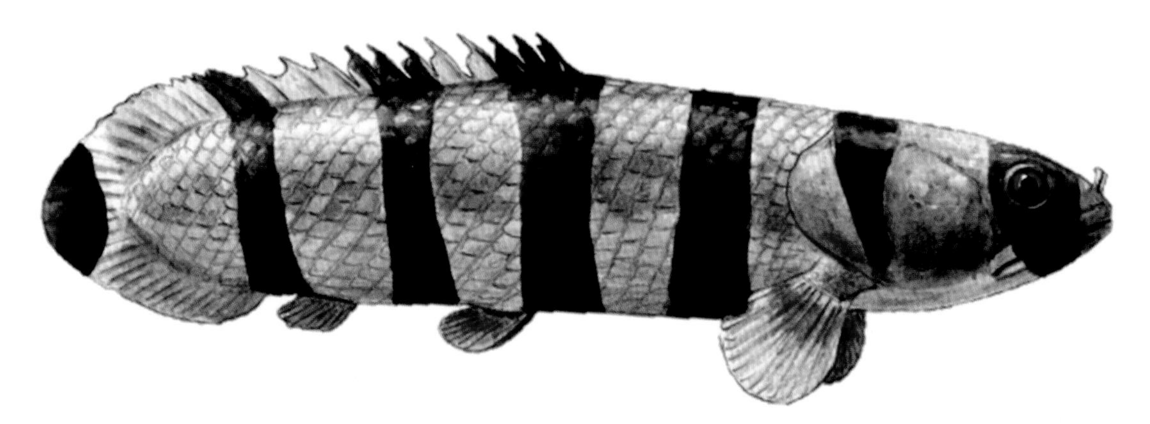

Fig 4.19. A reconstruction of *Serenoichthys kemkemensis*. Artwork by Joschua Knüppe.

In extant Polypteriformes, the number of scales on the lateral line is equal to the number of vertebrae (Daget and Desoutter 1983; Dutheil 1999a). If this is also true here, then *S. kemkemensis* would have had thirty vertebrae (Dutheil 1999a).

SORBINICHTHYS

In life, *Sorbinichthys* would have been a classical tropical fish in design. First discovered in the En Nammoura beds of Lebanon (Bannikov and Bacchia 2000), it proved distinct enough to warrant the creation of its own family, the Sorbinichthyidae. A North African species was later discovered in Morocco and named *S. africanus* (fig. 4.20) (Murray and Wilson 2011).

Sorbinichthys belongs to the Ellimmichthyiformes, a group closely related to the modern herring and sardines. *Sorbinichthys africanus* (fig. 4.21) is similar to the Lebanese *S. elusivo*; one of the differences between the two is that *S. africanus* has twelve or thirteen pleural ribs, while *S.*

Fig. 4.20. A reconstruction of *Sorbinichthys africanus*. Artwork by Joschua Knüppe.

Fig. 4.21. A specimen of *Sorbinichthys africanus*, with some mineral incrustations. Image provided by Thomas Bastelberge. See image at http://www.thefossilforum.com/index.php?/gallery/image/27344-sorbinichthys-africanus-murray-wilson-2011/.

elusivo has fifteen or sixteen (Murray and Wilson 2011). The maximum size of *S. elusivo* was 115 mm, while the *S. africanus* was smaller, ranging from 41.5 mm to 48.6 mm at most (Murray and Wilson 2011).

Sorbinichthys had large scales and a row of perhaps thirty scutes before the dorsal fin. Some of these scutes have spinelike projections (Murray and Wilson 2011). *Sorbinichthys* was also similar to some species of angelfishes, as both possess laterally compressed bodies with elongated fins (Murray and Wilson 2011). Behaviorally, *S. africanus* would have been a reef fish found in shallow waters and lagoons along the coastline. A predator with small, sharp teeth, *S. africanus* would have fed on small fish and invertebrates (Murray and Wilson 2011).

THORECTICHTHYS

One of the more recent additions to the North African biota is *Thorectichthys* from Morocco (fig. 4.22). While it's clearly an Ellimmichthyiform, a member of the family Paraclupeidae, Murray and Wilson (2013) chose to create the subfamily Thorectichthyinae to contain it.

Fig. 4.22. A reconstruction of *Thorectichthys marocensis*. Artwork by Joschua Knüppe.

The exact age of this genus is uncertain, as the beds are near the Cenomanian/Turonian border; it may postdate the Cenomanian and therefore not be a part of this ecosystem. After much hesitation, I've included it; hopefully, greater understanding of these beds will allow us to pin its age down with greater accuracy in the future.

Thorectichthys has two known species: *T. marocensis* and *T. rhadinus* (see fig. 4.23). The jaw of *Thorectichthys* was upturned, almost vertical in position. This gave *Thorectichthys* an extremely flat-faced appearance with the head being deeper than it is long (Murray and Wilson 2013).

Thorectichthys was such a remarkably deep-bodied genus that its body seems almost disproportionate. The body depth is usually between 57% and 65% of its standard length in *T. marocensis* and 40% to 48% in *T. rhadinus*. This makes *Thorectichthys* unique compared to other members

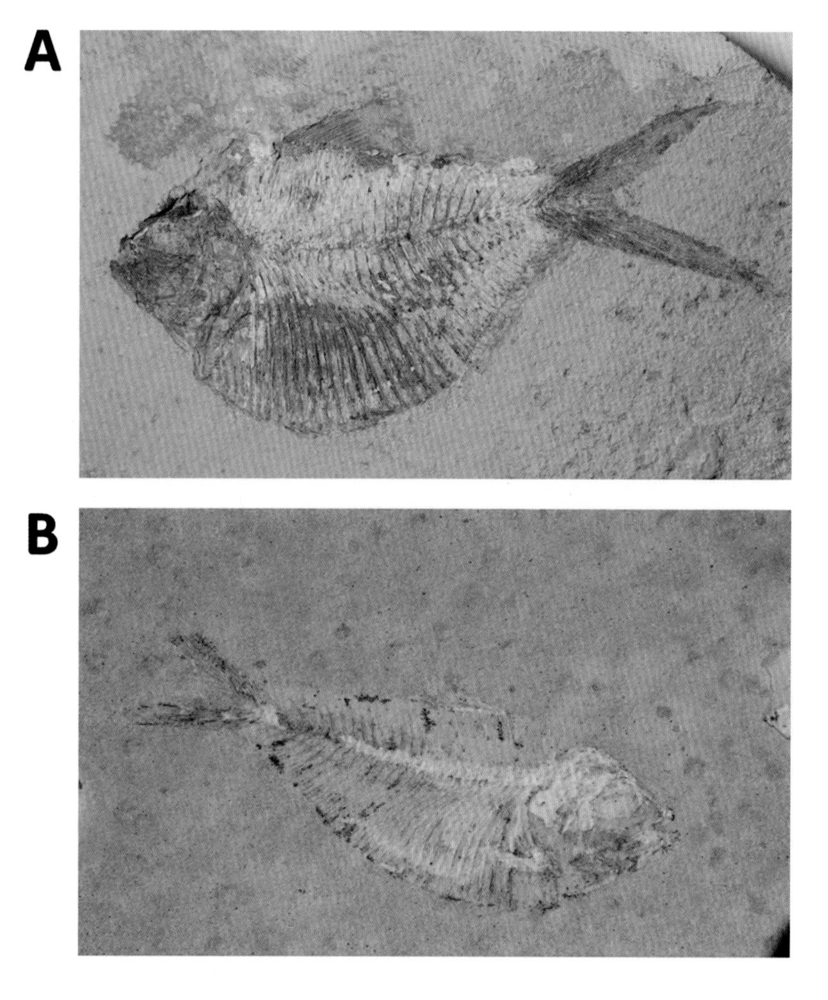

Fig. 4.23. The *Thorectichthys* species complex: *A*, *Thorectichthys marocensis*; and *B*, *Thorectichthys rhadinus*. Images provided by Thomas Bastelberge. See *A* at http://www.thefossilforum.com/index.php?/gallery/image/33316-thorectichthys-marocensis-murray-wilson-2013/; see *B* at http://www.thefossilforum.com/index.php?/gallery/image/33317-thorectichthys-rhadinus-murray-wilson-2013/.

of this family, where the body depth ratio is nowhere near as extreme. Only *Ezkutuberezi* and *Tycheroichthys* have higher ratios (Murray and Wilson 2013).

Thorectichthys marocensis has tall, fine, conical teeth. The *T. rhadinus* teeth are not well developed, but this may be a result of poor preservation (Murray and Wilson 2013). The conical structure of these teeth suggests that *Thorectichthys* was a specialized fish eater, and the upturned mouth suggests that it hunted at the surface.

PARANOGMIUS

Along with *Mawsonia* and *Onchopristis*, *Paranogmius doederleini* (fig. 4.24) would have been among the mightiest of the fish found in this region. First found in Bahariya, *P. doederleini* was diagnosed on the basis of vertebra along with fragmentary caudal and cranial remains (Weiler 1935). But the holotype has been subsequently lost.

Of course, this brings us to *Concavotectum moroccensis* (see fig. 4.25) from Morocco (Cavin and Forey 2008). Cavin and Forey consider the two genera to be possibly distinct and mention that there appear to be

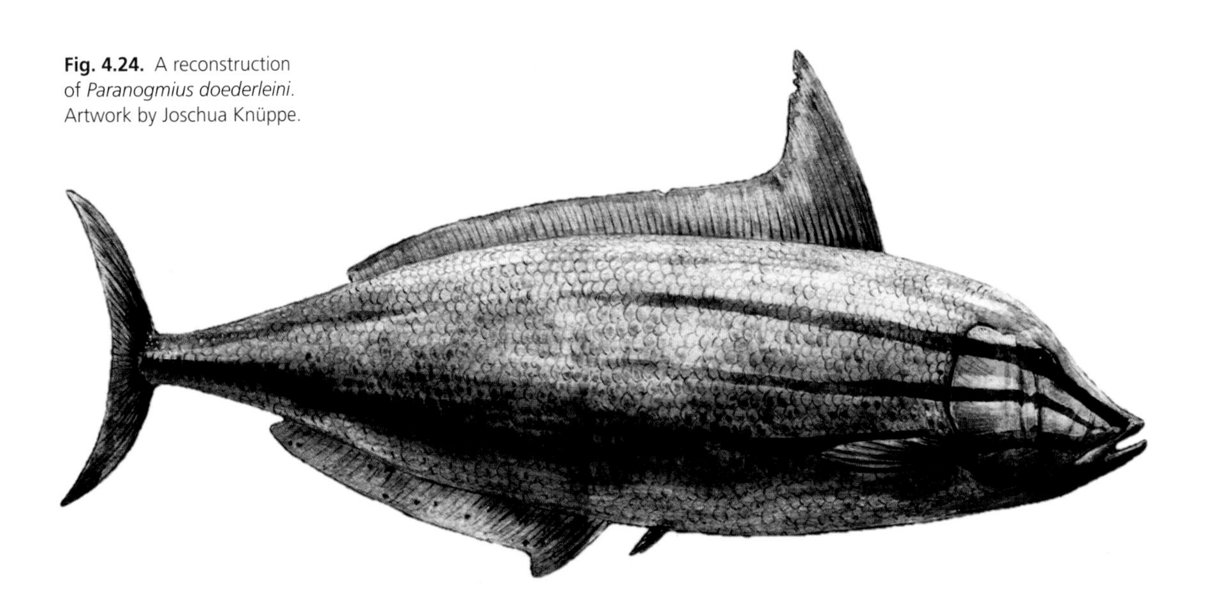

Fig. 4.24. A reconstruction of *Paranogmius doederleini*. Artwork by Joschua Knüppe.

some differences between *P. doederleini* and *C. moroccensis*, although it's impossible to be sure with the original *P. doederleini* material lost. So they were ultimately unable to reach any conclusion on the issue, although I suspect this is due to individual variation. Given that I believe that the two are synonymous, *Paranogmius* has priority as it was named first and based on clearly diagnosable material. So *C. moroccensis* is included herein under *Paranogmius*, although I admit that the synonymy is weak and might be invalidated in the future.

Paranogmius doederleini is likely a basal member of the Plethodidae, a family within the tselfatiiforms (Taverne and Gayet 2005). With its streamlined, muscular body, *P. doederleini* would have been the game fish of its day, although it would have been difficult to hook given its diet. Its dental plates suggest that *P. doederleini* was a filter feeder, feeding on small invertebrates (Cavin and Forey 2008; Cavin et al. 2015).

Everyone knows that fish breathe via gills. A fish gulps water in through the mouth, and that water then flows through the gill arches (those visible slits behind the head). These arches have a series of long filaments that form platelike structures called lamellae. The lamellae are where oxygen is absorbed from the water as it flows through the gills (Romer and Parsons 1977).

The lamellae are usually protected by a series of gill rakers, which protect the delicate gills from any solid objects that might be accidentally gulped in along with the water by filtering them out and diverting them toward the esophagus. Some fish, such as *P. doederleini*, have taken this one step further and use the gill rakers to deliberately filter edible material from the water and then eat it.

SPINOCAUDICHTHYS

Spinocaudichthys oumtkoutensis (fig. 4.26) is a strange beast in that we have multiple well-preserved specimens that have been of little help in

Fig. 4.25. The braincase of *Concavotectum*. *Concavotectum* and *Paranogmius* would have been closely related if not synonymous. Image provided by Olof Moleman. See image at http://www.thefossilforum.com/index.php?/collections database/chordata/bony-fishes/concavotectum-morocensis cavin-forey-2008-r1178/.

placing this species in the acanthomorph family tree. While clearly an acanthomorph (Johnson and Patterson 1993), it also has many plesiomorphic features previously unknown in this group, and some traits even appear to be homoplastic (Filleul and Dutheil 2001).

Recent analysis, while still unable to pin *S. oumtkoutensis* down precisely, suggest that it forms its own clade within the acanthomorphs, known informally as Clade B, which includes *Omosomopsis* as well as the "*Percopsiformes+Gadiformes+Zeiformes*" (Davesne et al. 2018).

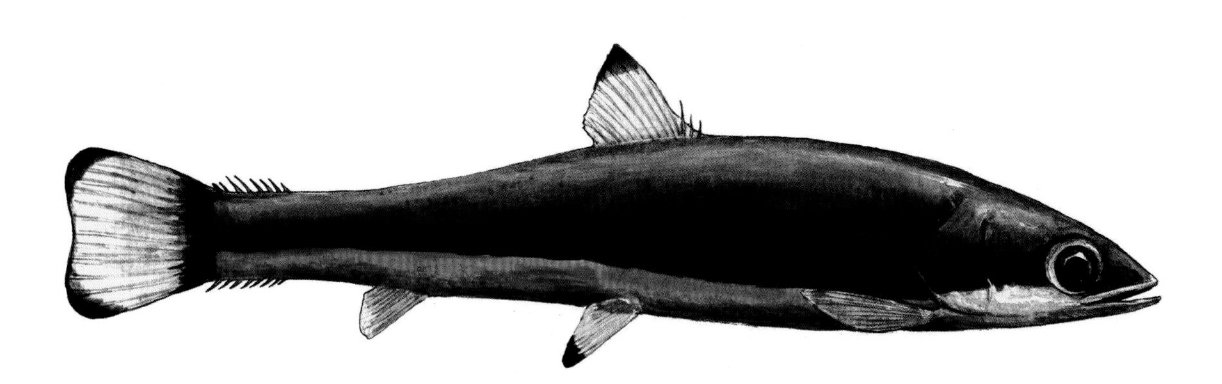

Fig. 4.26. A reconstruction of *Spinocaudichthys oumtkoutensis*. Artwork by Joschua Knüppe.

The original collection was composed of ten specimens, with Davesne and colleagues (2018) adding another three, all of which are juveniles. The fact that *S. oumtkoutensis* is known only from juvenile remains preserved together suggests that this species schooled together after hatching (Filleul and Dutheil 2001). Because of mineralization, much of the head is obscured in the holotype specimen, although parts of the skull are visible. *Spinocaudichthys oumtkoutensis* also has exceptionally large eyes (Filleul and Dutheil 2001). There may also be a second species of *Spinocaudichthys* present, as another specimen found in 2012 has less vertebrae than the holotype (Davesne et al. 2018).

Regardless, this possible second species does provide evidence of the diet of *Spinocaudichthys*. The intestine in *Spinocaudichthys* sp. is long, with an estimated six to eight coils (Davesne et al. 2018). This increase in intestinal length makes food take longer to pass through the system, a trait usually found in herbivores because plant material is difficult to digest and requires longer retention time for full absorption of any nutrients (Davesne et al. 2018).

LUSITANICHTHYS

Lusitanichthys africanus (fig. 4.27) seems to vary greatly in size, with the standard length ranging from 20 mm to 72 mm, and the majority measure in at 45 mm or less (Murray et al. 2013). Surprisingly, this is not due to ontogeny, as the bones of the smaller specimens are completely ossified, so these were clearly adults (Cavin 1999a). Yet there are no diagnostic

Fig. 4.27. A reconstruction of *Lusitanichthys africanus*. Artwork by Joschua Knüppe.

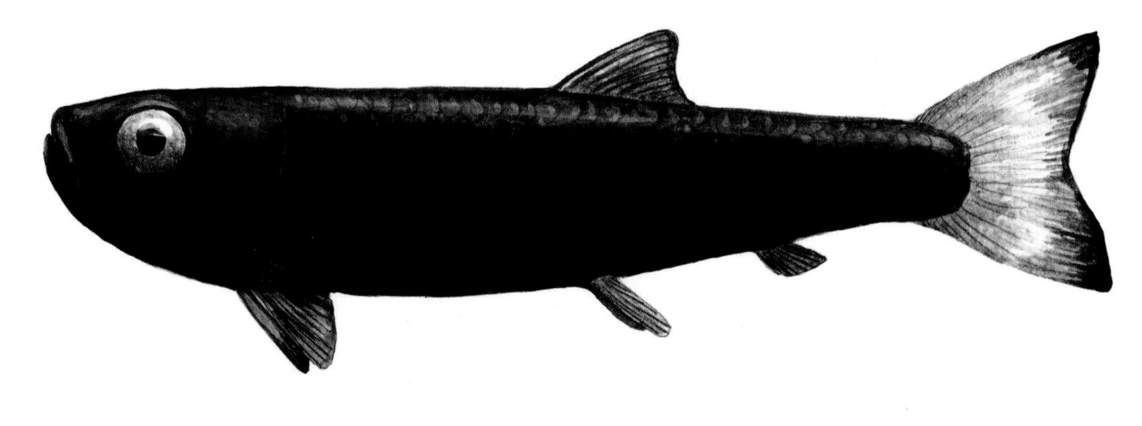

Fig. 4.28. A specimen of *Lusitanichthys africanus*. Image provided by Thomas Bastelberge. See image at http://www.thefossilforum.com/index.php?/gallery/image/33314-lusitanichthys-africanus-cavin-1999/.

differences between the various specimens (see fig. 4.28), so they belong to the same species (Murray et al. 2013).

What may be a second species of *Lusitanichthys* was reported by Martill and colleagues (2011). Murray and colleagues (2013) suggest that this specimen possibly has a lower vertebral count and a longer head, but they have not studied the specimen themselves. A full description of this specimen has never been published.

AIDACHAR

I remember reading *Whales and Giants of the Sea* by Rupert Oliver (1982) as a child. In it, there was a wonderful illustration by Bernard Long of the ichthyodectid *Xiphactinus*. That image still holds my imagination, even after all this time; the color pattern of the artwork (fig. 4.29) is an homage to that book, which helped shape my childhood.

North Africa was also home to at least one ichthyodectid species, *Aidachar pankowskii*. Forey and Cavin (2007) originally assigned this

Fig. 4.29. A reconstruction of *Aidachar pankowskii*. Artwork by Joschua Knüppe.

species, known from a braincase, to the genus *Cladocyclus*. Yet Martill and colleagues (2011) claimed that the lower jaw of the Kem Kem ichthyodectiform differed so significantly from *Cladocyclus* it should be given its own genus. This was a mistake, as Forey and Cavin never assigned a jaw to their *Cladocyclus*. The jaw in question, USNM PAL 521361, has been left in open nomenclature (Forey and Cavin 2007).

Yet *pankowskii* was ultimately transferred to a new genus, *Aidachar* (Cavin et al. 2015). On the basis of the revised description of *Aidachar paludalis* from Uzbekistan, it had been proved that *pankowskii* is actually a species of *Aidachar* (Mkhitaryan and Averianov 2011). Cavin and colleagues (2015) have subsequently assigned other jaw material to *A. pankowskii*.

The braincase of *A. pankowskii* measures 102 mm and is similar in proportion to *Cladocyclus gardneri* from Brazil (Forey and Cavin 2007). Like most ichthyodectiforms, *A. pankowskii* would have been an active predator. Its slim, elongated body was designed for traveling at speed in open water.

CLADOCYCLUS

The second species of ichthyodectiform is known from the Gara Sbaa beds of Morocco (Martill et al. 2011). This unnamed species apparently cannot be assigned to *A. pankowskii* and is supposedly closer to *Cladocyclus gardneri*. However, while Martill and colleagues (2011) claim that this species can be positively assigned to the genus *Cladocyclus* (fig. 4.30), they never give their reasons for this identification. Apart from noting that it's "considerably smaller" than *A. pankowskii*, they gave no further details, although others are studying the specimen.

Fig. 4.30. A reconstruction of *Cladocyclus* sp. Artwork by Joschua Knüppe.

HECKELICHTHYS

The third ichthyodectid, *Heckelichthys vexillifer*, from Morocco, has a complex history, having gone through multiple names before being given its own genus by Taverne (2008). Built for speed, *H. vexillifer* has tall dorsal and anal fins, coupled with an elongated caudal fin. Like the rest of the body, the head is elongated, albeit with short jaws (Taverne 2008).

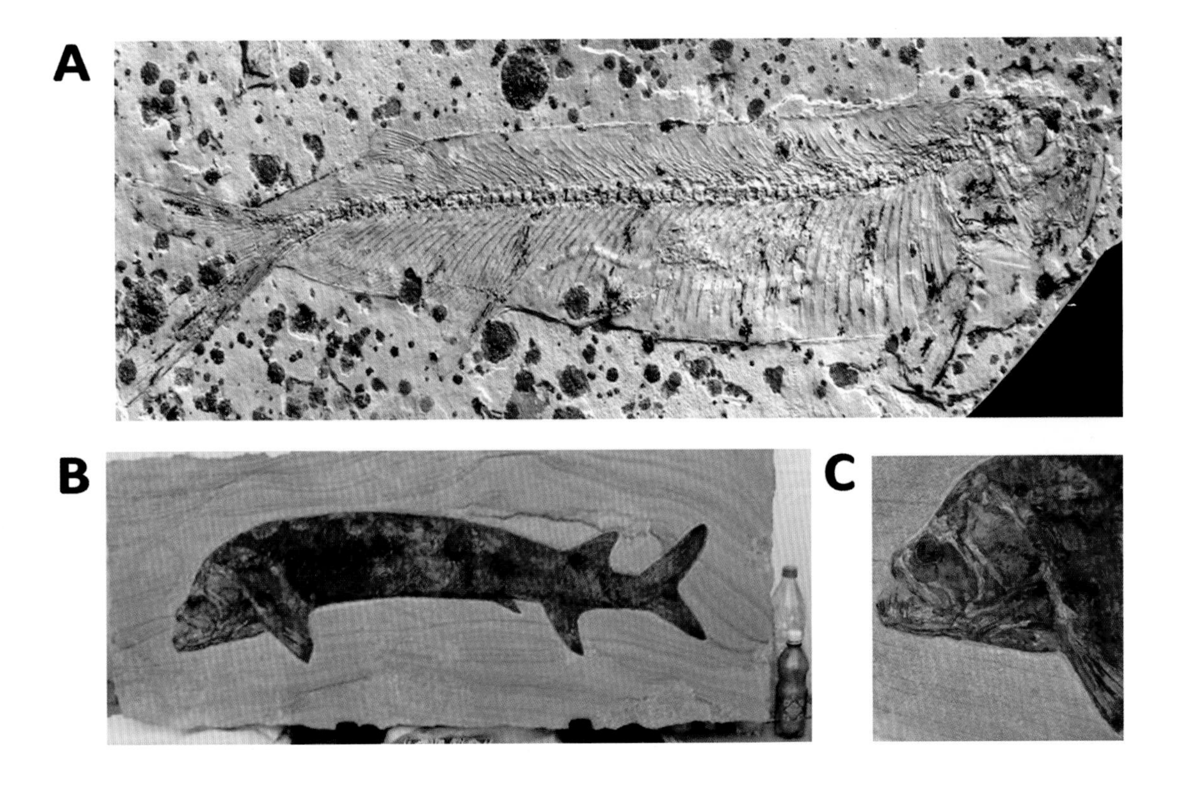

Fig. 4.31. A selection of ichthyodectiform fossils from Morocco, all currently held in private collections. A definitive identification is impossible as I have not personally examined any of these specimens, so all are currently left as *ichthyodectiform non det.* Images provided by Thomas Bastelberge and Jean-François Lhomme.

ERFOUDICHTHYS

Erfoudichthys rosae (fig. 4.32) was originally known from a solitary head with the first eight vertebrae (Pittet et al. 2010). Another specimen, MDE F-56, shows some features not seen in the holotype. For instance, the pectoral fins are preserved (Cavin et al. 2015). *Erfoudichthys rosae* also has a series of well-developed sensory canals (Pittet et al. 2010; Cavin et al. 2015).

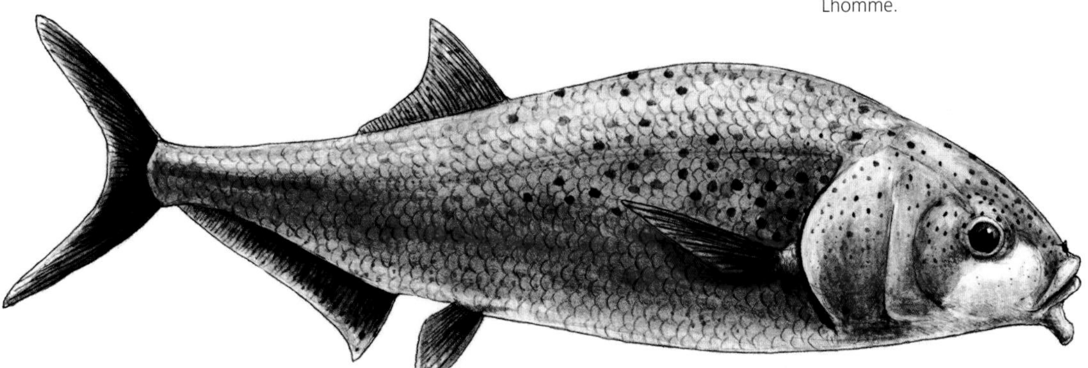

Fig. 4.32. A reconstruction of *Erfoudichthys rosae*. Artwork by Joschua Knüppe.

Fig. 4.33. A partial specimen of *Erfoudichthys rosae*. Scale bar is 10 mm. Image from Cavin et al. (2015, fig. 9A).

Without even a partial skeleton, it is nearly impossible to deduce habitat or behavior. The well-developed sensory canal system suggests that the detection of movement was very important to *E. rosae*. Perhaps *E. rosae* was a predator that specialized in hunting hidden prey, although it's also possible that it was a schooling fish and used these canals to orient itself within the group (Coombs et al. 2001). I consider the latter suggestion unlikely as *E. rosae* remains are so rare that I doubt it lived in large numbers.

DIPLOSPONDICHTHYS

Diplospondichthys moreaui (fig. 4.34) is another hard-to-place taxon, with Filleul and Dutheil (2004) considering it to be utterly unique among osteichthyans. The general form suggests an anguilliform, possibly related to *Enchelion* from Lebanon. Yet *D. moreaui* has many traits not seen in *Enchelion*, so the similarity seems to be more a result of convergence than common descent (Filleul and Dutheil 2004). Cavin (2017) suggests that *D. moreaui* may be a specialized halecomorph.

Fig. 4.34. A reconstruction of *Diplospondichthys moreaui*. Artwork by Joschua Knüppe.

While mostly complete, the skull is poorly preserved. The skeleton is also missing parts of the tail and fins (Filleul and Dutheil 2004). Scales are technically unknown for this genus, although strange ossifications are recorded. They run from the thirty-sixth vertebra down to the last dorsal and are shaped like a "three branched star" with a small spine (Filleul and Dutheil 2004). Their purpose is uncertain. They could have been buried within the muscles with only the spike protruding, as shown in the figure 3.34, or embedded in the skin like an actual scute.

Diplospondichthys moreaui was a carnivore, but its small eyes and slim, flexible body suggest that it was adapted for a burrowing lifestyle (Filleul and Dutheil 2004). It was probably a slow swimmer, as the tail vertebra lacked the attachment sites for large muscles (Filleul and Dutheil 2004).

PALAEONOTOPTERUS

Palaeonotopterus is a genus of notopterid (Greenwood and Wilson 1998). Known from Egypt and Morocco, *P. greenwoodi* (fig. 4.35) was originally known from just an isolated skull (Forey 1997), although more material has been found since then (Cavin and Forey 2001).

Fig. 4.35. A reconstruction of *Palaeonotopterus greenwoodi*. Artwork by Joschua Knüppe.

The tooth plate is also remarkably similar to those of *Plethodus*. Two species of *Plethodus* are known from the same region: *P. libycus* (Weiler 1935) and *P. tibniensis* (Schaal 1984). While the tooth plates of *Palaeonotopterus* show similarities to those of *P. libycus*, they lack the groves seen in *P. tibniensis*. So, while *P. libycus* is probably a synonym (Taverne 2000a; Cavin and Forey 2001), *P. tibniensis* might still be a valid species (Grandstaff 2018).

This creates a slight problem because both *Plethodus* species were named before *Palaeonotopterus*, which would make the latter the junior synonym. For this reason, it would perhaps be better if *P. libycus* were considered *nomen oblitum* and *P. greenwoodi* erected as a *nomen conservandum* since it's known from much better material and the holotype for *P. libycus* has been destroyed.

OMOSOMOPSIS

What follows next is a cavalcade of obscure taxa, many of which have not been re-described in decades. Others have only recently been named. The first we'll be discussing is *Omosomopsis simum* (Gaudant 1978). While most researchers consider it valid, little has been done with it since it was first named. Currently, *O. simum* is placed in the Polymixiidae

(Khalloufi et al. 2010). Regardless of its obscurity, *Omosomopsis* was a survivor, as the taxon survived the Cenomanian/Turonian extinction and made it as far as the Santonian age.

OMOSOMA

Just as obscure is *Omosoma tselfatensis* (Costa 1857), which was originally joined with *Omosomopsis*. Gaudant (1978) separated the two taxa, although the two are closely related (Khalloufi et al. 2010).

ELOPOPSIS

Elopopsis microdon was originally described by Heckel (1856) as being from Slovenia but is also known from Morocco (Arambourg 1954). According to Zouhri (2017), this taxon is characterized by a broad skull and teeth that are generally small except for the large ones on the premaxilla (the tip of the upper jaw).

TINGITANICHTHYS

Tingitanichthys heterodon is best known from Italy, but two specimens are also known from Morocco (Khalloufi et al. 2010). This species had a pronounced dorsal fin running down the back right to the base of the skull (Zouhri 2017).

Fig. 4.36. A complete specimen of *Errachidia pentaspinosa*. Image provided by Thomas Bastelberge. See image at http://www.thefossilforum.com/index.php?/gallery/image/33312-errachidia-pentaspinosa-murray-wilson-2014/.

ERRACHIDIA

Errachidia pentaspinosa (reconstructed on plate 5) is another of North Africa's deep-bodied taxa (Murray and Wilson 2014). Murray and colleagues (2013) initially published little about this taxon following its discovery, but we now know that this species has a body depth of 33 mm and a standard length of 50 mm, with the head representing about one-third of that length (Murray and Wilson 2014).

It was initially suggested that *E. pentaspinosa* (fig. 4.36) is a type of aipichthyoid (Murray et al. 2013), although it differs from the two known aipichthyoid families in the number of fin spines and thus cannot be included in either. As a result, this species was left as aipichthyoid *incertae sedis* (Murray et al. 2013).

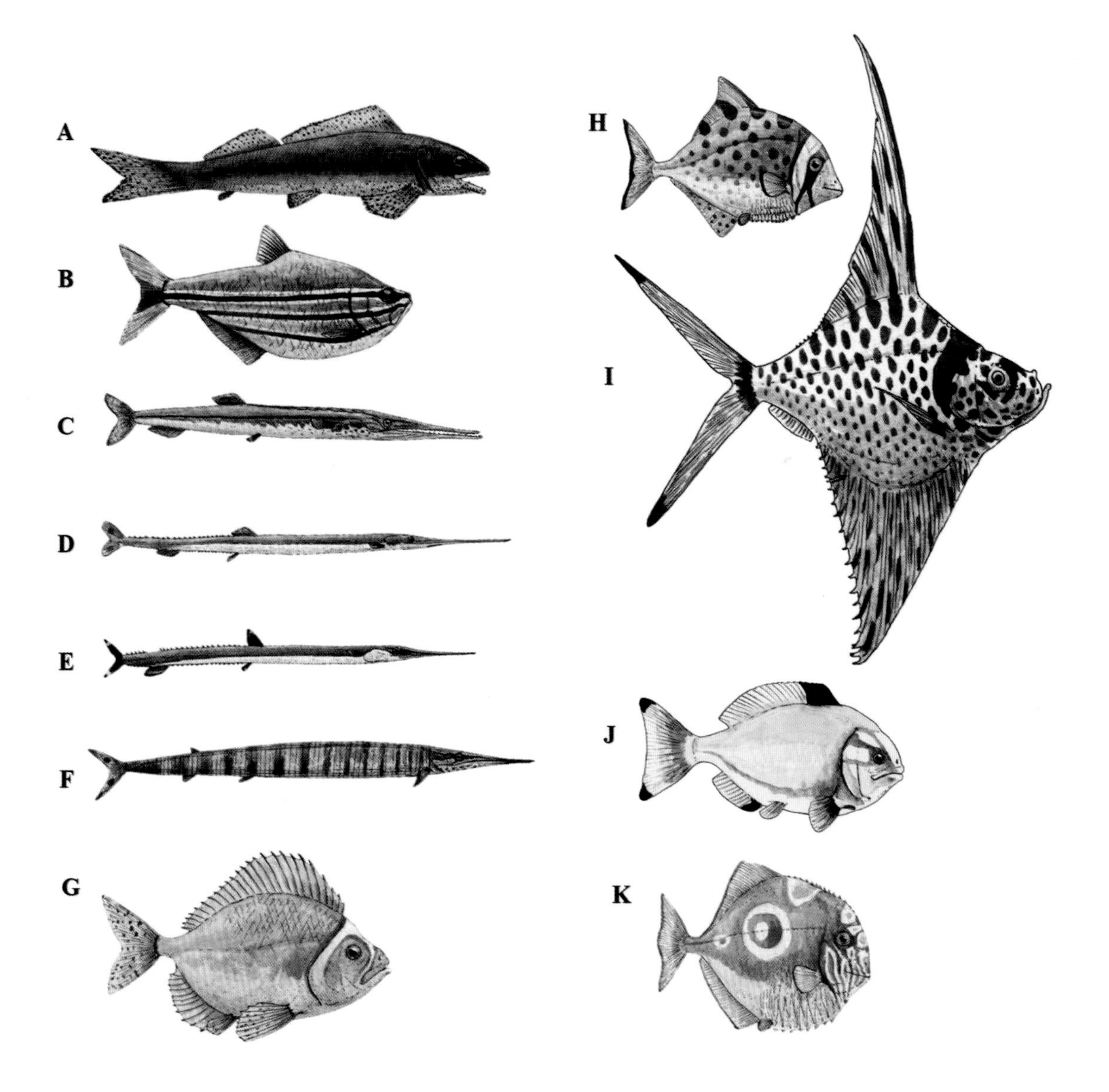

Plate 5. The fish of North Africa. *A, Agoultichthys; B, Armigatus; C, Saurorhamphus; D, Rhynchodercetis; E,* cf. *Dercetis; F, Belonostomus; G, Errachidia; H, Coelodus; I, Rhombichthys; J, Magrebichthys;* and *K, Paranursallia.* Images not to scale. Artwork by Joschua Knüppe.

MAGREBICHTHYS

Magrebichthys nelsoni (reconstructed on plate 5) was first noted by Murray et al. (2013) and named by Murray and Wilson (2014). Known from four specimens, it's smaller than the other acanthomorph species. The largest has a standard length of 24 mm, while the smallest has a standard length of just 17 mm. The body depth also varies from 47% to 58% of the standard length. This species appears to have changed during ontogeny, with its body becoming deeper as it grew (Murray and Wilson 2014).

PROTOSTOMIAS

Protostomias maroccanus (Arambourg 1943; Taverne 1991) is yet another taxon known from better-preserved specimens from Europe. A creature of the ocean abyss, *P. maroccanus* frequented the shallow coastal waters of North Africa as well.

Like many deep-sea fishes, *P. maroccanus* is suspected to have borne on its head a bioluminescent barbel used as a lure to draw in potential prey. As with *Enchodus*, the long, thin teeth on the lower jaw were evolved to pin struggling victims (Diedrich 2012).

Protostomias maroccanus also gives us a rare insight into the marine food chain in North Africa, as specimens have been preserved with *Clupavus*, *Paravinciguerria*, and juvenile *Ichthyotringa* in their stomachs (Arambourg 1943; Diedrich 2012).

TSELFATIA

Tselfatia formosa (Arambourg 1943) is another widespread genus, found in Africa, North America, and Europe. This species had small pectoral fins high up on the body, about level with the eyes. The impressive dorsal fin, the signature trait for this taxon, has an estimated 42 rays, with the fifth being the most robust. The skull is blunt but tall. The eyes are large, so sight was important to this species (Taverne 2000b; Maisch and Lehmann 2000).

DAVICHTHYS

Davichthys lacostei was originally known as *Holcolepis lacostei*, before being given its own genus by Forey (1973). Known from both Lebanon and North Africa, *D. lacostei* was a small species, 90 mm long (Arambourg 1954; Forey 1973). It can be differentiated from the other species of *Davichthys* by features such as the morphology of the median fin—a conjoined tail, dorsal, and anal fin (Forey 1973).

CLUPAVUS

Although it was named by Arambourg (1968), much of what we know of *Clupavus maroccanus* comes from later work by Taverne (1977, 1995) and Gayet (1981). Unlike many fish with elongated skulls, *C. maroccanus* also has a tall skull, with its height being 65% of its length. *Clupavus maroccanus* also has jaws superficially similar to those of the ichthyodectids (Taverne 1977).

Although Taverne (1977) originally placed *C. maroccanus* in the Clupeoidei, Gayet (1981) later pointed out that *C. maroccanus* possessed a Weberian apparatus (a joined swim bladder and auditory system). Taverne (1995) acknowledged this point, and as a result, the Clupavidae is now placed in the superorder Ostariophysi.

SYLVIENODUS

Initially, the pycnodont specimen of Martill and colleagues (2011)—fig. 8C in their publication—was referred to as *Pycnodus* sp., but they gave no reason for their identification. Murray and colleagues (2013) give only a slightly better analysis, noting that this specimen seems to represent a new genus of Pycnodontiform.

They would later be proved right, as Poyato-Ariza (2013) has identified this specimen as *Sylvienodus*. While a full description is still pending, Poyato-Ariza notes several features that unite fig. 8C with this taxon, including the simple morphology of arcocentra (part of the vertebrae), the lack of fenestra on the dermocranial bone (part of the skull covering the endocranium), and the number of vertebrae and dorsal ridge scales.

HOMALOPAGUS

Found in 2007, the first specimen of *Homalopagus multispinosus*, UALVP 47142, is missing both the entire caudal region and part of the skull (Murray et al. 2007). Another specimen, UALVP 51665, was discovered in 2013 and assigned to the same species, although the authors never give a reason (Murray et al. 2013). So, it's best to be cautious with this referral until a detailed study is conducted on UALVP 51665 and UALVP 47142.

TRIPLOMYSTUS

Triplomystus is an obscure genus, with five specimens known from Morocco. These will probably warrant their own species when properly described because they have larger ribs than other species of *Triplomystus* and the morphology of the scutes is also different (Khalloufi et al. 2017).

KERMICHTHYS

Kermichthys daguini remains an obscure taxon. Originally named *Thrissopater* by Arambourg (1954), it was later given its own genus by Taverne (1993). But this did little to resolve the taxonomic confusion surrounding the genus.

Partly this is because *K. daguini* is exceptionally rare. Currently, it's known from only two specimens, one from Morocco and the second from Italy. Given the poor preservation of both specimens, all we can say at the moment is that *K. daguini* has a slim elongated body and that the final three vertebrae were fused (Zouhri 2017).

COELODUS

Coelodus (reconstructed on plate 5) is another of those taxa known only from isolated remains, in this case a single tooth from Egypt (Slaughter and Thurmond 1974). Not much has been done with it since its discovery. Currently, it's left as *Coelodus* sp., although I suspect that it could be conspecific with, or at least closely related to, *C. syriacus* from the Middle East (Hussakof 1916).

Another damaged specimen, tentatively identified as cf. *Coelodus* sp., is known from Morocco. Its relationship to the Egypt material, and to the genus as a whole, is unclear, although it may warrant its own species in the future (Cooper and Martill 2020).

Like most other pycnodonts, *Coelodus* had a deep, laterally compressed body that gave it a flattened appearance. It's the teeth that stand out. The incisiform teeth grasped prey, and the back teeth then crushed it. This specialized dentition was useful for hard prey (Kriwet et al. 1999). *Coelodus* would have been another reef fish, dwelling in warm and shallow waters, although some species had a tolerance for fresh water (Kriwet et al. 1999).

NEOPROSCINETES

Neoproscinetes is a genus from Brazil with one species, *N. africanus*, native to Morocco (Cooper and Martill 2020). Like many African pycnodonts, *N. africanus* is known only from tooth and vomer fragments. In fish, the vomer, a facial bone, forms part of the roof of the mouth and often has teeth on it. So *N. africanus* is defined solely on dental autapomorphies.

AGASSIZILIA

Another recently named pycnodont is *Agassizilia erfoudina* (Cooper and Martill 2020). It too is known only from a fragmentary specimen, characterized by its distinctive tooth rows and the morphology of its medial teeth.

RHYNCHODERCETIS

One of the more outlandish creatures in this region was *Rhynchodercetis* (reconstructed on plate 5), known for its extremely elongated body—hence the unofficial family name for this group: the needlefish. *Rhynchodercetis* was a widespread genus, with species known from the Middle East to North America (Blanco and Alvarado-Ortega 2006). Currently, there is one species known from North Africa: *R. yovanovitchi* (Arambourg 1954; Taverne 1987). Cavin and Dutheil (1999) and Martill and colleagues (2011) also document *Rhynchodercetis* specimens.

The most obvious feature of this taxon is the elongated rostrum, with the lower jaw being slightly shorter than the upper jaw (Murray et al. 2013). Murray and colleagues (2013) described other specimens as *Rhynchodercetis* sp. These have triradiate scutes running down each flank, with the dorsal and pelvic fins located far down the body and the anal fin located near the caudal fin (Murray et al. 2013).

Murray and colleagues note that there is little to separate these specimens from *R. yovanovitchi*. The description above matches the characteristics of *R. yovanovitchi*. For that reason, I'm reasonably confident in assigning them to that species.

CF. *DERCETIS*

Another of the needlefish from this region is a species Murray and colleagues (2013) consider to be closely related to *Dercetis*. Known from Morocco, cf. *Dercetis* (reconstructed on plate 5) is a long, slim fish with an elongated rostrum, although not as elongated as with other members of the Dercetidae. This species also has triradiate scutes that form a single row along each side of the fish. Cf. *Dercetis* also has a third row of scutes, running along its dorsal midline (Murray et al. 2013).

One specimen even has its gut contents preserved. The prey item may also be a juvenile of the same species, which would constitute evidence of cannibalism (Murray et al. 2013).

BELONOSTOMUS

A reasonably successful genus, *Belonostomus* (reconstructed on plate 5) first evolved during the Jurassic and survived well into the Late Cretaceous. Arambourg (1954) assigned the Moroccan species of *Belonostomus* to *B. crassirostris*, which is also known from Italy. While this genus retains the slim body that's so characteristic of the needlefish (see fig. 4.37), this taxon is characterized by its broader build when compared to *Rhynchodercetis*.

Another specimen was mentioned by Martill and colleagues (2011), although it was never described. They did not assign it to any species,

Fig. 4.37. A *Belonostomus* specimen from Morocco. Image provided by Thomas Bastelberge.

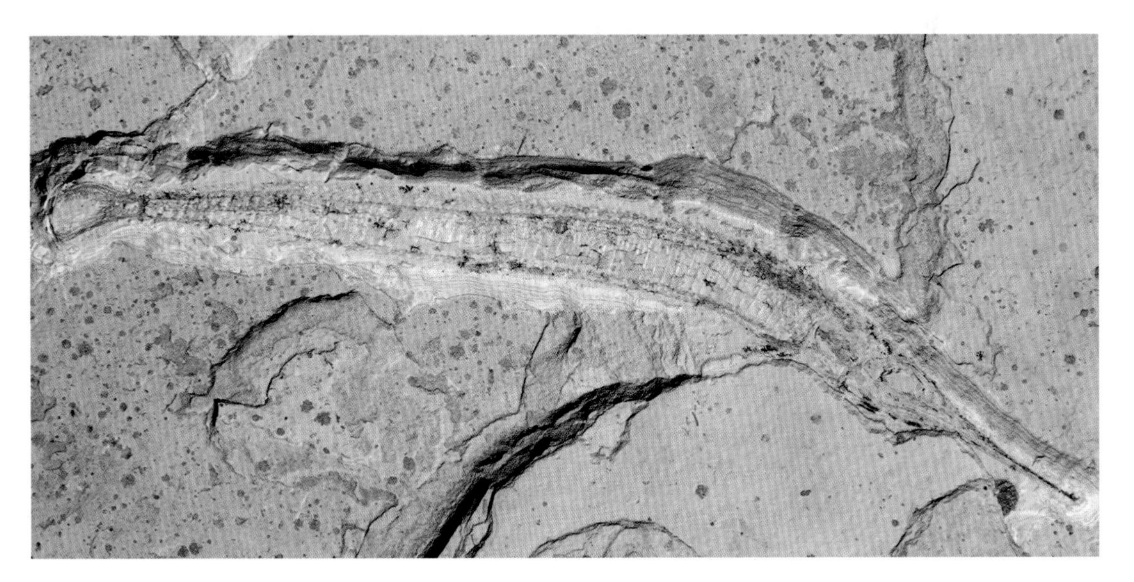

leaving it as *Belonostomus* sp., although they do comment that it appears to belong to the species group that includes *B. crassirostris* and is a closely related species.

SAURORHAMPHUS

Even more poorly known is *Saurorhamphus* (reconstructed on plate 5). Documented from Morocco by Martill and colleagues (2011), this specimen was never given a description. Currently, there are three named species in this genus: *S. freyeri* from Slovenia, *S. giorgiae* from Lebanon, and *S. judeaensis* from Israel. The Moroccan specimen is left as *Saurorhamphus* sp. It could belong to any of those species or be closely related to them.

RHOMBICHTHYS

Rhombichthys, as you can see from the reconstruction on plate 5, was shaped like a living boomerang, with its laterally compressed body accentuated by a tall dorsal fin and elongated belly scutes. The specimen has an estimated thirty-five belly scutes, with the last row having a "claw-shaped" structure on the end (Khalloufi et al. 2010).

Twice as tall as it is long, this specimen is currently left as *Rhombichthys* sp. While this Moroccan specimen shows "no significant differences" from *R. intoccabilis* from Palestine, the authors refrain from assigning it to that species until a full description is written (Khalloufi et al. 2010).

RHARBICHTHYS

Not to be confused with *Rhombichthys*, *Rharbichthys ferox* was named by Arambourg (1954), and its taxonomic position is still heavily debated to this day. Known from Morocco, Italy, and Brazil, this taxon is characterized by elongated jaws and a thin skull (Zouhri 2017).

AGOULTICHTHYS

Agoultichthys chattertoni (reconstructed on plate 5) is a macrosemiid from Morocco first described in detail by Murray and Wilson (2009).

Fig. 4.38. A specimen of *Agoultichthys* from Morocco. Image provided by Thomas Bastelberge. See image at http://www.thefossilforum.com/index.php?/gallery/image/8282-agoultichthys-chattertoni/.

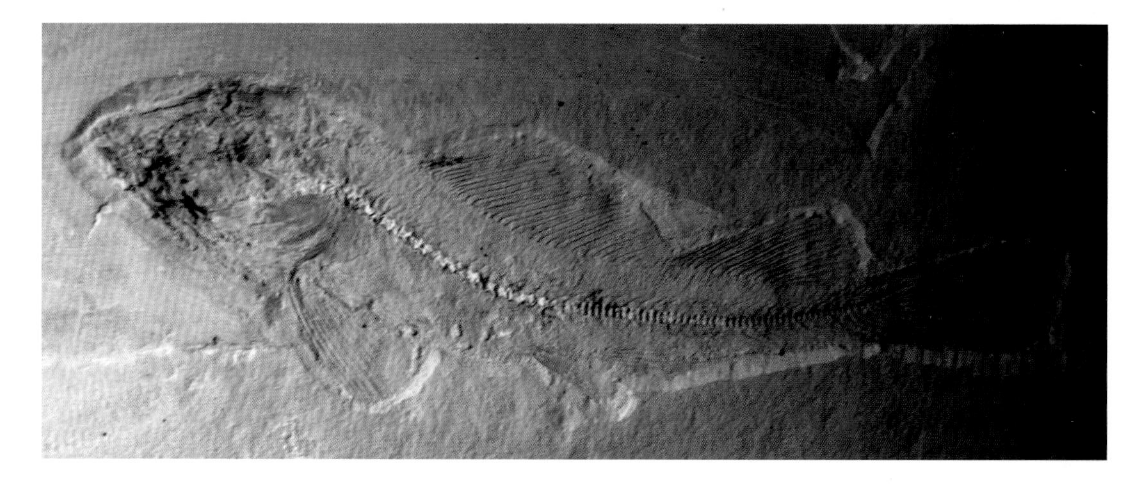

Since then, more specimens have been found (Martill et al. 2011), allowing for an amended diagnosis, although most details concerning taxonomic placement remain unchanged (Murray et al. 2013).

A complete description is given by Murray and Wilson (2009), who provided new information not visible on the holotype. These additional specimens (see fig. 4.38) confirm that the anal fin is not absent but is very small and is close behind the pelvic fins (Murray et al. 2013).

PARAVINCIGUERRIA

Paravinciguerria praecursor is a genus known from both Europe and North Africa. Since the Moroccan specimens are poorly preserved (Arambourg 1954), much of what we know comes from the European specimens.

Like many deep-sea species, it has small eyes and a huge mouth superficially similar to the extant taxa, *Cyclothone* (Carnevale and Rindone 2011). The gills are also known in this species, carrying an estimated six large gill rakers (Carnevale and Rindone 2011).

PARANURSALLIA

The story of *Paranursallia* (reconstructed on plate 5) begins in Morocco, with the discovery of a fish named *Palaeobalistum gutturosum* by Arambourg (1954). Poyato-Ariza and Wenz (2002) later referred the species to the genus *Nursallia*. They were close, as this species is indeed a member of the Nursalliinae, but they failed to realize they had a new genus.

The real answer was found thirty years ago when Mohsen Layeb discovered a wonderfully preserved specimen from the Cenomanian rocks of Tunisia, which was assigned to *Palaeobalistum gutturosum* by coworker Jean Gaudant—a shrewd analysis, as the Tunisian and Moroccan specimens are both members of the same new genus. But Layeb and Gaudant failed to realize that they'd found a new taxon (Taverne et al. 2015).

This mistake was rectified in 2015 when Taverne reexamined the Tunisian specimen. This specimen was given its own species, *spinosa*, and referred—along with the Moroccan species, *gutturosum*—to the new genus *Paranursallia* (Taverne et al. 2015). *Paranursallia gutturosa* has a larger number of scutes along its body than *P. spinosa*: twenty along its back and seventeen along its chest (Taverne et al. 2015).

ARMIGATUS

Armigatus is a small genus best known from Lebanon. Originally placed in the genus *Diplomystus*, it was given its own genus, *Armigatus*, by Grande (1982). Later the Moroccan specimens were given their own species by Vernygora and Murray (2016): *A. oligodentatus* (reconstructed on plate 5).

This taxon is named for its subtriangular armored scutes. Running in two rows, these stretch from the throat, along the flanks, and down to the anus, although there is a gap in the armor at the back of the head. They also change in size, with the smallest at the tail, and they get larger toward the pectoral fins (Grande 1982; Vernygora and Murray 2016).

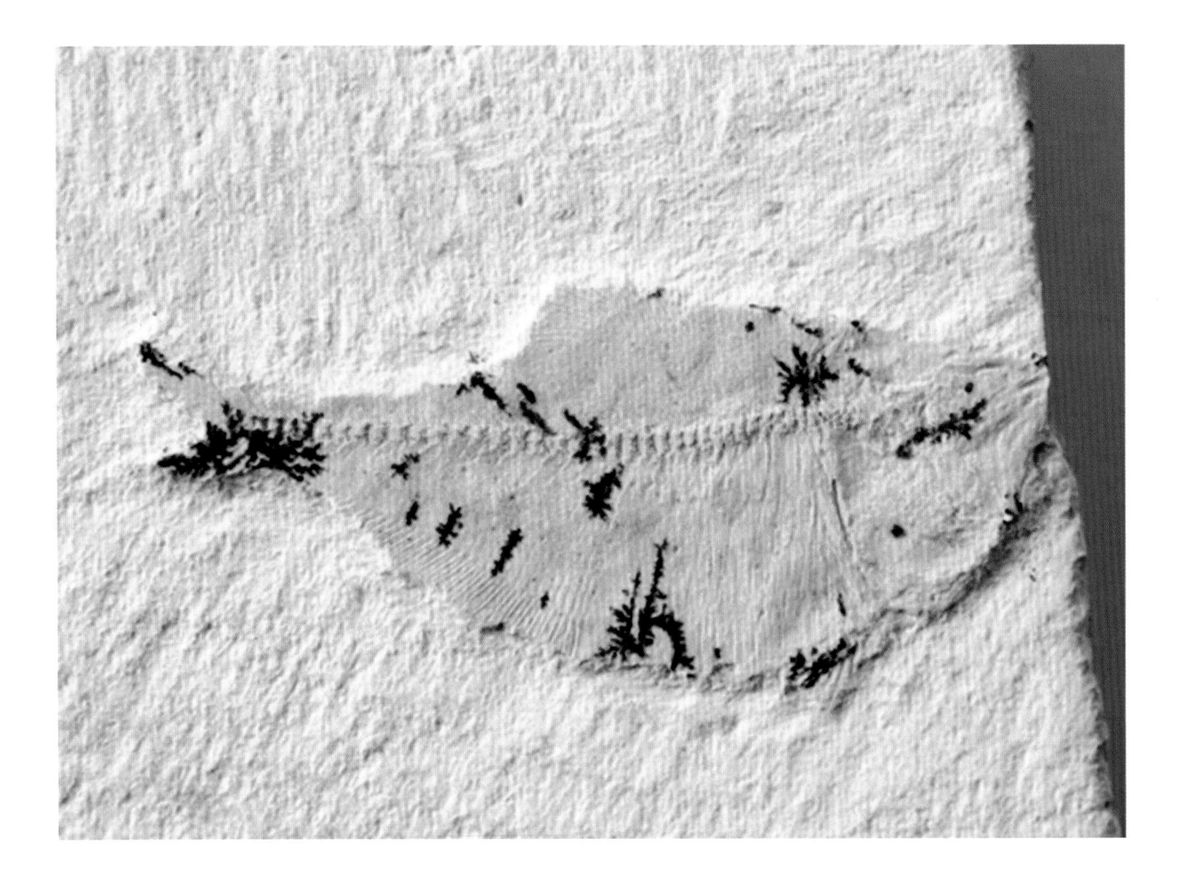

Fig. 4.39. A specimen assigned to *"Diplomystus"* from Morocco. Image provided by Aaron Miller of ancientearth-tradingco.com.

Surprisingly, none of the dozens of specimens preserves the anal fin in any detail. There appear to be fewer bones in this fin than most other Ellimmichthyiformes have, so the anal fin was probably small (Vernygora and Murray 2016).

Other specimens of "*Diplomystus*" are known from Morocco (Dutheil 1999b; Martill et al. 2011), but it's unclear whether they belong to *A. oligodentatus* or truly are a species of *Diplomystus* (see fig. 4.39). Khalloufi and colleagues (2017) left these specimens as Clupeomorph *incertae sedis*.

Actinistia

MAWSONIA

The Actinistia, or lobe-finned fish, are a group of osteichthyans that are more closely related to the lungfish and terrestrial vertebrates than they are to the other osteichthyans. While only two species are alive today, they were more diverse in the past, and of all the fish of Bahariya, *Mawsonia* is undoubtedly the star (Weiler 1935; Grandstaff 2006).

Of the multiple species of *Mawsonia* known, only one is native to North Africa: *M. libyca* (fig. 4.40). Other African species include *M. tegamensis* from Niger and *M. ubangiana* from the Congo. As we are only discussing North Africa, *M. tegamensis* and *M. ubangiana* will not be discussed here. The taxonomy for *Mawsonia* is far from settled, a

Fig. 4.40. A reconstruction of *Mawsonia libyca*. Artwork by Joschua Knüppe.

matter complicated by the lack of neotype for *M. libyca* (the holotype was destroyed during World War II).

The Moroccan mawsoniid specimens are assigned their own species, *M. lavocati* (Tabaste 1963), which is now known to be a species of *Axelrodichthys* (Cavin and Longbottom 2008). The problem with differentiating mawsoniid specimens is due to the large amount of individual variability seen in *Mawsonia* and the difficulty of finding complete specimens. This confusion has reached the point that the validity of *M. libyca* has been questioned (Carvalho and Maisey 2008).

Fig. 4.41. A reconstruction of *Mawsonia* sp. Image provided by Hirokazu Tokugawa and the special exhibition of Gunma Museum of Natural History.

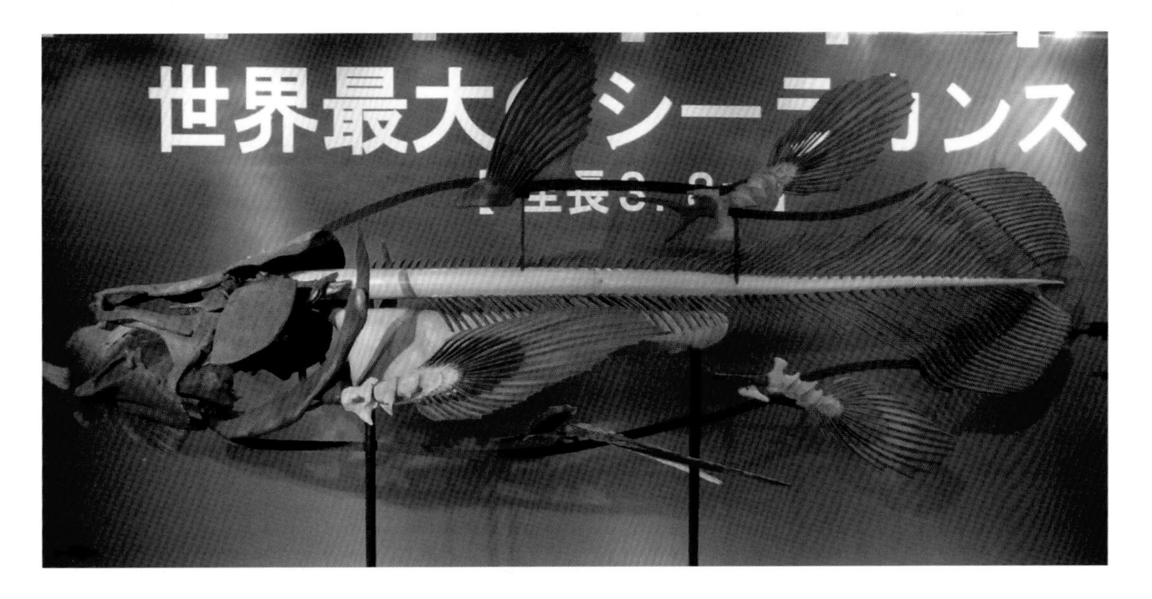

Grandstaff (2011) identified two new specimens of *M. libyca*. One of the *M. libyca* specimens found in 2000 possesses features that overlap with those of the lost holotype, allowing direct comparison (Grandstaff et al. 2002). Grandstaff also cautions that the two species, *M. libyca* and *M. gigas*, are separated chronologically (by 50 million years) as well as geographically (by the Atlantic Ocean and Trans-Saharan Seaway). So it's unlikely that these populations would not have been separated from their last common ancestor during all that time.

On this basis, *M. libyca* is retained here as a valid species. For all the quibbling over taxonomy, *Mawsonia* has a distinguished evolutionary history. It's a member of the Coelacanthiformes, a type of lobe-finned fish that evolved between 416 and 397 mya and are closely related to the rhipidistians, which would give rise to all land vertebrates (Gordon et al. 2004).

Mawsonia was a huge fish, with sizes ranging from 3.5 to 6.3 meters (11 to 20 feet) long (Wenz 1981; Carvalho and Maisey 2008; Medeiros et al. 2014; Dutel et al. 2014). Dutel and colleagues (2012) observe that a trend toward gigantism repeatedly appeared among coelacanths, presumably as a result of competition in predator-rich environments; the size would have allowed coelacanths to overpower prey and avoid being preyed on themselves.

The tactic worked, and *Mawsonia* was a highly successful genus, with *M. libyca* known from many disarticulated remains (see fig. 4.41). This suggests that these waters were host to a large population. While no neotype has yet been described from this wealth of material, hopefully it won't be too long until Bahariya's signature fish taxon gets the attention it deserves.

AXELRODICHTHYS

Axelrodichthys lavocati (fig. 4.42) was once believed to be a species of *Mawsonia* (Tabaste 1963). Cavin and Longbottom (2008) suggest that

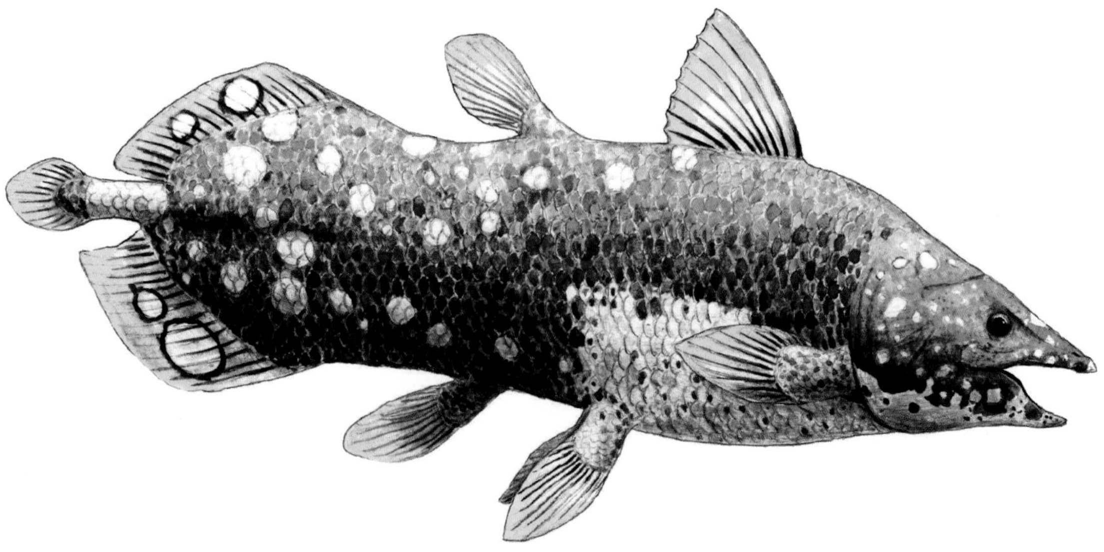

Fig. 4.42. A reconstruction of *Axelrodichthys* sp. Artwork by Joschua Knüppe.

M. lavocati was actually a species of *Axelrodichthys*, and Fragoso and colleagues (2018) finally transferred *lavocati* to that genus.

Axelrodichthys can be clearly distinguished from *Mawsonia* by a suite of characteristics, such as the concave dorsal margin of the parietonasal shield (part of the skull) (Maisey 1986; Forey 1998; Yabumoto and Brito 2013). *Axelrodichthys* also has three pairs of nasals, unlike *Mawsonia*, which has two (Fragoso et al. 2018).

Carvalho and Maisey (2008) consider A. *lavocati* to be a synonym of *Mawsonia gigas* (a species of *Mawsonia* from South America). However, Yabumoto and Uyeno (2005) were able to compare a specimen of A. *lavocati* (then known as M. *lavocati*) at the National Science Museum in Tokyo with the *lavocati* holotype described by Tabaste (1963) and Wenz (1981). Some of the features seen in the Tokyo specimen are almost identical to those of the holotype.

So *lavocati* is a valid species. It is also considered a species of *Axelrodichthys*, as the only real difference between A. *lavocati* and the South American *Axelrodichthys araripensis* is that the lachrymojugal (cheekbone) in A. *lavocati* is straight and not curved (Fragoso et al. 2018).

This brings us to another mawsoniid specimen mentioned by Cavin and Forey (2004). It may belong to A. *lavocati*, as MDE F36 shares

A

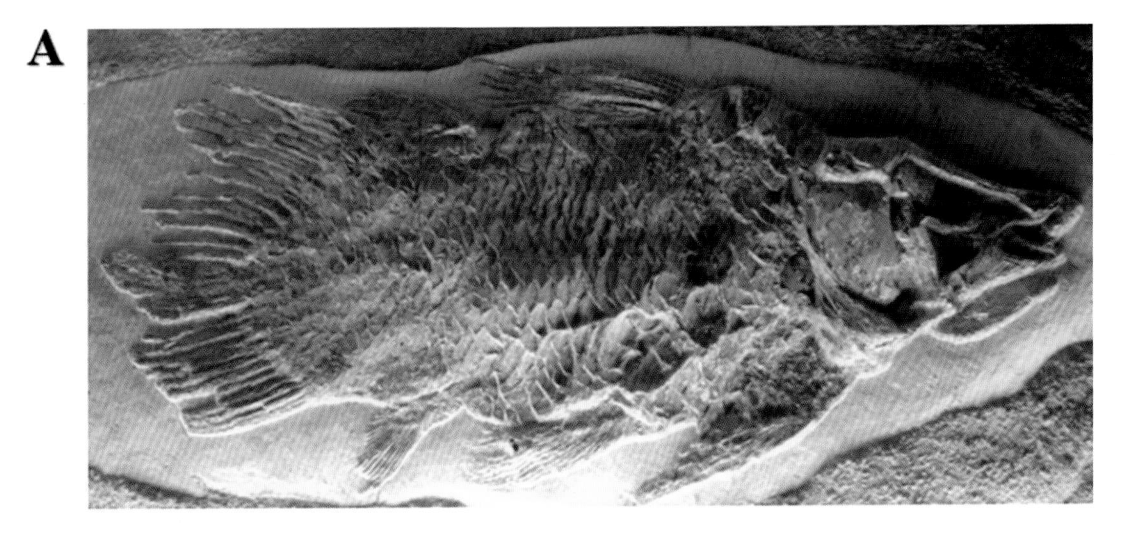

B

Fig. 4.43. *A*, a complete *Axelrodichthys* specimen from Brazil; and *B*, an isolated principal coronoid bone from Morocco. Images provided by Daderot / CC-BY-SA-1.0. See Wikimedia Commons and Cavin et al. (2015, fig. 3.C1).

features with A. *araripensis* (Marise 1986). Another mawsoniid specimen from Morocco is UMI-30 (see fig. 4.43), a right principal coronoid (a jawbone) that is clearly from a mawsoniid as it is saddle shaped, with a posterior end rising higher than the anterior end (Forey 1998; Cavin et al. 2015). The difference is that the angular process is smaller than in *Mawsonia*, leading to a cautious referral to *Axelrodichthys* (Cavin et al. 2015).

A lot of what we know of A. *lavocati* must come from comparisons with A. *araripensis*. Many of the A. *araripensis* specimens from Brazil are juveniles (Yabumoto and Brito 2013). This has led to suggestions that this marine species traveled to quiet lagoons to reproduce en masse (Yabumoto and Brito 2013) and maybe even gave birth to live young, like the modern coelacanth *Latimeria* (Grandstaff 2018). The young would then remain in these lagoons until they were capable of surviving in the open ocean.

Dipnoi

NEOCERATODUS

The Dipnoi, or lungfish, possess a host of primitive features compared to other osteichthyans, yet they also possess the ability to breathe air. As such, their relatives, the rhipidistians, are considered the ancestors of all land vertebrates (Gordon et al. 2004).

Fig. 4.44. A reconstruction of *Neoceratodus africanus*. Artwork by Joschua Knüppe.

One such lungfish from North Africa is *Neoceratodus africanus* (fig. 4.44). Since we have living examples of this genus (N. *forster* of Australia) to work with, one might think that we'd have a great idea of this animal's anatomy and lifestyle. Not so. The tale of N. *africanus* and other African lungfish species is a classic example of how mistakes in the literature can go unnoticed for years until we have a tangled mess to unravel.

Haug (1905) discovered the fragmentary remains of a lungfish from North Africa. These he provisionally assigned to the genus *Ceratodus*. The fact that these belonged to *Neoceratodus* went unnoticed until the 1980s (Martin 1982a). More examples were found by Stromer (1914), who followed Haug and assigned them to the genus *Ceratodus*. This was hardly an unreasonable decision, as he had only tooth plates to work with. Tooth plates are a thick mass of dentine and petrodentine attached to the palate and lower jaw, with projecting enamel-covered ridges that function like individual teeth.

We must also remember that Stromer's lungfish collection was huge, with more than 350 specimens to catalog (Peyer 1925). The task would have been overwhelming, especially as he was facing a series of mounting financial and personal problems at the time (Nothdurft et al. 2002). Ultimately, most of this collection would go undescribed, except for those specimens analyzed by Peyer (1925) and Weiler (1930).

Peyer (1925) accurately noted that there was a degree of variation in the tooth plates of N. *africanus* (or C. *africanus* as it was then known) but concluded that there could be any number of ridges on the tooth plate in this species. This trend of accepting any degree of variation would be repeated by Arambourg and Joleaud (1943) and Tabaste (1963). Then Martin (1984a) ultimately just made things worse. While he was right on the money in placing *Ceratodus africanus* in the genus *Neoceratodus* (Martin 1982a), he later synonymized C. *pectinatus* and C. *brasiliensis* with N. *africanus* (Martin 1984a).

In his defense, it is difficult to tell the difference between *Ceratodus* and *Neoceratodus*, and Martin did base this decision on the results of biometric analysis. However, Churcher and De Iuliis (2001) note that the analysis itself was flawed because the sample size for C. *pectinatus* was too small (n = 7) to produce a robust analysis. Thus, I've decided to follow the description given by Tabaste (1963) as amended by Churcher and De Iuliis (2001).

Cavin and colleagues (2015) estimate that N. *africanus* would have been less than 1 meter (3 feet) long. Like the surviving member of its genus, N. *africanus* would have been primarily carnivorous, feeding on both invertebrates and vertebrates alike (Lake 1978).

RETODUS

Churcher and De Iuliis (2001) raised the possibility that more than one species of lungfish was present in Bahariya. This prediction came to pass when Churcher and colleagues (2006) announced the discovery of a new genus, *Retodus tuberculatus* (fig. 4.45), from Egypt's Kharga and Dakhleh Oases. Other specimens were found in Algeria and one tooth plate from Bahariya.

As with *Neoceratodus africanus*, *Retodus tuberculatus* was originally considered to be a species of *Ceratodus* (Tabaste 1963). Martin (1981,

Fig. 4.45. A reconstruction of *Retodus tuberculatus*. Artwork by Joschua Knüppe.

1982a, 1982b, 1984a, 1984b) later synonymized it with N. *africanus*. However, a tooth plate from the Kharga Oasis in 1998 showed features that are distinct from both *Ceratodus* and *Neoceratodus*, including characteristics previously unknown in the tooth plates previously assigned to the species *tuberculatus*.

This distinct morphology led to the erection of a new genus, *Retodus*. The main difference between *Neoceratodus* and *Retodus* is that *R. tuberculatus* has only four ridges on its tooth plate. These are also linked by a series of smaller cross ridges and hollows.

In behavior and overall form, *R. tuberculatus* was presumably similar to N. *africanus*. However, the *R. tuberculatus* tooth plates are huge by lungfish standards, with the largest being 100 mm long by 36 mm wide (Churcher et al. 2006). This suggests that *R. tuberculatus* was a giant, several meters long (Churcher 1995).

Another critical difference is how *R. tuberculatus* used its tooth plates to process food. Most lungfish crush their prey between the plates. In *R. tuberculatus*, the evenly spaced ridges run transversely across the plate, which makes the usual rotary motion impossible. Instead, prey would be sliced by ridges sliding past one another as the jaws closed, and then crushed when the ridges of the upper tooth plate hit the bottoms of the furrows on the lower plate, and vice versa (Churcher et al. 2006). This unique tooth plate morphology suggests that *R. tuberculatus* may have possessed a highly modified jaw. But as no *R. tuberculatus* jaw has ever been found, this cannot be confirmed.

The tooth plates and the large size of this animal would have given it a unique place in the North African ecosystem. The unusual tooth plate morphology shows that *R. tuberculatus* was feeding differently, or possibly on different prey, from all other African lungfish. Also, given that lungfish are opportunistic predators feeding on a wide range of prey (Lake 1978), its larger size would have only broadened its feeding envelope and allowed it to attack prey of reasonable size.

ARGANODUS

Another species of North African lungfish is *Arganodus tiguidiensis* (fig. 4.46), a reasonably widespread species known from Algeria, Morocco, Niger, and Brazil (Tabaste 1963; de Broin et al. 1974; Castro et al. 2004; Candeiro et al. 2011; Cavin et al. 2015).

Fig. 4.46. A reconstruction of *Arganodus tiguidiensis*. Artwork by Joschua Knüppe.

The Moroccan specimen is known from an isolated and damaged tooth plate (Cavin et al. 2015). It can confidently be assigned to *A. tiguidiensis* because it's twice as long as it is wide and has ridges running to the posterior edge; these are diagnostic traits for *A. tiguidiensis* (Tabaste 1963).

"*CERATODUS*"

Given the tangled history of African lungfish, you can imagine my suspicion when dealing with tooth plates assigned to the genus *Ceratodus*, from Egypt and Morocco (Stromer 1936; Weiler 1935; Dutheil 1999b; Nothdurft et al. 2002). The fact that some of these identifications were only mentioned in passing made me wary, although Schaal (1984) gave a full description and Cavin and colleagues (2015) reported new material from Morocco with identifiable features that are diagnostic for the species *C. humei*.

However, experience has shown that remains previously assigned to *Ceratodus* have ultimately proved distinct on a generic level. The most likely reason for this is that the traits traditionally used to identify tooth plates are nondiagnostic. So while *Ceratodus* is retained herein, I would very much like to see the results of a Bayesian analysis on the Moroccan and Egyptian dipnoans, as I suspect that identification could change.

Fish *Incertae Sedis*

"*Idrissia* cf. *jubae*"

One taxon I don't know what to make of is *Idrissia* cf. *jubae* (fig. 4.47). Originally described by Arambourg (1954), this taxon is a common enough sight in online fossil stores, but I can find little published information on this genus. And the original paper naming the taxon is almost impossible to find.

Surprisingly, this doesn't stop people arguing over its taxonomic placement. The latest view is that it can't be placed in the teleost group and remains a wildcard taxon (Khalloufi 2010).

Fig. 4.47. A specimen assigned to *Idrissia* cf. *jubae*. Image provided by Thomas Bastelberge. See image at http://www.thefossilforum.com/index.php?/gallery/image/10989-idrissia-cf-jubae/.

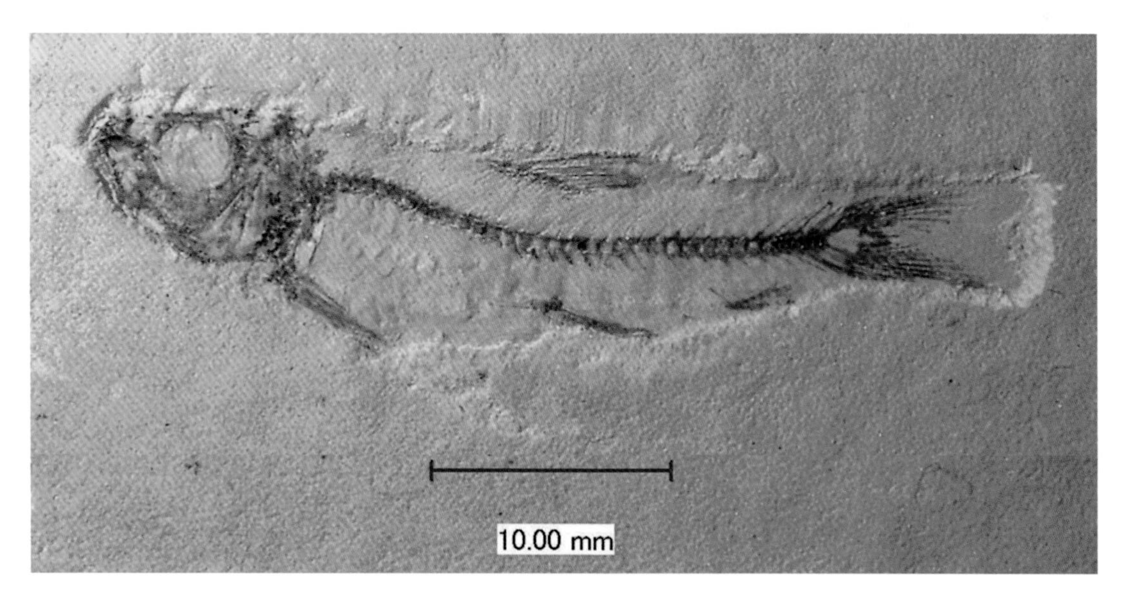

Fig. 4.48. A specimen assigned to the same taxa as UALVP 51599. Image provided by Thomas Bastelberge.

Fig. 4.49. A pycnodontid specimen tentatively assigned to the same taxa as UALVP 43595. Image provided by Thomas Bastelberge. See image at http://www.thefossilforum.com/index.php?/gallery/image/11004-pycnodont-non-det/.

"*Scapanorhynchus*"

The shark *Scapanorhynchus subulatus* was identified from Bahariya by Stromer (1927). Werner (1989) would later synonymize it with *Carcharoides planidens*; however, as *Carcharoides* is a taxon found in the Cenozoic era (Cappetta 1987), her conclusion has been thrown into doubt. As a result, the identity of this specimen is unclear.

"*Macromesodon*"

Cooper and Martill (2020) tentatively identify a damaged pycnodont prearticular bone (part of the jaw) as belonging to cf. *Macromesodon*. The origin of this specimen is unknown, although it came from somewhere near Erfoud, Morocco. While the tooth morphology and number of tooth rows match those of *M. daviesi* from Britain (which itself is a synonym of *M. macropterus*, according to Poyato-Ariza and Wenz [2002]), more specimens are needed before we can be confident in this referral.

Specimen 19 11 X II

Stromer (1936) identified a partial ichyodectoid tooth from Egypt; he referred to *Saurodon*. However, this referral is unlikely as *Saurodon* did not evolve until after the Cenomanian and is restricted in its range to Europe and North America (Everhart 2014). So, while there are similarities to *Saurodon*, the Egyptian specimen is best left as Ichyodectoidei *incertae sedis*.

Specimen UALVP 51599

Unfortunately, the affinities of UALVP 51599 are nowhere near as settled (see fig. 4.48). This species is 102 mm long, and the depth from the base of the dorsal fin to the underside is only 15% of the fish's standard length (Murray et al. 2013). The possibly modified anterior neural arches (the bony protuberances projecting from the tops of the vertebrae) may indicate that it belongs with the Ostariophysi (Murray et al. 2013), but the neural arches are too badly preserved to be certain.

Specimen UALVP 51625

While this species is known from an almost complete specimen, a large number of the bones have been damaged during the preservation process (Murray et al. 2013). The specimen was 81 mm long, with the head accounting for 29 mm, or 36%, of that length. This was a reasonably deep-bodied fish, 23 mm from the base of the dorsal fin to the underside (Murray et al. 2013).

Murray and colleagues (2013) have tentatively assigned this species to the Chanoidei. While I agree with this provisional assignment, we will need better-preserved specimens before it can be named.

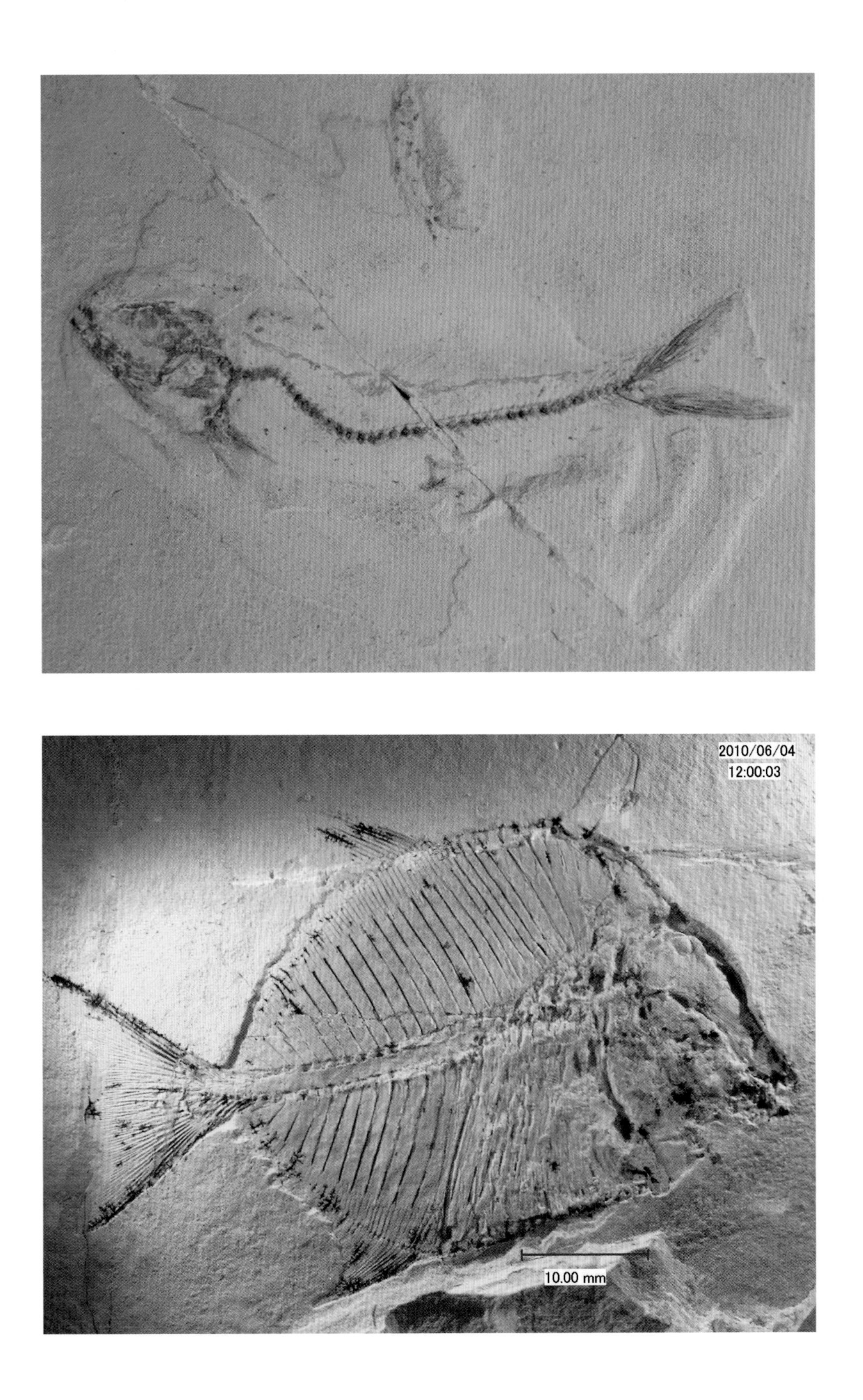
2010/06/04
12:00:03
10.00 mm

Specimens UALVP 43595, UALVP 51608, UALVP 51667, UALVP 51713, and UALVP 51669

This taxon (see fig. 4.49) is a member of the Pycnodontidae because it possesses a parietal peniculus (a bony strut attached to the parietal bone), which is an apomorphy of this family (Nursall 1996; Murray et al. 2013). While the parietal peniculus is not visible in all of the specimens, this might be a result of poor preservation.

Another specimen discovered by Cavin and colleagues (2010)—referred to as fig. 8A in their publication—was assigned to this taxon, on the basis of its general appearance (Murray et al. 2013). However, this referral was extremely tentative as they did not see the specimen themselves and the specimen was poorly illustrated (Cavin et al. 2010); thus it is only provisionally referred to the same genus as the others herein.

Unnamed Ellimmichthyiform

The most recent addition to the ellimmichthyiforms was found in Morocco by Khalloufi and his team. Still unnamed, this new taxon is known from seven specimens and can be identified by its triple rows of armored scutes, some of which have spines and other bony projections (Khalloufi et al. 2017).

Historic Taxa

Stromerichthys aethiopicus was named on the basis of material found by Stromer (Weiler 1935). Torices and colleagues (2010) view *Stromerichthys* as problematic, and Cavin and colleagues (2015) agree, as they wrote that *Stromerichthys* is a chimaera: the operculum (facial bones that also protect the gills) belongs to *Mawsonia*, the maxilla is from *Bawitius*, and the scales are from *Obaichthys*. This means that *Stromerichthys* is now a *nomen nudum*. The name exists but has no material assigned to it. For this reason, *Stromerichthys* is included herein only for those interested in the history of African paleontology.

Fig. 5.1. A reconstruction of *Oumrkoutia anae*. Artwork by Joschua Knüppe.

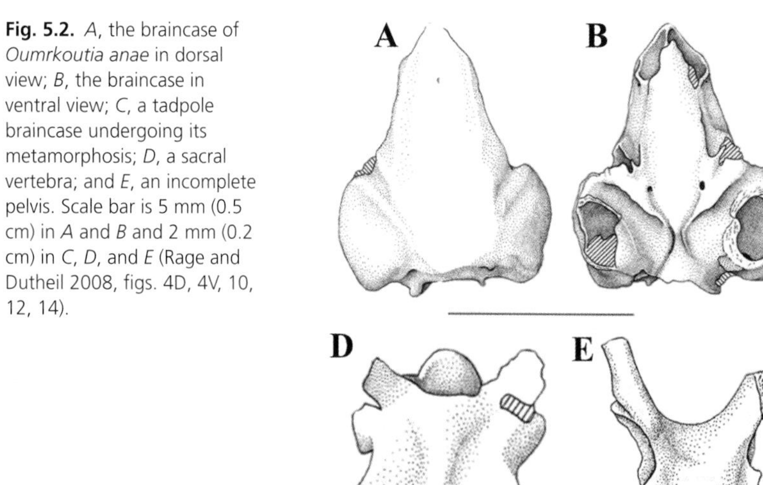

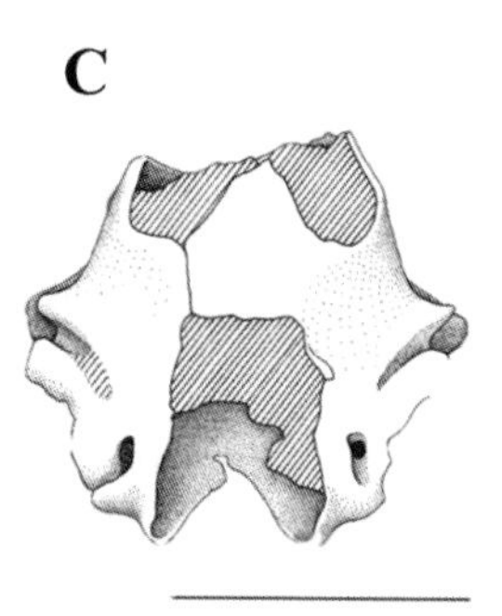

Fig. 5.2. *A*, the braincase of *Oumrkoutia anae* in dorsal view; *B*, the braincase in ventral view; *C*, a tadpole braincase undergoing its metamorphosis; *D*, a sacral vertebra; and *E*, an incomplete pelvis. Scale bar is 5 mm (0.5 cm) in *A* and *B* and 2 mm (0.2 cm) in *C*, *D*, and *E* (Rage and Dutheil 2008, figs. 4D, 4V, 10, 12, 14).

The Fauna of North Africa: Vertebrates (Tetrapoda)

5

The Tetrapoda is the group that includes all four-limbed vertebrates: reptiles, mammals, birds, and amphibians. Even animals that have lost some or all of their limbs, such as snakes, belong to this group because they still evolved from a four-limbed ancestor.

Amphibians

Amphibians can be easily recognized and defined by most people. Frogs and their kin are omnipresent in most habitats, despite their current decline in numbers (McCallum 2007). Excluding the extinct groups, the amphibians are neatly divided into three groups: the Anura (frogs), Gymophiona (caecilians), and Caudata (salamanders).

Oumrkoutia

As a group, the Anura hit the evolutionary ground running. One of the earliest proto-frogs, *Triadobatrachus*, which lived 250 mya, already had the basic bauplan: large head, shortened torso, and large hind limbs (Piveteau 1936). Cenomanian frogs were no different in their refusal to deviate from the classic design.

The most well-known species of amphibian from this region is *Oumrkoutia anae* (Rage and Dutheil 2008). *Oumrkoutia anae* (fig. 5.1) is known from twenty-one individuals encompassing both cranial and noncranial elements. Some of the specimens are from juveniles, giving us a good picture of the changes this species went through during ontogeny, although Rage and Dutheil are unclear about whether these specimens are tadpoles or individuals that have developed legs. Their use of the word *larval* may suggest the former, but they also mention "immature individuals," which may suggest the latter (see fig. 5.2).

Oumrkoutia anae can be identified as a pipid frog, sharing eleven features with this group (Rage and Dutheil 2008). The Pipidae are a strange clade. They have flattened bodies, are almost totally aquatic, and lack tongues (Cannatella and Trueb 1988). Other pipoids are known from the Cretaceous of Gondwana (Báez 1981), and members of this group still exist in South America and Africa (Ford and Cannatella 1993).

However, *O. anae* possesses features that make the genus unique within this group—so many that it seems to have more differences than similarities. These features include separate paired nasals and a skull that is flat, but not wedge shaped, in lateral view, unlike those of other pipoids (Rage and Dutheil 2008); adults even retain their *fissura metotica*—an opening in the skull—a trait that is usually found only in tadpoles (Rage and Dutheil 2008).

More typical pipoid traits are to be found in the vertebrae, although even there *O. anae* differs slightly from other pipoids by possessing longer neural arches. I wonder if *O. anae* represents a subgroup within the pipoids that was unique to Mesozoic Africa, but this cannot be verified because we lack the data, which makes the exact placement of *O. anae* within the Pipidae impossible (Rage and Dutheil 2008).

Kababisha

The salamander *Kababisha* (fig. 5.3), from Morocco, is known from thirty-eight vertebrae and two partial braincases (Rage and Dutheil 2008). A member of the Sirenidae, these salamanders are distinguished by their elongated, eellike bodies. They retain gills even as adults and only possess forelimbs; the hind limbs are completely atrophied (Farrand 1982; Lanza et al. 1998).

Fig. 5.3. A reconstruction of *Kababisha* sp. Artwork by Joschua Knüppe.

Rage and Dutheil (2008) note that the sirenid remains from Morocco (see fig. 5.4) are similar to *Kababisha sudanensis*, a species from the Cenomanian of Sudan (Evans et al. 1996). The differences are that the occipital condyle (the knob at the back of the skull, where it connects to the vertebrae) of the Moroccan species is longer than that of *K. sudanensis* and has a marked depression that is absent in *K. sudanensis*. The two Moroccan braincases also differ from each other, indicating that there were

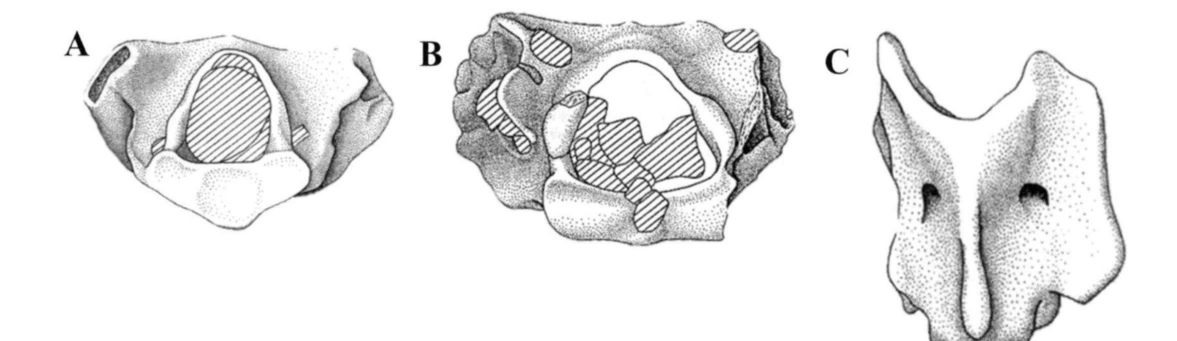

Fig. 5.4. *A* and *B*, a partial braincase of *Kababisha*; and *C*, vertebra, in ventral view (Rage and Dutheil 2008, figs. 1[p], 2[p], 3[v]).

either two different species present or just one species with pronounced sexual dimorphism (Rage and Dutheil 2008).

While Rage and Dutheil (2008) decided to be cautious and refer the remains to *Kababisha* sp., Titus and Loewen (2013) question this assignment. Rage and Dutheil (2008) themselves note some features that are reminiscent of the microsaurids (a group of Lepospondyl amphibians). Regardless, the Sudanese and Moroccan *Kababisha* are both related, whatever their taxonomy placement may be.

Amphibia *incertae sedis*

There were at least two other species of frog in the Cenomanian of Morocco (see fig. 5.5); the morphology of these isolated vertebrae and cranial material differs not only from those of O. *anae* but also from each other (Rage and Dutheil 2008). Beyond that, it's impossible to assign these remains to any Anuran family.

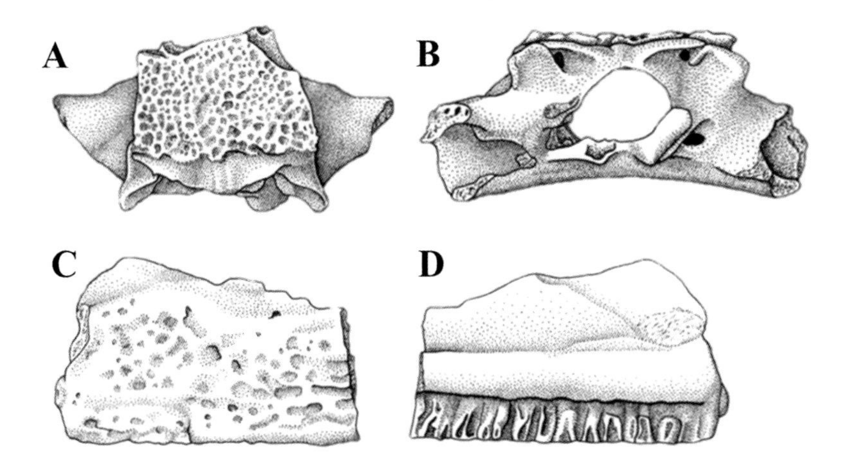

Fig. 5.5. *Anuran incertae sedis* from Morocco. Partial braincase in *A*, dorsal view; and *B*, posterior view. Partial maxilla in *C*, lateral view; and *D*, mesial view (Rage and Dutheil 2008, figs. 15–16).

Mammalia

While dinosaurs have become more and more popular in recent years, mammals are still well represented in many Mesozoic locations (Meng et al. 2006; Foster 2007). Unfortunately, this does not apply to the Cenomanian of North Africa, where the mammalian record is sparse.

Mammalia *incertae sedis*

SPECIMEN ZIN PC 1/31

The only mammal fossil we have is one indeterminate caudal vertebra from Libya (see fig. 5.6). The vertebra lacks a neural arch, which suggests that it was from the distal part of the tail (Nessov et al. 1998).

The height and width of the specimen, which is 4.8 mm long, are similar to those of the morganucodontid mammal *Eozostrodon* from Great Britain. The vertebra tapers more toward its posterior end than the vertebrae of *Eozostrodon*, suggesting that this species had a shorter tail (Nessov et al. 1998).

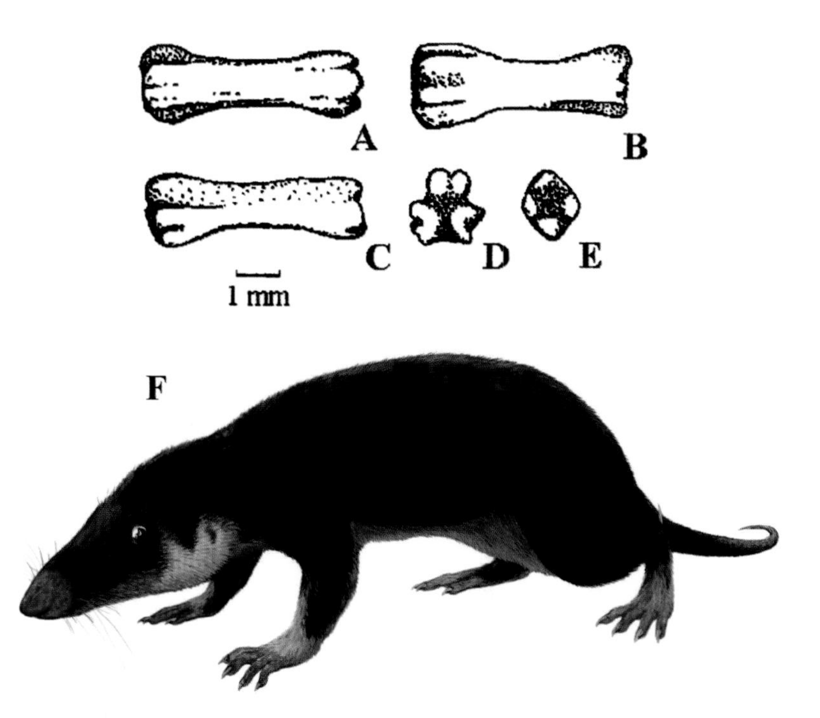

Fig. 5.6. Specimen Zin PC. Caudal vertebrae in *A*, dorsal; *B*, ventral; *C*, lateral; *D*, anterior; and *E*, posterior views. A reconstruction of the *morganucodontid* mammal, *Morganucodon* (F). The North African genus would probably have been similar in general appearance. Copyright 1998. From Nessov et al. (1998). Reproduced by permission of the Society of Vertebrate Paleontology, http://www.vertpaleo.org. Artwork by Michael B. H. / CC-BY-SA-3.0. See Wikimedia Commons.

Mammalia Trace Fossils

Mammal footprints have been found in Tunisia (Contessi 2013a), although the limited number of specimens prevented Contessi from assigning them to any ichnotaxon. The form is known from two footprints, each about 30 mm long, which were almost certainly made by the same individual (Contessi 2013a).

Reptilia: Testudines

Testudines, also known as the Chelonia, contains turtles, terrapins, and tortoises. This group is characterized by a shell composed of the dermal bones, ribs, and backbone.

Apertotemporalis

Nobody seems to have given *Apertotemporalis* much thought since it was first described (Stromer 1934b). The reason for that is the destruction of the holotype. I was under the impression that *A. baharijensis* (reconstructed on plate 6) was known from a partial skeleton that included the skull. So imagine my horror when I finally got hold of a copy of Stromer's original paper and discovered that this genus is known only from fragmentary pieces: a partial cranium with only the ear region preserved in detail.

Part of the left side of the skull is even misshapen, which Stromer (1934b) believed to be a pathological condition. The braincase itself is 95 mm wide. While the jaw joints were narrow and slightly convex, *A. baharijensis* had a pronounced processus ectopterygoideus (a bony plate connecting to the upper jaw), suggesting that it had a powerful bite (Stromer 1934b).

While only estimating, Stromer suspected that A. *baharijensis* had a shell approximately half a meter long. Ecologically, A. *baharijensis* would have been a generalist omnivore, feeding on both plant and animal matter.

Galianemys

Galianemys from Morocco is much better known. Currently, two species of *Galianemys* are known—G. *whitei* and G. *emringeri* (reconstructed on plate 6)—and each of those species is known from six skulls. A small, pointed skull characterizes the genus (see fig. 5.7), which has traits that are unique among the bothremydid turtles (Gaffney et al. 2002).

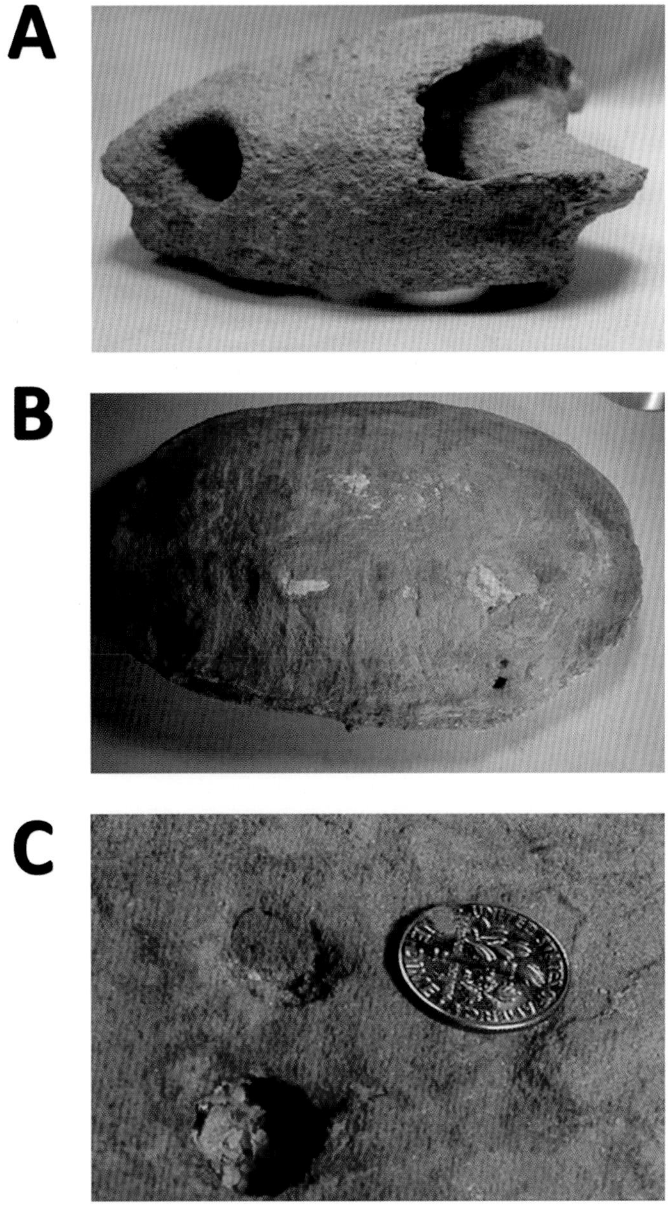

Fig. 5.7. *A*, a complete cranium of *Galianemys* sp.; and *B* and *C*, a shell assigned to *Galianemys*. Note the two puncture wounds on the shell; it's difficult to deduce what creature made them but left the rest of the shell undamaged, although the owner suggests a crocodilian. Images courtesy of John P. Adamek, Fossilmall.com. See images *B* and *C* at http://www.fossilmall.com/Stonerelic/vertebrates/MTU2/MTU2.htm.

Hamadachelys

Hamadachelys escuilliei (reconstructed on plate 6) is also known only from its skull. It was originally described by Tong and Buffetaut (1996), and Gaffney and colleagues (2006) have given a revised diagnosis for this turtle. Although the skull is mostly complete, it's been suggested that some missing features, such as the hyoid bones and branchial horns, were mistakenly destroyed during preparation (Gaffney 2001).

Dirqadim

Once again, the holotype of *Dirqadim schaefferi* is only a skull. *Dirqadim schaefferi* is yet another side-necked turtle (reconstructed on plate 6), albeit a member of the Euraxemydidae. This species is characterized by its heavyset skull (Gaffney et al. 2006). This taxon lived in both rivers and deltas, suggesting that it was tolerant of brackish water (Gaffney et al. 2006).

Testudines *incertae sedis*

SPECIMEN 911 XII 33

1911 XII 33 is made up of separate bones—a humerus, femur, and ilium—that were found together and that almost certainly belonged to the same animal (Stromer 1934b). On the basis of the femur, 911 XII 33 was assigned to the Pelomedusidae, although that's impossible to prove without more material.

SPECIMENS 1911 XII 55, 1912 VIII 94, AND 1912 VIII 95

Stromer (1934b) also reported heavily ornamented shell fragments, probably a plastron, about 150 mm long and 65 mm thick. On the inside of one of the fragments is what appears to be a partial rib (Stromer 1934b). Stromer assigned these fragments to the side-necked turtles.

SPECIMEN 1922 X 53

1922 X 53 is a partial cervical vertebra, 550 mm long and 310 mm wide. The neural arch is pronounced and slanted (Stromer 1934b). Nothing else was published about this vertebra. However, it's possible that some or all of this postcranial material reported by Stromer belongs to *Apertotemporalis*. There is no way to be sure until we have another specimen of *Apertotemporalis*, preferably with postcranial material, to compare against.

SPECIMENS GZG.V.22.664 AND GZG.V.22.665

GZG.V.22.664 is an anterior plastron from Morocco, which has large intergular scutes and smaller gular scutes (Hans-Volker 2010). Hans-Volker thinks these shells belong to *Galianemys* sp., but it's impossible to be sure given that *Galianemys* is only known from skulls, so we have no basis for comparison. This is further complicated by the fact that many shell fragments show morphological differences from one another (some are narrow and others broad, with sutures that are either angular or straight) that indicate they belong to two separate species.

Other Testudines

Recently a whole host of turtle postcranial material has been gathered from the Cenomanian Maghrabi Formation of Egypt (Abuelkheir and Abdelgawad 2017). But these specimens have yet to be described.

Plate 6. The turtles of North Africa. *A*, *Hamadachelys*; *B*, *Dirqadim*; *C*, *Galianemys whitei*; *D*, *Galianemys emringeri*; and *E*, *Apertotemporalis*. Images not to scale.

Testudines Trace Fossils

Numerous turtle footprints are known from Morocco and Libya (Belvedere et al. 2013). Belvedere and colleagues (2013) claim that turtle tracks are the second most common footprint found in Morocco. Ibrahim and colleagues (2014a), however, did not find a single specimen, although they admit that this could be because footprints were preserved only in certain areas.

And some of these tracks are huge, with some footprints being 100 mm or larger (Belvedere et al. 2013). This suggests that either the known species of turtle in this region grew larger than the current specimens suggest or there is another species of turtle out there that we have yet to find in the fossil record (Ibrahim et al. 2014a).

Squamata

Until relatively recently, the Squamata was neatly divided into two groups: the snakes (Ophidia) and the lizards and amphisbaenians (Lacertilia). But life is not that simple, and recent research has shown that the Lacertilia is paraphyletic (Estes et al. 1988). Some infraorders such as the Iguania (iguanas, chameleons, etc.) belong in the same group as snakes, yet other types of lizard (such as geckos) don't. The new clade that unites the snakes and their closest relatives among the lizards is called the Toxicofera (Fry et al. 2009).

Simoliophis

The oldest snakes known date to roughly 112 mya (Vidal et al. 2009), although their origin probably goes back further than that. Hsiang and colleagues (2015) estimate that they evolved 128.5 mya. The origin of the snakes is important to our discussion because the Cenomanian of North Africa was home to a primitive genus named *Simoliophis* (Sauvage 1880; Nessov et al. 1998; Nothdurft et al. 2002).

Currently, there are three known species of *Simoliophis*: *S. rochebrunei*, *S. libycus*, and an unnamed species from Bahariya (fig. 5.8). Nopcsa

Fig. 5.8. A reconstruction of *Simoliophis* sp. from Bahariya. Artwork by Joschua Knüppe.

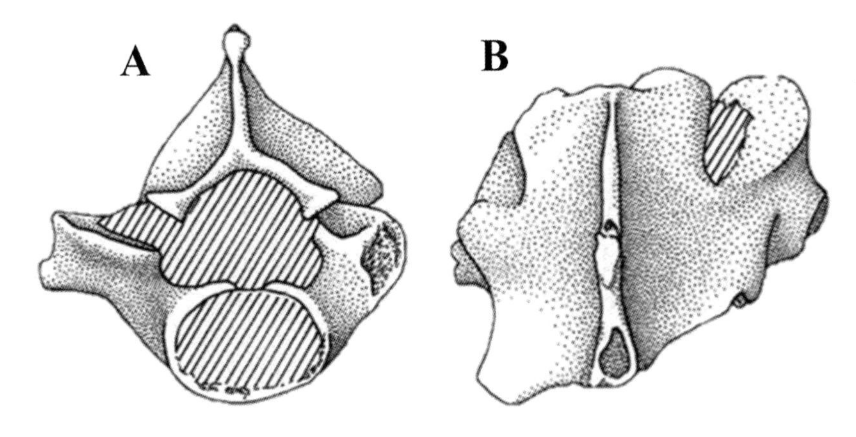

Fig. 5.9. Unidentified vertebra left as *Simoliophiidae incertae sedis* in *A*, anterior; and *B*, dorsal views (Rage and Dutheil 2008, fig. 18). Reproduced by permission of https://www.schweizerbart.de.

(1925) assigned all of these to *S. rochebrunei*; however, Rage and Escuillie (2003) point out that Nopcsa's reconstruction was chimeric and consider the three species to be distinct. That opinion is followed herein. Since this means that *S. rochebrunei* is now known only from Europe, it will not be considered further. The remaining two species are both from Africa.

One reason that the *Simoliophis* from Bahariya has not been assigned a name is that it's supposedly chimeric: the bones from two snake species mixed together (Rage and Escuillie 2003; Rage and Dutheil 2008). However, a later review of this genus has concluded that such variation in the vertebral column is normal for the genus, so all the Bahariya material may belong to *Simoliophis* after all (Rage et al. 2016).

Rage and Dutheil (2008) also raise the possibility of a fourth species, noting that the Moroccan *Simoliophis* differs from the others. However, they could not deduce whether these features are taxonomically important. Surprisingly, Rage and colleagues (2016) make no further comment on that issue in their otherwise detailed review. So, while currently left as *S.* cf. *libycus*, the Moroccan *Simoliophis* may be given its own species in the future.

One trait found in the vertebrae and ribs of *Simoliophis* is pachyostosis, a thickening of the bones. Hoffstetter (1955) suggests that increased bone density evolved to provide the animal with natural ballast while swimming. This is supported by Scanlon and colleagues (1999), whose calculations show that the density of the animal's body tissues must have been near that needed to maintain neutral buoyancy.

We can also fill in some of the blanks by looking at other members of the Simoliophiidae, such as *Eupodophis* from Lebanon and *Pachyrhachis* from Palestine (Haas 1979; Lee and Caldwell 1998). Both species retained the hind limbs, and as all three taxa are closely related, we can be confident that *Simoliophis* also possessed hind limbs (Haas 1979; Rage and Escuillie 2003). Indeed, *Simoliophis* and *Pachyrhachis* are so closely related that they may even be synonyms (Rieppel and Head 2004), although that cannot be proved at this time (Rage et al. 2016).

The Simoliophiidae stand out because of the huge sizes some of their members attained. While no size estimate has been given for *S. libycus* as far as I know, Knüppe (2015) estimates that the Bahariyan *Simoliophis* could reach 4 meters (13 feet) in length—the size of the modern African rock python.

Simoliophis was an obligate marine genus; its fossils are only found in marine or near-marine sediments. Buffrénil and Rage (1993), Scanlon and colleagues (1999), and Rage and colleagues (2016) have shown that *Simoliophis* was capable of long dives and was adapted for life on the bottom of shallow reefs. *Simoliophis* was a slow swimmer, so it's likely that it hunted by ambush (Rage et al. 2016).

Norisophis

Known only from isolated vertebra, *Norisophis begaa* is special in that it may be a true snake. This suggests that while the Ophidia may have evolved in Laurasia, the crown Serpentes may be of Gondwanan origin (Klein et al. 2016).

Norisophis begaa is characterized by its incredibly reduced neural spine. The vertebra itself is 7.46 mm wide and 7.02 mm long (Klein et al. 2016). Given the limited material, no reconstruction of *N. begaa* has been attempted. Grandstaff (2018) suggests that *N. begaa* may have been about a meter long, on the basis of comparison with modern snakes.

Ecologically, *N. begaa* was probably a burrower since the small neural spine is a feature found in burrowing snakes and amphisbaenians.

Lapparentophis

The Lapparentophiidae are a poorly known family of snakes found only in Africa during the Albian to Cenomanian periods, with the possible exception of *Pouitella*, from the Cenomanian of France. The namer of this family, *Lapparentophis defrennei*, hails from Algeria, near the border with Libya (Hoffstetter 1959).

Known from three thoracic vertebrae, *L. defrennei* has such features as a narrow neural arch, two deep fossae at the bottom of the zygantrum (the socket where the zygosphene of the following vertebra fits, forming a joint), and a small neural canal and zygosphene (part of the neural arch). Some of the other features found in *Lapparentophis* are rare or unknown in extant snakes, as you would expect given that *Lapparentophis* is one of the earliest snakes known (Hoffstetter 1959).

Another specimen, consisting of two vertebrae, was found in Morocco and assigned to a new species, *L. ragei*. This species can be differentiated from *L. defrennei* because of several features, such as the longer length of its vertebrae, its overall smaller size, and the presence of parazygosphenal foramina—although this varies between the specimens, as it's present on the holotype but small or absent on the paratype. *Lapparentophis defrennei* seems to lack this feature entirely (Vullo 2019).

Ecologically, both species were terrestrial, living on floodplains, although the presence of marine fish alongside *L. defrennei* suggests that the coastline was nearby (Hoffstetter 1959).

Fig. 5.10. A reconstruction of *Jeddaherdan aleadonta*. Artwork by Joschua Knüppe.

Jeddaherdan

Iguanas are some of my favorite extant animals, so *Jeddaherdan aleadonta* (fig. 5.10) was a welcome discovery. This taxon is known from just a partial jaw with four teeth (Apesteguía et al. 2016). The amount of tooth wear and the lack of new teeth erupting from the back of the jaw suggest that it was almost fully grown, while the fact that the back teeth hadn't fully fused with the labial (lip) wall shows that it was still not entirely mature yet (Apesteguía et al. 2016). The teeth of *J. aleadonta* suggest that this taxon was an omnivore, with a predilection for soft plant matter (Apesteguía et al. 2016).

Bicuspidon

The latest addition to the lizard fauna is *Bicuspidon hogreli*, which is known from a single right dentary purchased from a local collector (Vullo and Rage 2018). *Bicuspidon* belongs to the Polyglyphanodontia, a once incredibly diverse group that became extinct along with the dinosaurs.

The specimen is 23.8 mm long and has teeth that are not as tightly spaced and are much larger than those of the other species of *Bicuspidon*. As the genus name suggests, the taxon has teeth of different shapes and sizes, ranging from conical to bicuspid in morphology (Vullo and Rage 2018). Based on this, *B. hogreli* was most likely an omnivore.

Squamata *incertae sedis*

MADTSOIIDAE

The madtsoiid snakes are a varied group whose members constricted their prey as does the modern python. Twenty madtsoiid vertebrae, from different sites, are known from Morocco (Rage and Dutheil 2008). This species was reasonably small. The biggest vertebra in the collection has a centrum less than 5 mm in length. Rage and Dutheil (2008) mention the possibility that two madtsoiid species were present, citing the variation between the anterior and posterior vertebrae, but that is unlikely because it would mean that one is only known from anterior vertebrae and the other from only posterior vertebrae.

NIGEROPHIIDAE

Nigerophiid snakes are known from eighteen trunk vertebrae and four caudal vertebrae from Morocco (Rage and Dutheil 2008). These vertebrae are remarkably elongated with a narrow centrum. While they show the characteristics typical of the Nigerophiidae, Rage and Dutheil also mention unique features, so the referral to the Nigerophiidae is tentative as a result.

SPECIMEN NUMBER UNKNOWN

Currently, there are two species of unnamed lizard known from North Africa. The best specimen comes from Gara Sbaa and looks to be an almost complete skeleton. It was sold to an unknown buyer, so it may never be seen again.

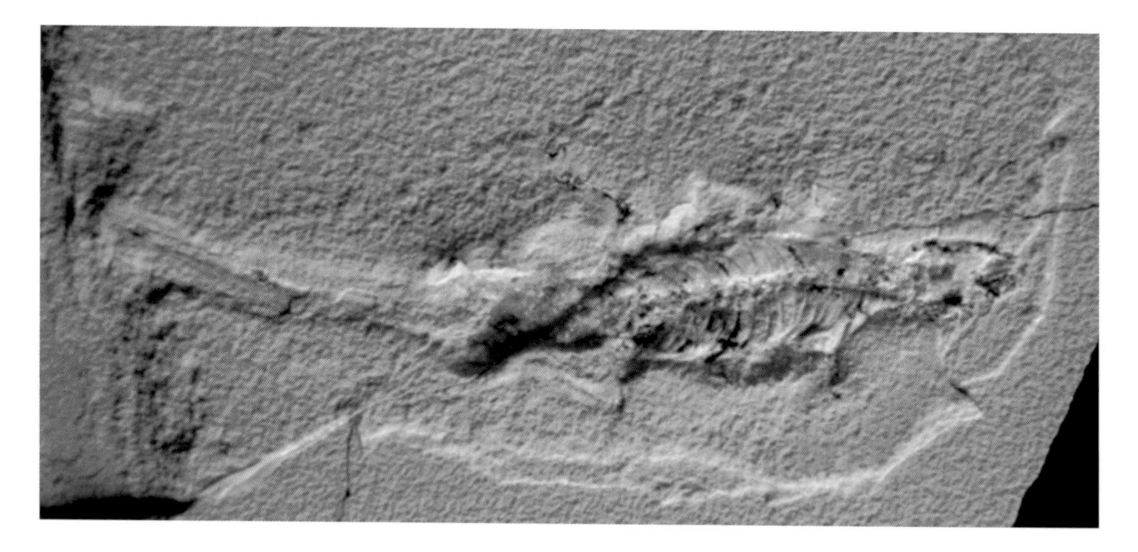

Fig. 5.11. The skeleton of an unnamed lizard from Morocco. The specimen's current location is unknown, and the only published photograph is too low in quality to make out any anatomical details. Reproduced from Martill et al. (2011, fig. 12). Copyright 2011. Published by Elsevier Masson SAS. All rights reserved.

What we know of this species comes from this single photograph (fig. 5.11) published by Martill and colleagues (2011). Pasini (personal observation 2013, in Garassino and Pasini 2018) has also seen a lizard fossil at Gara Sbaa, but it's unclear what specimen it was.

SPECIMEN UCRC PV115

The second specimen is known from a single vertebra, 4 mm in length, from the Tafilalt and Guir region (Rage and Dutheil 2008). The specimen probably possesses a zygosphene-zygantrum system (joints between the vertebrae), which, while rare in modern lizards, is common among ancient forms (Rage and Dutheil 2008). Rage and Dutheil (2008) suggest that this vertebra belongs to a limbless anguid lizard. No illustrations were provided.

Plesiosauria

While there were many types of marine reptile during the Mesozoic, the ones we're interested in are members of the Plesiosauria. The Plesiosauria can be split into two families: the well-known Plesiosauroidea, with their long necks, and the Pliosauroidea, with their short necks and large crocodile-like heads.

Polycotylide *incertae sedis*

Pliosaurs are known from the fragmentary remains of several individuals found in Bahariya (fig. 5.12), including vertebrae and part of a jaw. They were originally named *Peyerus capensis* by Stromer (1933, 1935) and were later referred to the genus *Leptocleidus*.

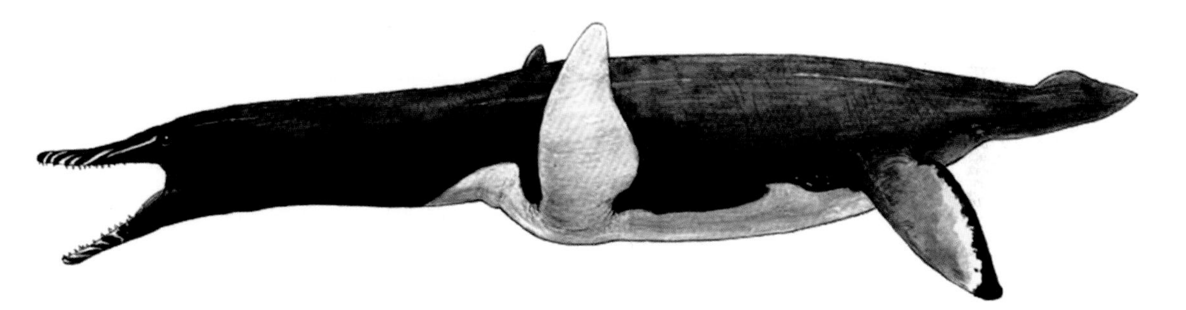

Fig. 5.12. A reconstruction of the Egyptian polycotylid. Artwork by Joschua Knüppe.

But was this Stromer's original intention? Sachs (2014) has untangled the history of *Leptocleidus capensis* and discovered that all is not what it seems. The holotype of *Leptocleidus capensis*, from South Africa, was originally named *Plesiosaurus capensis* by Andrews (1911). Stromer (1935), during his work on the Bahariyan pliosaurid, believed that the South African material warranted its own genus and so coined the name *Peyerus*.

Later, Persson (1963) realized that *P. capensis* belonged to *Leptocleidus* of Great Britain, and because the name *Leptocleidus* was created first, it has priority over *Peyerus*. Thus, it seems that the name *Peyerus* was only ever intended for the South African species; Stromer never named the specimens from North Africa, referring to them only as specimens A to D (Stromer 1936; Sachs 2014).

Indeed, the North African pliosaurid lacks many of the diagnostic traits seen in *Leptocleidus*. The shape of the mandible and vertebrae, for instance, are very different (Stromer 1935; Cruickshank 1997; Kear and Barrett 2011).

Sachs (2014) has mentioned his involvement with a planned annotated edition of Stromer's original manuscript on the subject, but until it's published, the Bahariyan pliosaurid is herein left as Polycotylidae *incertae sedis* pending any further revisions.

Archosauria: Crocodylomorpha

The Crocodylomorpha evolved roughly 230 mya, yet the first *true* crocodilians—the Eusuchia—only evolved 83.5 mya. Before that there was a whole array of strange creatures, sharing a common ancestor with the Eusuchia, which had gone off on their own evolutionary paths.

Elosuchus

Perhaps symptomatic of the confusion in crocodylomorph taxonomy is *Elosuchus* (Lavocat 1955a; de Broin 2002), as we can't seem to find a family for it. De Broin (2002) believed that it warranted its own group, the Elosuchidae. Later, others would place it within the Pholidosauridae (Fortier et al. 2011) or within a Pholidosauridae complex (de Andrade et al. 2011). More recent work seems to confirm that close relationship, placing the Elosuchidae in the clade Tethysuchia, although the validity of the Elosuchidae and Pholidosauridae is still not clear (Meunier and Larsson 2017).

Fig. 5.13. A reconstruction of *Elosuchus cherifiensis*. Artwork by Joschua Knüppe.

Elosuchus is traditionally separated into two species: *E. cherifiensis* from North Africa (fig. 5.13) and *E. felixi* from Niger. Although one Society of Vertebrate Palaeontology (SVP) abstract suggests that *E. felixi* is a synonym of *E. cherifiensis* (Meunier and Larsson 2015), Young and

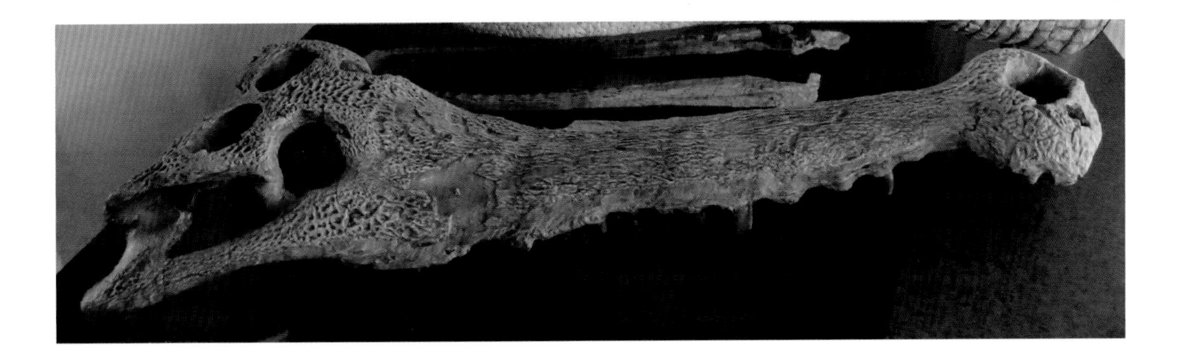

Fig. 5.14. The skull of *Elosuchus*. Credit Ghedoghedo / CC-BY-SA-4.0. See Wikimedia Commons.

colleagues (2016) disagree and give *E. felixi* its own genus, *Fortignathus*. Once Meunier and Larsson had time to conduct a phylogenetic analysis, they agreed that *E. felixi* lacked the apomorphies seen in *Elosuchus* (Meunier and Larsson 2017). But while *Elosuchus* lost one species, it gained another, *E. broinae* from Algeria (Meunier and Larsson 2017).

This taxon is best known from a collection of skulls (see fig. 5.14); at least fifty partial specimens and several hundred skull fragments are known from both species (de Broin 2002). Among this material are cervical and dorsal vertebrae, designated as paratypes for *E. cherifiensis* (de Broin 2002). However, this wealth of material needs to be reviewed to ensure that all these specimens do indeed belong to *E. cherifiensis*, given the removal of *E. felixi* to *Fortignathus*—especially as there could be yet another taxon present (Meunier and Larsson 2017; Ibrahim et al. 2020a).

Elosuchus was also described as possessing narrow, rectangular osteoderms that run in two rows down the back, as well as polygonal armored plates on its chest and stomach (de Broin 2002). Young and colleagues (2016) now question whether these osteoderms belong to *Elosuchus*, preferring to leave them as Crocodyliforms indeterminate.

Ecologically, *Elosuchus* would have been a giant fish eater, with some skulls measuring over 1 meter (3 feet) long. But like the modern False Gharial, it was large and robust enough to have also preyed upon terrestrial animals.

Aegyptosuchus

Now we come to the Aegyptosuchidae and the Stomatosuchidae. Both groups are limited to the Cenomanian of North Africa, and both evolved extremely platyrostral—elongated and flat—skulls.

Aegyptosuchus peyeri (fig. 5.15) is known primarily from a partial braincase and skull roof, 249 mm long (Stromer 1933). Interestingly, the skull of *A. peyeri* has a much smaller bony boss than its sister taxon *Aegisuchus witmeri* (Holliday and Gardner 2012).

The rest of the holotype includes the left scapula, the right coracoid (part of the shoulder girdle that connects with the scapula), part of the jaw, and a series of isolated vertebrae, from the cervicals down to the sacrals. Other skull fragments and a humerus found in the same area were not included. The difference in size between those specimens and the

Fig. 5.15. A reconstruction of *Aegyptosuchus peyeri*. Artwork by Joschua Knüppe.

holotype shows that they could not have come from the same individual (Stromer 1933).

At one time, the aegyptosuchids were considered members of the Stomatosuchidae because of their similar appearance (Carroll 1988). Even as late as 2020, some have attempted to resurrect the idea that *A. peyeri* was synonymous with *Stomatosuchus* (Ibrahim et al. 2020a), but I find the arguments put forward to be unremarkable, especially as neither appear to be closely related; the aegyptosuchids are eusuchians while the stomatosuchids are neosuchians, a different branch of the family tree altogether (Holliday and Gardner 2012).

Some attempt to get around this by suggesting that the Aegyptosuchidae are actually neosuchians as well, basing this argument on some vertebrae found in association with *A. peyeri* (Stromer 1933), but offer no re-description of the material. This also contradicts the phylogenetic analyses by Holliday and Gardner (2012). More likely any neosuchian traits were a result of this taxon's basal position in the eusuchian family tree.

While only estimating, Stromer (1933) suggested that *A. peyeri* would have equaled the Nile crocodile in size. The teeth of *A. peyeri* were rather short and blunt (Stromer 1933), suggesting that it fed upon large, possibly armored prey.

Aegisuchus

Following World War II, the Aegyptosuchidae were all but forgotten until *Aegisuchus witmeri* (fig. 5.16) was found. Known from a braincase, *A. witmeri* is diagnosed by autapomorphies that include a prominent rugose boss on the top of the skull. The morphology of the braincase also demonstrates that *A. witmeri* had a broad and flat skull like that of its sister taxon, *A. peyeri* (Holliday and Gardner 2012).

This taxon stands out because of its cranial ornamentation, the aforementioned enlarged bony boss on the center of the skull. This is interpreted as being a display structure. Extant crocodilians often show off their profiles during territorial displays and fights over mating rights. *Aegisuchus witmeri* might have used similar displays, raising and swinging its head to show off the thick tissue and skin supported by its skull boss (Holliday and Gardner 2012).

It's not often appreciated that *A. witmeri* was one of the largest crocodilians in history. This species possibly measured 15 meters (49 feet)

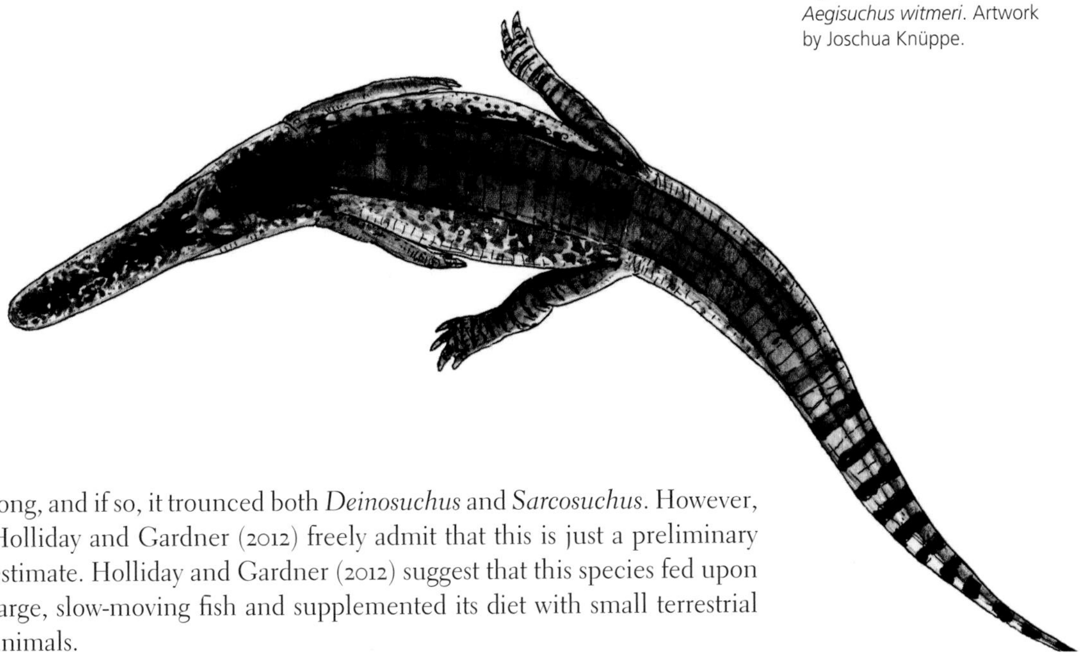

Fig. 5.16. A reconstruction of *Aegisuchus witmeri*. Artwork by Joschua Knüppe.

long, and if so, it trounced both *Deinosuchus* and *Sarcosuchus*. However, Holliday and Gardner (2012) freely admit that this is just a preliminary estimate. Holliday and Gardner (2012) suggest that this species fed upon large, slow-moving fish and supplemented its diet with small terrestrial animals.

Stomatosuchus

Many crocodilians from this region have platyrostral skulls, but *Stomatosuchus inermis* (fig. 5.17) took this trend to the extreme (Stromer 1925; Nopcsa 1926). So what was *S. inermis* doing with such an extensively modified set of jaws? One theory that gained early traction is that it was a filter feeder. Nopcsa (1926) based this theory on the fact that the species supposedly lacked teeth and suggests that *S. inermis* had a baggy throat pouch or gular sack.

While *S. inermis* may have had a gular sack, given its edentulous mandibles, the mandibular symphysis isn't preserved, making it difficult to say for sure (Motani et al. 2015), and the idea that it was used to filter feed is unlikely. I suspect that S. *inermis* was more like a pelican, using its throat pouch to scoop up fish (Marchant and Higgins 1990). Sereno and Larsson (2009) put forward an alternative hypothesis, suggesting that the stomatosuchids hunted by sitting motionless on the riverbed until prey swam into their bear-trap jaws.

Fig. 5.17. A reconstruction of *Stomatosuchus inermis*. Artwork by Joschua Knüppe.

Even the idea that S. *inermis* lacks teeth is questionable. No teeth were found with the holotype and only known specimen, but Stromer (1925) points out that the preservation was too poor to say for sure that they were absent, especially given that the skull had tooth sockets on the lower jaw. The fact that its closest relative, *Laganosuchus*, has teeth also supports that position.

Laganosuchus

Following World War II, no further Stomatosuchidae remains were discovered until 2009, when another member of the family was found. *Laganosuchus* is known from two species: *L. maghrebensis* from Morocco (fig. 5.18) and *L. thaumastos* of Niger. As the latter hails from a country to the south of North Africa, it will not be discussed further.

Ibrahim and colleagues (2020a) have suggested that *Laganosuchus* might be a senior synonym of *Aegisuchus*. I see no particular reason to believe this, beyond the fact that there is no overlapping material between the two taxa, especially as a phylogenetic analysis shows that making *Aegyptosuchus*, *Aegisuchus*, and *Laganosuchus* sister taxa would require an extra 281 steps (Holliday and Gardner 2012).

Fig. 5.18. A reconstruction of *Laganosuchus maghrebensis*. Artwork by Joschua Knüppe.

The teeth of *L. maghrebensis* are cone shaped with flattened crowns. The dentary symphysis (a cartilaginous joint) of *L. maghrebensis* also seems to have fused with age (Sereno and Larsson 2009). No size estimate was given for this species, although the Nigerien species could reach lengths of 6 meters (20 feet).

Libycosuchus

The third of Stromer's crocodyliforms is *Libycosuchus brevirostris* (Stromer 1914; Buffetaut 1976). Although the holotype specimen was destroyed, further material has been discovered over the years by Nothdurft and colleagues (2002), Tumarkin-Deratzian and colleagues (2004), and Buffetaut (1976). However, some of these specimens may be referable to *Hamadasuchus* (Larsson and Sues 2007).

Fig. 5.19. A reconstruction of *Libycosuchus brevirostris*. Artwork by Joschua Knüppe.

With its short snout and orbits so large they're almost froglike, *L. brevirostris* (see fig. 5.19) cuts an easily recognizable figure. Some compare this species to the hyena, based on its short snout and powerful jaw muscles (Buffetaut 1982). While I agree that *L. brevirostris* had a powerful bite, its teeth don't appear suited to bone crushing (Stromer 1914). I suspect that *L. brevirostris* was more like a reptilian honey badger—a nasty little predator that was opportunistic in its diet.

Lavocatchampsa

Like *Libycosuchus*, *Lavocatchampsa sigogneaurussellae* (fig. 5.20) was a Notosuchian and basal member of the family Candidodontidae (Martin and de Broin 2016). The holotype specimen is a partial skull, 38.5 mm long. Some parts are too badly crushed to make out much detail (Martin and de Broin 2016).

The teeth make this taxon stand out. This species has heterodont multicuspid teeth that range from subconical to molariform in shape. The fourth and fifth maxillary teeth are also the largest, with the sixth being the widest (Martin and de Broin 2016). The dentary teeth are subconical, with accessory cusps (Martin and de Broin 2016).

Ecologically, *L. sigogneaurussellae* would have fed on hard foodstuffs, such as crustaceans, gastropods, and insects.

Fig. 5.20. A reconstruction of *Lavocatchampsa sigogneau-russellae*. Artwork by Joschua Knüppe.

Hamadasuchus

At least superficially, *Hamadasuchus rebouli* (fig. 5.21) is a return to form for those expecting traditional crocodilians. While the exact taxonomic placement of *H. rebouli* has varied over the years (Buffetaut 1991, 1994), Larsson and Sues (2007) proposed a clade called the Sebecia for the Sebecidae and the Peirosauridae. The latter family includes *H. rebouli*.

Fig. 5.21. A reconstruction of *Hamadasuchus rebouli*. Artwork by Joschua Knüppe.

This species is well known, with both adult and immature specimens. A recent review suggests that there may have been a second species of *Hamadasuchus* in this region, although it's difficult to separate them, given the fragmentary nature of all the existing specimens (Figueredo et al. 2017). Ecologically, *H. rebouli* was a carnivore. However, it would have been a mostly terrestrial, doglike one.

Antaeusuchus

The newest crocodyliform from Morocco is *Antaeusuchus taouzensis*. NHMUK PV R36829 is an isolated mandible with eighteen teeth (Hunt 2018). Originally given only a brief description, the specimen ended up in a polytomy with several other unnamed notosuchians. A full description, which allowed us to better resolve the taxonomy, came out a few years later.

Now known from two mandibles, *A. taouzensis* is a sister taxon to Hamadasuchus, with conical teeth that become more compressed from the tenth tooth onward (Nicholl et al. 2021). This taxon just highlights how badly we need a partial notosuchian skeleton from this formation rather than bits and pieces. Once we have a Rosetta Stone to work with, then we can really begin to get a handle on their diversity.

Crocodyliform *incertae sedis*

"*ARARIPESUCHUS*"

Araripesuchus rattoides from Morocco is known only from parts of the dentary. It is characterized by an enlarged first dentary tooth, followed

Fig. 5.22. A reconstruction of the Moroccan "*Araripesuchus*." Artwork by Joschua Knüppe.

by a smaller second dentary tooth. The fourth dentary tooth is also twice the size of the others and caniniform in shape. These procumbent teeth gave this species its name: the "rat-like" *Araripesuchus* (Sereno and Larsson 2009).

Other specimens, mostly skull material, are known from Morocco. These individuals had skull lengths between 100 and 150 mm long, with one dentary being twice the size of the holotype specimen. Sereno and Larsson (2009) suggest that this material represents a northern species of *Araripesuchus*.

Figueredo and Alexander (2015) came to the opposite conclusion: A. *rattoides* has none of the diagnostic features of *Araripesuchus*, according to their revised diagnosis for this taxon, and so warrants its own genus. As the taxonomy is up in the air at the moment, nothing else can be stated about A. *rattoides* beyond the fact that it belonged to a Uruguaysuchidae clade (Figueiredo and Alexander 2015).

SPECIMEN JP CR683

While still unpublished, JP Cr683 is a left coracoid with an unusual "axe shaped" morphology (see fig. 5.23), once thought to belong to a subfamily of carnivorous dinosaurs called the Unenlagiinae (Singer 2015).

Cau (2015d) deduced that JP Cr683 is actually from a crocodilian. The specimen shows similarities to *Simosuchus*, although this comparison may be deceiving as that was the only taxa Cau (2015d) compared it with.

It's possible that JP Cr683 is conspecific with *Libycosuchus*, *Lavocatchampsa*, or the unnamed taxa discussed later. However, these possibilities are impossible to prove or disprove at the moment, given the lack of overlapping material for comparison.

SPECIMEN FSAC-KK 07

FSAC-KK 07 is a large jugal from a neosuchian, possibly from a sebecid. Some features suggest a close relationship with *Hamadasuchus*, although other features are unique. It's approximately 150% larger than those known for *Hamadasuchus*, measuring over 15 cm long. The posterior

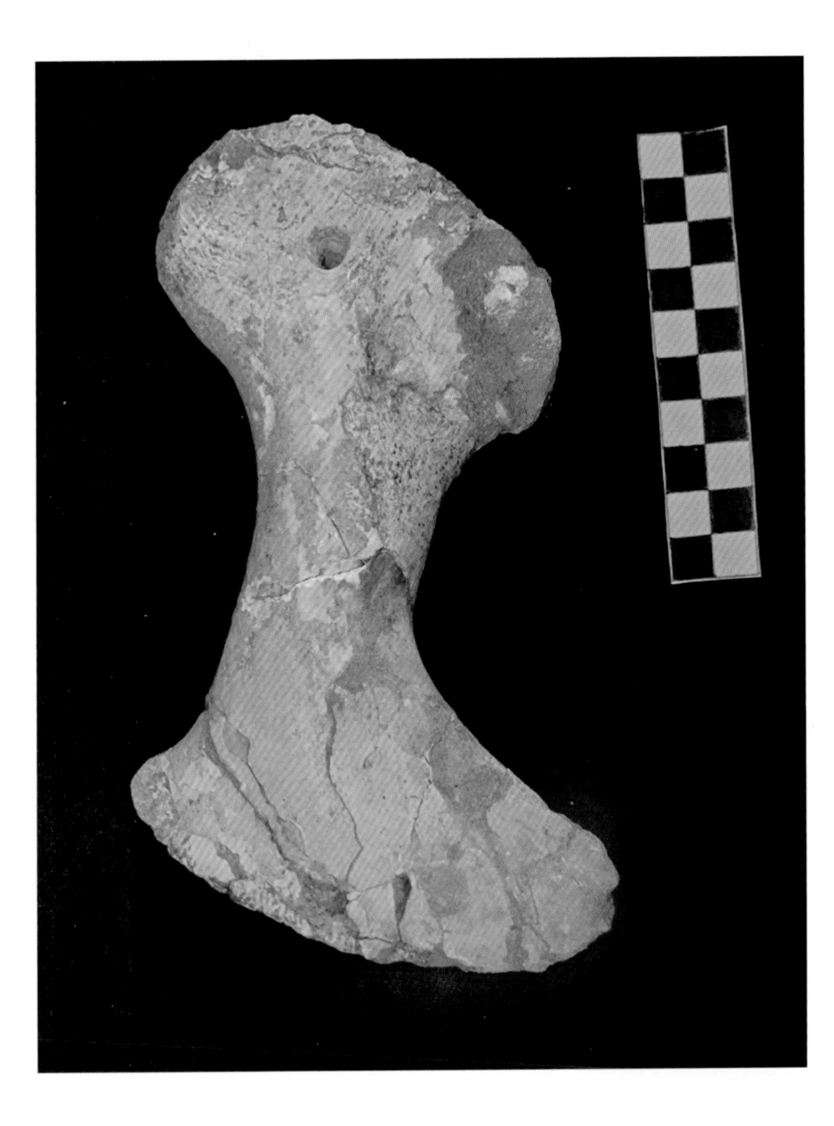

Fig. 5.23. JP Cr683, the holotype of "*Maroccanoraptor.*" Hopefully, this specimen will be officially described in the future. Image provided by JuraPark.

process also tapers in this specimen, whereas it expands in *Hamadasuchus* (Ibrahim et al. 2020a).

UNNAMED NEOSUCHIAN

Found in both Morocco and Niger, this new species is identified as a long-jawed sebecosuchid, on the basis of unspecified features in the nostril region. This species has tusklike first dentary teeth, which slip into large sockets on the opposite jaw. It's also claimed that this taxon has a vomeronasal organ (used to detect odor particles), which would make it unique among modern crocodyliforms (Larsson et al. 2015).

This species also ranks among the largest crocodyliforms to have ever lived. Larsson and colleagues (2015) state that the skull alone was 2 m (6 ft) long. What makes this taxon truly special are its similarities to *Spinosaurus*. Both taxa, despite not being related, have evolved a terminal rosette (long, interlocking teeth at the end of the snout) and laterally compressed jaws. Just how this river basin managed to support two titans in the same ecological niche will be discussed later.

UNNAMED NOTOSUCHIAN

Another recent addition to the North African biota is a notosuchian, known from a lower jaw. What makes this specimen stand out is its teeth. The crowns are big, with large cusps in the center giving way to smaller cusps on the edges. The tenth tooth forms a large tusk.

It's suspected that this new genus is related to the *Simosuchus*, although it could also be conspecific with the lone coracoid specimen JP Cr683. If it truly is related to *Simosuchus*, then this species was probably heavily armored. Regardless, judging by its teeth, it was probably herbivorous (Rego and Evans 2017).

Crocodilian Trace Fossils

A solitary crocodilian footprint from Morocco was discovered by Belvedere and colleagues (2013). At 284 mm wide and 221 mm long, this track was made by a large individual. The track itself is preserved in reasonable detail. The toes are elongated, with digit III being the longest, at 22.9 inches.

Historic Taxa

Stromerosuchus aegyptiacus (Kuhn 1936) was named on the basis of material found by Stromer in 1911. The specimens were impounded by Egyptian authorities, and when finally mailed to Stromer, they were badly damaged during shipment. It's likely that *S. aegyptiacus* is actually composed of bits and pieces of *Aegyptosuchus* and *Stomatosuchus* (Nothdurft et al. 2002). For this reason, and because of the poor quality of the remains, which are now lost, *S. aegyptiacus* is now regarded as a *nomen dubium*.

Baharijodon carnosauroides (Kuhn 1936) was also based on damaged and scrappy material from Egypt, which was also destroyed. As a result, it too is now considered a *nomen dubium* (Nothdurft et al. 2002).

Kemkemia auditorei was named by Cau and Maganuco (2009) on the basis of an isolated caudal vertebra. This specimen, which is 60.48 mm long and 33.81 mm tall, lay forgotten at the Museo di Storia Naturale di Milano (Cau 2009) before anyone realized that it has a suit of autapomorphies (Cau and Maganuco 2009).

Cau and Maganuco (2009) initially claimed that *K. auditorei* was theropod dinosaur. However, this is where things began to go wrong for *Kemkemia*. In 2011, Andrea Cau was contacted by Federico Agnolin and Gabriel Lio, who had read his paper and come to a different view. They suggested that *K. auditorei* was a crocodilian (Cau 2012a).

Their timing was fortuitous, as Cau and Maganuco were also having qualms about their identification (Cau 2012a). This communication resulted in a review of *K. auditorei*, which supported the idea that *K.*

auditorei was a crocodilian (Lio et al. 2012; Cau 2012b). As if that weren't complex enough, yet another paper was published suggesting that *Kemkemia* might belong to a spinosaurid (Chiarenza and Cau 2016). This saga is not over yet.

Pterosauria

Here's a group to fire the imagination, even if some people insist on calling them flying dinosaurs no matter how often we in the scientific community correct them. Remarkably diverse, they ranged from the size of a small bird to the size of a small airplane (Witton 2013a). In recent years, many new taxa have been named from North Africa, usually on the basis of the most fragmentary specimens. Personally, I'm not so sure that's wise, as it makes the referral of other specimens to that genus, and accurate reconstructions of it, difficult.

Siroccopteryx

Siroccopteryx moroccensis (fig. 5.24) goes down in history as the first pterosaur to be named from Morocco. A pity then that it's not known from better material, as S. *moroccensis* is known only from a partial jaw with attached teeth (Mader and Kellner 1999), although Ibrahim and colleagues (2010) mention a second, undescribed jaw in storage in Casablanca.

Fig. 5.24. A reconstruction of *Siroccopteryx moroccensis*. Artwork by Joschua Knüppe.

Mader and Kellner (1999) believed *S. moroccensis* to be a member of the Anhangueridae. Unwin (2001), however, not only transferred it to the Ornithocheiridae but also considered *Siroccopteryx* to be just a new species of *Coloborhynchus*, which was named first and which therefore has priority.

That did not go unopposed. Fastnacht (2001) argued that this taxon has more in common with *Anhanguera*. Rodrigues and Kellner (2008) pointed out that the synonymy between *Siroccopteryx* and *Coloborhynchus* was based on the resemblance *Siroccopteryx* has with another pterosaur species, *Uktenadactylus*. When Unwin (2001) synonymized *Siroccopteryx* with *Coloborhynchus*, *Uktenadactylus wadleighi* was considered to be a species of *Coloborhynchus*. When *U. wadleighi* was given its own genus by Rodrigues and Kellner (2008), that rendered Unwin's synonymy moot.

Rodrigues and Kellner (2008) go on to point out that *Siroccopteryx* differs from both *Uktenadactylus* and *Coloborhynchus* in the shape and angle of its crest as well as the shape of the tip of the beak. For that reason, *Siroccopteryx* is herein considered a valid genus. As to its family, Ibrahim and colleagues (2010) consider it to be an ornithocheirid, as does Witton (2013a).

Brink and colleagues (2015) have studied isolated ornithocheirid teeth from the Kem Kem beds and identified two distinct tooth morphologies based on the microstructure, showing that there was a second ornithocheirid taxon present besides *S. moroccensis*. However, this assumes that one of the two morphologies identified belongs to *S. moroccensis*. That cannot be demonstrated at present, and the taxonomy of the Kem Kem ornithocheirids has become a lot more complex since then (Jacobs et al. 2020).

Nicorhynchus

Originally identified as species of *Coloborhynchus*, *Nicorhynchus fluviferox* is known from just a partial premaxilla from Morocco. The specimen preserves four broken teeth and one pair of empty tooth sockets (Jacobs et al. 2018). While *N. fluviferox* shows some similarities to *Siroccopteryx*, a suite of unique features separates the two taxa (Jacobs et al. 2018; Holgado and Pêgas 2020). In diet and ecology, *N. fluviferox* would have been a piscivore similar to *Siroccopteryx*.

Ornithocheirus

Ornithocheirus is a genus with a tangled history. Known from fragmentary remains, it soon became a wastebasket taxon for many British pterosaurs. It was also caught up in a tug-of-war between Harry Seeley and Richard Owen, who had very different views on *Ornithocheirus* taxonomy. Regardless, only the type species, *O. simus*, is now considered valid (Witton 2013a).

A portion of premaxillae belonging to *Ornithocheirus* was identified in 2020. This is the first confirmed occurrence of this genus outside of

Great Britain, with the specimen assigned to *O.* cf. *simus* (Jacobs et al. 2020).

Several morphological features of the Moroccan specimen are almost identical to *O. simus*. However, there are also differences, such as the presence of small depressions on the surface, the shape of the anterodorsal margin, and the size and position of the tooth sockets (Jacobs et al. 2020). With *O. simus* being so poorly known, it's unclear whether these features are taxonomically important. So the Moroccan *Ornithocheirus* may get its own species or genus at some point.

Anhanguera

The first evidence of this Brazilian pterosaur in North Africa is a partial dentary, FSAC-KK 5005. The referral to this genus was based on the shape of the mandibular crest and the position of the tooth sockets. The Moroccan species of *Anhanguera* is currently left as *A.* cf. *piscator* (Jacobs et al. 2020).

However, many of the other traits seen in this specimen are not unique to *Anhanguera*, or even to the Anhangueridae in general, as many are also found in the Ornithocheiridae (Jacobs et al. 2020). Given this and the fragmentary nature of the specimen, I suspect that FSAC-KK 5005 may also warrant its own species or genus in the future.

Leptostomia

One of the latest pterosaurs to be named is *Leptostomia begaaensis* (Smith et al. 2020). It's known only from the most fragmentary remains, consisting of just a section of upper beak. However, the specimen does suggest intriguing possibilities; the beak is both flattened and thickened, and exceptionally long.

Such morphology is also seen in kiwis and wading birds that probe the vegetation or mudflats for buried invertebrates. This suggests that *L. begaaensis* was a specialized invertivore, feeding along riverbanks and beaches at low tide (Smith et al. 2020).

Beyond that, there is little else to say. Given the paltry nature of the holotype, its taxonomic placement is uncertain. It may belong to the Azhdarchidae, the Chaoyangopteridae, or a new azhdarchoid clade altogether (Smith et al. 2020).

Apatorhamphus

This genus name refers to the difficulty its authors had in identifying the family to which the specimen belonged; *apatos* means "deceptive" in ancient Greek (McPhee et al. 2020). The holotype specimen, FSAC-KK 5010, is a highly fractured premaxilla lacking the anterior tip (McPhee et al. 2020). Various other specimens have also been referred to this genus: BSP 1993 IX 338, CMN 50859, FSAC-KK 5011, FSAC-KK 5012, FSAC-KK 5013, and FSAC-KK 5014 (McPhee et al. 2020).

Apatorhamphus gyrostega is tentatively assigned to the Chaoyangopteridae, although it would have been one of the earliest chaoyangopterids known if that identification is accurate. Ecologically, A. *gyrostega* would have occupied a similar ecological niche to that of the azhdarchid pterosaurs, hunting terrestrial prey like a stork. However, most chaoyangopterids are known to have delicate jaws, which suggests that they did not feed on large prey (Witton 2013a).

Alanqa

The best-known pterosaur from this region, *Alanqa saharica* (fig. 5.25), was first discovered by Wellnhofer and Buffetaut (1999). More material was subsequently found and described by Ibrahim and colleagues (2010), who gave this pterosaur its name.

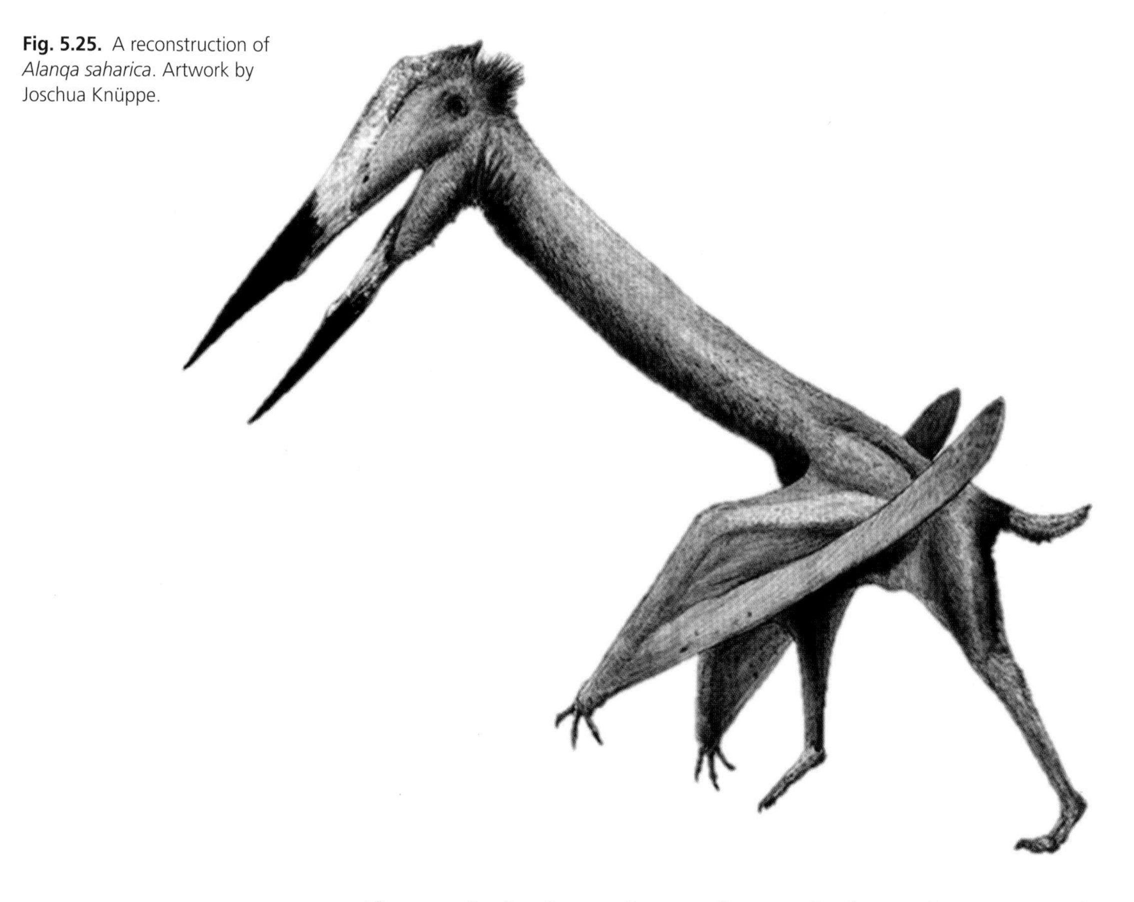

Fig. 5.25. A reconstruction of *Alanqa saharica*. Artwork by Joschua Knüppe.

Alanqa saharica is now known from multiple jaw fragments and vertebrae (see fig. 5.26). However, not everyone agrees on the identity of that material. Rodrigues and colleagues (2011), while agreeing that the jaw fragment BSP 1996 I 36 belongs to A. *saharica*, argue that specimens BSP 1993 and CMN 50859 cannot belong to that genus because the jawbone cross section is different, despite the overall similarities in shape.

Averianov (2014), however, assigned all azhdarchid material from the Kem Kem beds to A. *saharica*. He believed that the differences observed in the morphology were due to the individual animals being of different

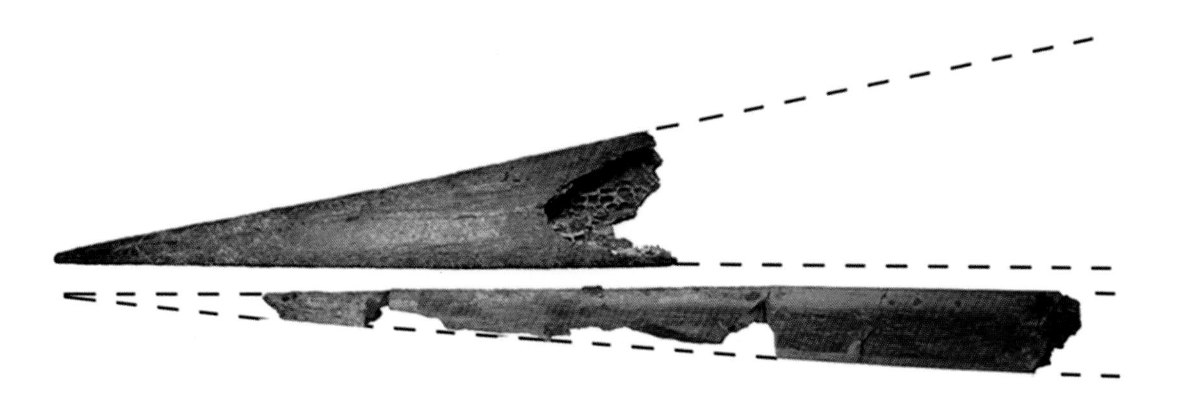

Fig. 5.26. The beak of *Alanqa*. Image from Ibrahim et al. (2010, fig. 4A).

ages. Averianov's growth sequence is as follows: specimen BSP 1997.I.67 represents the youngest known individual, followed by specimens MNU-FRJ 7054-V, BSP 1993.I.338 and BSP 1996.I.36 (both from animals of a similar age), CMN 50859 (a subadult), and FSAC-KK 26, which represents an adult specimen.

In young animals, the sagittal crest runs almost to the tip of the beak, but as they get older, the jaws get longer and the crest occupies less and less of the beak (Averianov 2014). Averianov also mentions that it's possible that only males of this species possessed sagittal crests as adults.

I'm inclined to agree with Averianov, at least in part, as his idea does solve the problem posed by Rodrigues and colleagues (2011), whose specimens have different cross sections yet are similar in most other ways. Specimens BSP 1997.I.67, MNUFRJ 7054-V, and BSP 1993.I.338 form a natural grade, as shown in Averianov's figure 1. Only BSP 1996.I.36 and FSAC-KK 26 seem to lack a smooth progression from the others.

Ibrahim and colleagues (2010) estimate that the wingspan of the holotype specimen was 4 meters (13 feet). However, one vertebra from another individual suggests the existence of larger specimens, measuring 6 meters (20 feet). The jaws made it truly stand out. Martill and Ibrahim (2015) note that the most recent specimen possesses two protuberances on the upper jaw, with a median eminence on the lower jaw fitting perfectly between them.

Martill and Ibrahim (2015) suggest that this trait most likely evolved as a feeding mechanism, allowing A. *saharica* to crush hard foodstuffs in its beak. Hard-shelled crustaceans and other invertebrates are quite common in this environment (Garassino et al. 2008a; Guinot et al. 2008; Garassino et al. 2011). Morocco also has an abundance of turtles.

Although I agree that feeding in the water and probing in the sediments for prey is unlikely, as azhdarchid necks were rather inflexible and their feet unsuited for wading (Witton and Naish 2008; Witton 2013a; Martill and Ibrahim 2015), A. *saharica* could have fed at low tide while such prey was active and out of the burrows. Regardless, A. *saharica* fed on hard prey items, along with small prey capable of being swallowed whole without the need for oral processing.

Xericeps

Xericeps curvirostris is another azhdarchoid, known only from a partial lower jaw. It can be differentiated from *A. saharica* on the basis of the anatomy of the bony ridges on the lower jaw and the fact that the jaw itself curves upward (Martill et al. 2018).

These bony ridges are similar to those of *Alanqa* and it was once thought there was a close taxonomic relationship between the two (Martill et al. 2018). *Xericeps* presumably also fed on hard prey items, although the shape of the beak suggests niche partitioning between the two.

Afrotapejara

The presence of tapejarids in North Africa has long been debated. Wellnhofer and Buffetaut (1999) identified what appeared to be a partial anterior jaw from such a creature—BSP 1997 I 67. Ibrahim and colleagues (2010) supported this identification, although they note that it could belong to a thalassodromid instead. Averianov (2014), however, rejected both identifications and identified this specimen as belonging to a juvenile *A. saharica*.

That brings us back to the tapejarids again. It's difficult to tell whether BSP 1997 I 67 came from the upper or lower jaw, but the depth shows that it would have had a large crest in life, atypical for an azhdarchid (Thomas 2017a).

The North African tapejarid was finally named in 2020. The holotype, FSAC-KK 5004, is a fragmentary upper jaw, missing both the tip of the beak and the crest. *Afrotapejara zouhri* has a row of small foramina on the side of the jaw, running along the edge of the palate. The palate itself also has a single, boss-like protuberance near the back (Martill et al. 2020).

Two other jaw fragments, FSAC-KK 5006 and FSAC-KK 5007, have also been assigned to *A. zouhri*. BSP 1997 I 67 may also belong to this genus, assuming that it is indeed a lower jaw. But it's impossible to say for sure given the lack of overlapping specimens for comparison (Martill et al. 2020).

It's impossible to deduce size on the basis of such fragmentary specimens, but if BSP 1997 I 67 is proportioned like other tapejarids, it would have ranked among the largest (Thomas 2017a). Ecologically, tapejarids are mostly considered to be herbivores/frugivores or possibly seed eaters (Meng 2008; Wu et al. 2017), but Martill et al. (2020) suggest that *A. zouhri* was an omnivore.

Pterosaur *incertae sedis*

SPECIMEN MN 7054-V

A jaw fragment, MN 7054-V, was originally described as belonging to a pteranodontid (Kellner et al. 2007). Ibrahim and colleagues (2010) considered this unlikely, and I tended to dismiss the specimen out of hand as being too fragmentary to be of any value.

However, my conversations with Henry Thomas (2017b) have led me to reconsider that view, as MN 7054-V does have some diagnostic features. Thomas has very cautiously assigned MN 7054-V to the Chaoyangopteridae on the basis of its similarities to *Shenzhoupterus* and *Lacusovagus*. It may possibly belong to *Apatorhamphus*, although a confident referral is impossible at this time.

UNNAMED PTERANODONTIA

This isolated wing phalanx represents the first pterosaur from Bahariya. While still undescribed, it appears to be from a pteranodontian pterosaur rather than the azhdarchids or ornithocheirids (Sallam et al. 2021).

DINOSAURIA

Dinosauria is the group containing the most recent common ancestor of *Megalosaurus* and *Iguanodon*, as well as all their descendants (Currie and Padian 1997). While the outline of the family tree is generally stable, the taxonomic ground at the base is shifting, and it will take a while for a new consensus to appear (Baron et al. 2017 vs. Mortimer 2017).

Spinosaurus

Spinosaurus is the most iconic of all African dinosaurs, with only *Brachiosaurus* (the African variety now renamed *Giraffititan*) giving it a run for its money in the popular media. And from late 2014 onward, there's been a flurry of re-descriptions, to the point where it's now arguably overexposed. The tribalism, hostility, and passive-aggressiveness from dinosaur aficionados, and from some professional researchers who should know better, is draining the enjoyment out of it.

Fig. 5.27. A reconstruction of *Spinosaurus aegyptiacus*. Artwork by Joschua Knüppe.

And there's plenty to disagree over, so the next few sections will probably be controversial. *Spinosaurus* has become a paradox in that the more data we collect, the less anyone seems to agree on anything. The first contentious feature is those short legs, which led to the reconstruction of this taxon as a reptilian dachshund.

The story began with Stromer, who discovered a specimen of *Spinosaurus* he did not quite know what to make of. The limb proportions were all wrong, so instead of assigning the material to *S. aegyptiacus* (fig. 5.27), he left it as "*Spinosaurus B*" (Stromer 1934a). He should have trusted his initial instincts, as another specimen (*Spinosaurus* C), this time from Morocco, shows almost identical proportions. The animal to which it belonged really did have a disproportionally small pelvis and hind limbs. As a result, Ibrahim and colleagues (2014b) concluded that *Spinosaurus* must have been a knuckle-walking quadruped.

Doubts were raised almost immediately, especially as others were unable to replicate the results. Following criticisms that there was a scaling error (Hartman 2014; Witton 2014a), Ibrahim and Maganuco responded by providing landmarks to measure against (in Witton 2014b). Using these new landmarks when he tried again, Witton got almost the same result as Ibrahim, leading him to announce his renewed confidence in the published proportions (Witton 2014b).

But while I personally accept that this taxon had surprisingly short limbs for a theropod, I agree with Headden (2014) in that I'll need to see a detailed reconstruction of the arm, wrist, and shoulder morphology before I accept that *Spinosaurus* was a quadruped. And while the hindlimbs were short, the limb musculature shows no sign of a similar reduction (Witton 2020). Even Ibrahim and his team (2020b) have now quietly backed away from that argument, insisting that it could walk quadrupedally but was not obligated to do so.

The second issue is the amalgamation of all spinosaurid material into *S. aegyptiacus*. Previously, Russell (1996) named two new taxa from Morocco: *Spinosaurus maroccanus* and *Sigilmassasaurus brevicollis*. McFeeters and colleagues (2013) concluded that *S. maroccanus* and *S. brevicollis* were the same taxon, with *S. brevicollis* having priority. Then Ibrahim and colleagues (2014b, 2020a) and Smyth, Ibrahim, and Martill (2020b) decided that all of this material belongs to *S. aegyptiacus*.

While the question of *Sigilmassasaurus* will be discussed later, there are at least several differences between the Moroccan *Spinosaurus* and *S. aegyptiacus* that the previous authors gloss over, many of which have proven phylogenetic value in other theropods (Stromer 1915, 1934a; Evers et al. 2015; Cau 2015a; Maganuco and Dal Sasso 2018; Witton 2020). As a result, I'm personally convinced that there were at least two *Spinosaurus* species in North Africa: *S. aegyptiacus* from Egypt and *Spinosaurus* sp. from Morocco.

Spinosaurus is noted for its tetradactyl (four-toed) feet. Ibrahim and colleagues (2014b) believe that these are an adaptation for swimming, given the similarity with the webbed feet of extant shorebirds. Witton (2014b), however, suggests that this toe morphology could also be used for spreading the animal's weight on soft substrate. Interestingly, the pedal ungual (the bone at the end of the toe that supports the claws) from immature specimens has the same flattened morphology as that of the adults (Maganuco and Dal Sasso 2018; Lakin and Longrich 2019).

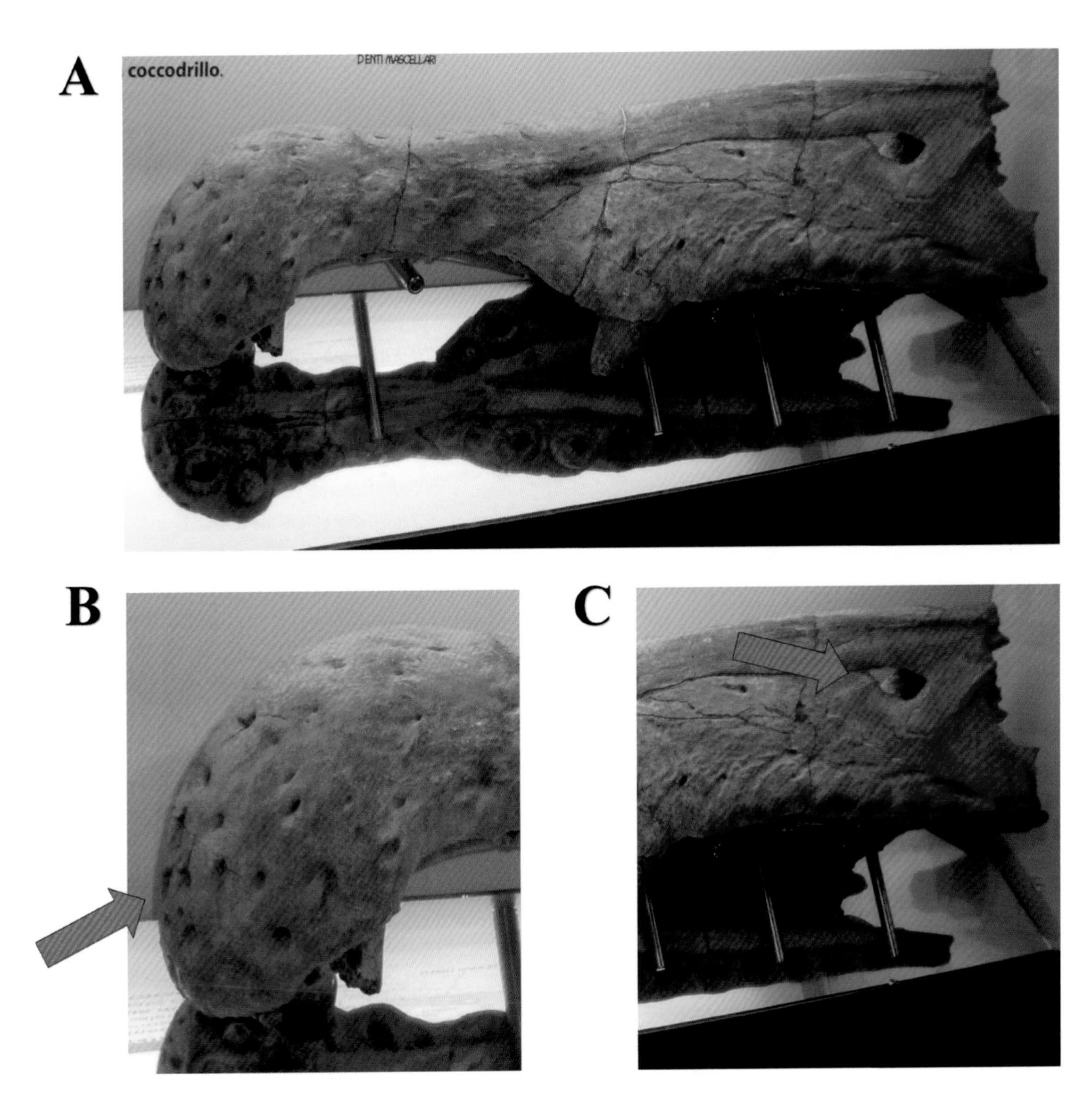

Fig. 5.28. *A*, a fragmentary skull of a *spinosaurid*. Previously assigned to *Spinosaurus*, it could also belong to *Sigilmassasaurus*. *B*, most of the upper jaw is preserved, allowing us to see the pitlike structures along the tip of the snout; usually believed to be pressure receptors, they may instead have housed the nerves and blood vessels from the lips. And *C*, the retracted nostrils. Credit Ghedoghedo / CC-BY-SA-3.0. See Wikimedia Commons.

This naturally raises an interesting question about parenting in this genus, given that the young could already swim and were possibly already self-sufficient. However, the smallest specimen known was already an estimated 1.78 meters (5.83 feet) long (Maganuco and dal Sasso 2018), so we'll need remains of an actual *Spinosaurus* hatchling before we can even begin to answer that question.

That brings us to one of the defining characteristics of this species: the elongated neural spines. Stromer (1915) reconstructed the sail as being covered in skin, which he thought was for either display or thermoregulation. Bailey (1997) proposed an alternative, suggesting that the sail was a muscular hump or fat deposit as seen in a bison. That idea has not gained widespread acceptance, because the similarities between the neural spines of *Spinosaurus* and the modern bison are more superficial than anything else.

Gimsa and colleagues (2015) proposed another model: the sail was used for maneuverability during swimming, much as in the modern sailfish. I personally see very few similarities between the flexible, retractable sail of a sailfish and the immobile sail of *Spinosaurus*.

But maybe Gimsa and his team were on to something when they commented on "*Spinosaurus*' muscular chest and neck" (Gimsa et al. 2015, 2), as Cau (2014) suggests that the morphology of the cervical series allowed the neck to be held sub-perpendicular to the rest of the spine, creating the pronounced U-shaped bend seen in the neck of some birds. This would have allowed the animal to remain upright, despite its reduced hind limbs, as folding the neck backward would help support an elongated head (Cau 2014).

Supporting the head in this posture would have required a series of elastic nuchal ligaments, which in turn would require large attachment sites, hence the enlarged neural spines. As these ligaments are passive, constantly pulling the head and neck backward, the animal would require powerful flexor muscles to move the head and neck, and the spinosaurids have large pectoral regions perfect for anchoring large neck musculature (Cau 2014).

Cau (2017) took this idea further with the suggestion that caudofemoralis, iliofibularis, and epaxial muscles would have had the same combined force as the neck extensor muscles. These effectively acted as a counterweight and gave the animal the leverage needed to hoist its front end upright. In this scenario, the tall spines over the hips would have acted as an attachment site, much like the towers on a suspension bridge.

One of the more recent changes is the discovery of a nearly complete tail that shows features no one ever expected. *Spinosaurus* had a large fin running the full length of the tail, making it look superficially like a newt. This fin is composed of tall, reclining neural spines and chevrons and was presumably used for swimming (Ibrahim et al. 2020b).

While the notion of *Spinosaurus* as an awesome, aquatic superpredator was quickly embraced by dinosaur fans, to the point where it's becoming a new dogma, professional opinions remain mixed. Hone (in Chang 2020) remains "extremely unconvinced" by this idea, as am I. I suspect that people are making the mistake Bailey (1997) made and assuming that superficial similarities in appearance equate to similar functions.

The fact that spines overlap as they bend backward means that they would all project at different angles when the tail curved in an arch, which would potentially snap them. Defenders of this idea have pointed out that a series of muscles and ligaments down the length of the fin would have helped brace the elongated neural spines and chevrons during bending (Witton 2020). While this is perfectly plausible, it's also an empty assertion, as there is no research yet to demonstrate that this network of muscles and ligaments even existed. It also contradicts the reconstruction offered by Ibrahim and colleagues (2020b).

Crocodiles also have wide transverse processes down almost the full length of the tail. These provide anchor points for the thick muscles needed to flex the tail into tight arcs, as well as the overall strength needed to drive their bodies through the water (Grigg and Kirshner 2015; Witton 2020). *Spinosaurus* lacks that. Contra Pierce (in Chang 2020), the tail was actually typical for a theropod in this respect; the transverse processes are broad at the base of the tail but shorten almost immediately, demonstrating a weak musculature down most of its length. So I seriously doubt that this puny tail could provide the power to propel a multiton bulk through the water (Henderson, in Chang 2020).

So without a more detailed reconstruction and accompanying biomechanical studies, I remain unconvinced that this tail was used for swimming. A far more parsimonious explanation, in my mind, is that it was used for display purposes.

Henderson (2018) has also provided computer modeling suggesting that *Spinosaurus* couldn't dive or even swim well at all. This was because of its pneumatized skeleton, the large sail on its back, and its raised center of mass, which left it laterally unstable and liable to topple over onto its side when it was floating on the surface. At best it could only have waded in shallow water.

I admit I was dubious about that at first. Although 3D modeling is an irreplaceable tool, I can't help but remember how the DinoMorph models of *Diplodocus* (Stevens and Parrish 1999; Stevens 2002) went from being the hottest thing ever to discredited (Upchurch 2000; Taylor et al. 2009).

While Henderson's methodology is sound (his program produced accurate results for the emperor penguin, which was used as a test case), his model is based on the Ibrahim and colleagues (2014b) restoration, the full description of which has yet to be published and with which others have had problems in the past (Hartman 2014; Witton 2014b). Given the assumptions and unpublished data that went into that restoration, it was hard for me to view Henderson's hypothesis as credible.

Yet it was difficult to sustain my objections in light of recent developments. *Spinosaurus* has become so well known that only the skull, neck, and arms are uncertain (Ibrahim et al. 2020b). Could a wider chest and that new enlarged tail provide enough weight to overcome Henderson's calculations? I doubt it, but we need the experiment to be redone with the new reconstruction before we can say anything for certain.

Sereno and colleagues' (2015) studies of *Spinosaurus* swimming speed also concluded that this taxon was capable of just 3 meters/second—better than a human swimmer but appallingly slow compared to other semiaquatic taxa. While this obviously needs to be reviewed in light of new data, it does tie into Henderson's work, as well as suggestions by Nash (2015c), who thinks that *Spinosaurus* was more akin to a hippopotamus than a primitive whale. Contrary to popular belief, hippopotamuses usually don't swim; they walk along the riverbed and use their momentum to temporarily glide through the water before touching down again (Coughlin and Fish 2009).

While it's still early days, with many claims and ideas yet to be tested, I personally support the minority view that *Spinosaurus* was adapted for walking along shallow riverbeds instead of for swimming in the water column (Nash 2015c; Hone and Holtz 2019, 2021; Henderson 2018).

It was also once thought that *Spinosaurus* had pressure receptors inside the snout, much like a crocodile, for targeting prey in murky water (dal Sasso et al. 2009; Ibrahim et al. 2014b). However, it's been shown that the foramina, pits on the skull that contain nerves, are no different from those of any other large theropod. *Spinosaurus* had an estimated 125 foramina across four of its jawbones (Ibrahim et al. 2014b), which is pitiful compared to a crocodilian, which can have hundreds more on just one. So it's likely that the foramina were not pressure receptors but merely housed nerves and blood vessels for its lips (Witton 2020).

Despite this, the idea that *Spinosaurus* was a piscivore is no longer in doubt. The aforementioned adaptations for an existence primarily around water (Amiot et al. 2010a; Sereno et al. 2015; Arden et al. 2019), as well as carbon isotope studies (Hassler et al. 2018), have all but settled the debate.

Spinosaurus was a giant, with size estimates running from 15 meters (49 feet) long and 6 tonnes (Glut 1982); 15 meters (49 feet) long and 4 tonnes (Paul 1988b); 16 to 18 meters (52 to 59 feet) long (dal Sasso et al. 2009); and 12.6 to 14.3 meters (41 to 47 feet) long and 12 to 20.9 tonnes in weight (Therrien and Henderson 2007). At the moment, because of uncertainty with our current reconstructions, it's impossible to get a valid estimate, beyond saying that it's large (Persons et al. 2020).

Sigilmassasaurus

Sigilmassasaurus was named by Russell (1996) on the basis of neck vertebrae. On top of this, he also suspected that *Spinosaurus B* could also be *S. brevicollis*, or at least an Egyptian species of *Sigilmassasaurus*. The low neural arches also suggest that the sail would be smaller or located farther down the torso (see fig. 5.29), which is not what we see in *Spinosaurus* (Headden 2018).

Fig. 5.29. A reconstruction of *Sigilmassasaurus brevicollis*. Artwork by Joschua Knüppe.

Sereno and his team (1996, 1998) later claimed that *S. brevicollis* was a junior synonym of *Carcharodontosaurus*. Novas, Dalla Vecchia, and D. Pais (2005), while disputing the synonymy with *Carcharodontosaurus*, suggested that at least some of the material belonged to an iguanodontian. Then *S. brevicollis* was mostly forgotten, until McFeeters and colleagues (2013) reevaluated the genus and concluded, on the basis of a suite of diagnostic characteristics, that it was indeed a distinct taxon.

Then Ibrahim and colleagues (2014b) decided that *Sigilmassasaurus* was a junior synonym of *Spinosaurus*. Evers and team (2015) soon retaliated. Their study suggested not only that *S. brevicollis* was still valid but that *Spinosaurus B* had dorsal vertebrae similar to those of *Sigilmassasaurus* and so probably belongs to *Sigilmassasaurus*, whereas *Spinosaurus C* had dorsal vertebrae like *Spinosaurus* and so probably belongs to *Spinosaurus*. They may very well be right about *Spinosaurus B* being a species of *Sigilmassasaurus*, but if so, this means that *Sigilmassasaurus* also has reduced hind limbs (Cau 2015a).

Smyth, Ibrahim, and Martill (2020) would later support the claim that *Sigilmassasaurus* is a junior synonym of *Spinosaurus*, on the basis of comparisons with *Suchomimus*, which shows clear morphological change along its axial column. So all the differences and possible autapomorphies used to define *Sigilmassasaurus* can instead be explained by the position of the vertebrae in the neck.

While it was published far too recently to be the subject of any reviews, the general reception to the Smyth, Ibrahim, and Martill (2020) paper appeared poor, and I too am skeptical. Never has the often-subjective nature of the taxonomy been laid bare. For instance, the absence of centroprezygapophyseal fossae and interzygapophyseal laminae in the confirmed *Sigilmassasaurus* vertebrae is seen as important by Evers and colleagues (2015), while Smyth, Ibrahim, and Martill (2020) dismiss these features as taxonomically irrelevant!

Then there is also the misinterpretation of features in Ibrahim and colleagues (2020a) to consider. For instance, one photograph of a *Sigilmassasaurus* vertebra that supposedly shares features with *Spinosaurus* was actually photographed from a poor angle, making it look as though it had features it didn't have. In actuality, the specimen was just badly crushed (Evers 2020).

A lot of Smyth, Ibrahim, and Martill's (2020) work is also based on assumptions about the position of the vertebra in the series. For instance, does BSPG 367 2006 I 57 represent C5 or C6? A lot of interpretations hinge on its position (Smyth, Ibrahim, and Martill 2020), yet without a complete cervical series from Morocco, we can't be sure, no matter the comparisons we make with related taxa.

Others have noted the presence of two spinosaurid morphotypes in Morocco, with differences seen in the frontals and quadrates, for instance (Russell 1996; Richter et al. 2012; Hendrick et al. 2016; Lakin and Longrich 2019; Arden et al. 2019). Smyth, Ibrahim, and Martill (2020) also dismiss these as intraspecific or ontogenetic variations or argue against the spinosaurid identity of some of the material (Ibrahim et al. 2014b), even though Hendrick and colleagues (2016) reviewed the degree of intraspecific variation seen in the quadrates of *Suchomimus*, *Baryonyx*, and *Allosaurus* and found it negligible.

Frankly, I'm also suspicious of their methodology, since if we apply it elsewhere, most if not all spinosaurid taxa would become synonyms of *Baryonyx* or *Spinosaurus*. *Ichthyovenator*, for instance, if we looked solely at its cervical vertebra and ignored the rest of the skeleton, would probably become *Spinosaurus* as well, given the marked similarities it too shares with *Sigilmassasaurus* (Allain et al. 2012). *Oxalaia* has already been sunk into *Spinosaurus*, even though it exists on another continent (Smyth, Ibrahim, and Martill 2020). Smyth and his team believe they've sunk *Sigilmassasaurus*, but I fear that they've actually sunk the entire Spinosauridae—an intolerable concept, as it almost certainly underestimates the diversity of this family.

This of course reawakens the long-standing arguments on spinosaurid diversity and on drawing the line between them, given the similarities seen among all the various spinosaurid taxa (Evers et al. 2015), as well as the disagreements between researchers over what features are genuine autapomorphies (Sues et al. 2001; Milner 2003; Hutt and Newbery 2004). Much of which, like the debate over *Spinosaurus/Sigilmassasaurus*, seem to have become battles of semantics rather than hard science (Candeiro et al. 2017).

Regardless, even if I accepted all the arguments made by Smyth and his team, it still avails them nothing. *Sigilmassasaurus* then becomes Spinosauridae *incertae sedis*; it does not become *Spinosaurus*, let alone *S. aegyptiacus*. There's no overlapping material for comparison, and they can't say it has no defining autapomorphies and then somehow positively identify it on a species level.

So I personally believe that there were two different spinosaurid taxa in North Africa, and *Sigilmassasaurus* is retained herein as a valid taxon. It's the most parsimonious explanation for the evidence, rather than invoking a *Spinosaurus* that has more variability than the domestic dog. This insistence on accepting any degree of variation in that genus will, I fear, result in a wastebasket taxon without any proper definition at all.

Ultimately, contra Smyth, Ibrahim, and Martill (2020), the synonymy between *Sigilmassasaurus* and *Spinosaurus* can only ever be verified by the discovery of considerable overlapping material that would allow us to compare them to each other rather than to their relatives. Until that hypothetical *Spinosaurus* skeleton, with a complete cervical series, comes to light, we have no hard evidence in favor of their conclusion, so I believe that it's best to maintain the original taxonomic position.

Carcharodontosaurus

While *Spinosaurus* might have been odd for a theropod, *Carcharodontosaurus* (fig. 5.30), was true to type. It was originally called *Megalosaurus*

Fig. 5.30. A reconstruction of *Carcharodontosaurus saharicus*. Artwork by Joschua Knüppe.

saharicus (Deparet and Savornin 1925), but Stromer (1931) later found a partial skeleton and gave it its own genus.

Stromer's holotype specimen was known from a partial braincase, rib, and maxilla, four vertebrae (one caudal and three cervical), fibula, femora, and pubes (which were damaged during shipping), and a left ischium (Stromer 1931; Rauhut 1995; Nothdurft et al. 2002). Stromer also included an ilium, but that referral has been called into doubt (Rauhut 1995). Brusatte and Sereno (2007) later added a partial skull, SGM-DIN 1 (see fig. 5.31).

Fig. 5.31. The skull of *Carcharodontosaurus*. This reconstruction is based on specimens from both Egypt and Morocco and may prove chimeric in the future. Credit Bramfab / CC-BY-SA-4.0. See Wikimedia Commons.

Once the debate over *Spinosaurus* taxonomy began, I half expected someone to say that the Moroccan and Egyptian specimens come from different species of *Carcharodontosaurus*. And as if on cue, Chiarenza and Cau (2016) noted that the Moroccan and Egyptian specimens have differently shaped maxillary interdental plates (the bony region between the teeth). For the moment I will keep the Moroccan and Egyptian specimens united as *C. saharicus*, but I suspect that will change should *Carcharodontosaurus* get a systemic review.

Studies of the inner ear and endocranium of *C. saharicus* have shown that its brain was similar to that of *Allosaurus*. Given the size of its cerebrum, this was probably not a particularly smart animal, but the optic nerve was large, suggesting that vision was important in the species (Larsson 2001).

Current size estimates for this taxon run from 12 to 13.3 meters (39 to 44 feet) and an estimated 6.1 to 15.1 tons in weight. The skull alone would have been 1.6 meters (5.2 feet) long (Sereno et al. 1996; Seebacher 2001; Therrien and Henderson 2007).

Yet some have suggested that this taxon was a bit of a wimp. Henderson and Nicholls (2015) suggest that *C. saharicus* was only capable of carrying 850 kg at most in its jaws, effectively deadlifting less than the world's strongest man! That comment was made by Nash (2015a), who points out that these calculations fail to consider the muscle groups beyond the neck. He suggests that the leg, abdominal, and tail muscles would also play a role.

I tend to agree with Nash, especially as beefy hind limbs were the rule in carcharodontosaurids (Novas 2009). I expect that *C. saharicus* would have been more than capable of attacking sauropods and large ornithopods.

It makes little difference in the end, as *C. saharicus* would undoubtedly have just eaten its victim where it fell if it couldn't carry it away. Few would have bothered it, as *C. saharicus* would have been at the top of the North African food chain. Given that the pillarlike hind limbs were designed to hold its immense weight (Novas 2009), it's doubtful that *C. saharicus* was a fast runner. So it was almost certainly an ambush predator.

Sauroniops

Kudos to Cau and colleagues (2012a, 2012b), as *Sauroniops pachytholus* (fig. 5.32) is probably one of the best dinosaur names I've ever heard. The "Dark Lord" of North Africa is all the more mysterious since it's known from only one isolated specimen (see fig. 5.33), a left frontal bone from the skull (Cau et al. 2012a, 2012b). This has led some to claim that it's a synonym of *Carcharodontosaurus* (Ibrahim et al. 2020a), but this ignores all the autapomorphies documented for this taxon (Cau et al. 2012a, 2012b). So this taxon does appear valid.

The thickened skull roof shown in this specimen makes *S. pachytholus* superficially similar to the abelisaurids (Bonaparte et al. 1990;

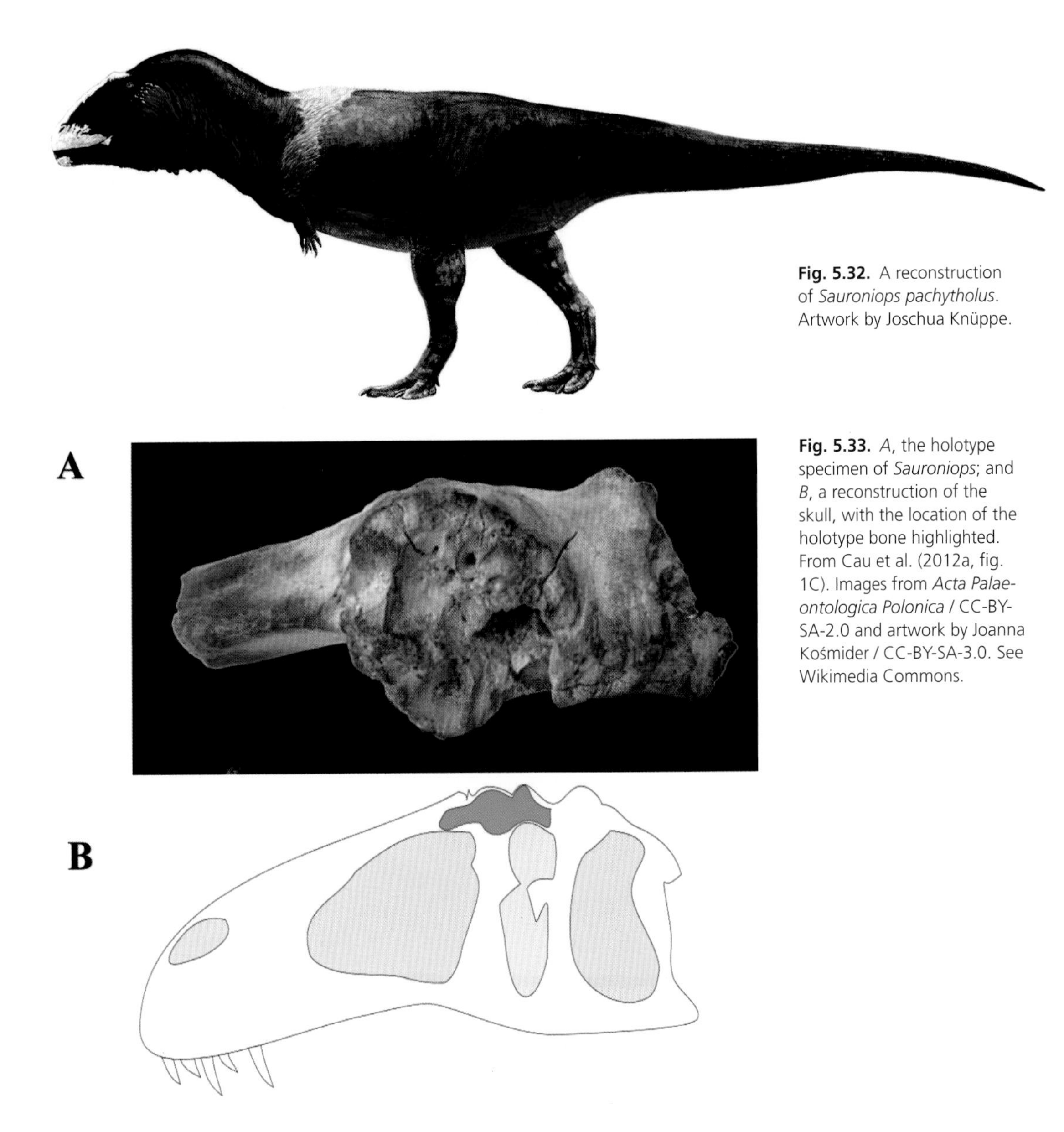

Fig. 5.32. A reconstruction of *Sauroniops pachytholus*. Artwork by Joschua Knüppe.

Fig. 5.33. *A*, the holotype specimen of *Sauroniops*; and *B*, a reconstruction of the skull, with the location of the holotype bone highlighted. From Cau et al. (2012a, fig. 1C). Images from *Acta Palaeontologica Polonica* / CC-BY-SA-2.0 and artwork by Joanna Kośmider / CC-BY-SA-3.0. See Wikimedia Commons.

Sampson and Witmer 2007). The exact function of this headgear is unknown, although display and intraspecific combat are the obvious choices (Cau et al. 2012b).

Paulina-Carabajal and Coria (2015) describe a similar-looking specimen from Argentina and suggest that this specimen belongs to the Tyrannosauroidea. If accurate, this would mean that S. *pachytholus* was the first tyrannosaurid ever found in Africa. Personally, I suspect that the features demonstrated by the Argentinian specimen are homoplastic. Cau (2015b) also adds that Paulina-Carabajal and Coria used an older megamatrix in their analysis and not the updated one provided in Cau and colleagues (2012b).

Personally, I think it more likely that *S. pachytholus* occupied an unranked basal position in the Carcharodontosauridae (Cau et al. 2012b). The exact size of *S. pachytholus* is impossible to determine, but if proportioned like *Carcharodontosaurus*, it would have ranked near the top of the food chain.

Bahariasaurus

Unlike its compatriots, *Bahariasaurus ingens* has been almost completely forgotten. Of course, there is a reason for this lack of reconstruction. *B. ingens*, as originally described by Stromer, is, in fact, a hodgepodge of thirty-two specimens that were found in the same general area but in different beds (Stromer 1934a; Rauhut 1995, Mortimer 2014). It's possible that this has resulted in a chimaera.

Stromer was well aware of this possibility, and the remainder of the material—eighteen more vertebrae, a femur, fibula, scapula, coracoid, fragmentary ilium, ischium, and other pubes specimens—was referred only with great caution. Contrary to popular belief, the tibia was never referred to *B. ingens* (Stromer 1934a; Rauhut 1995; Mortimer 2014). Indeed, I suspect that the tibia is not from the same individual animal, as the rest of the limb is far too large in comparison.

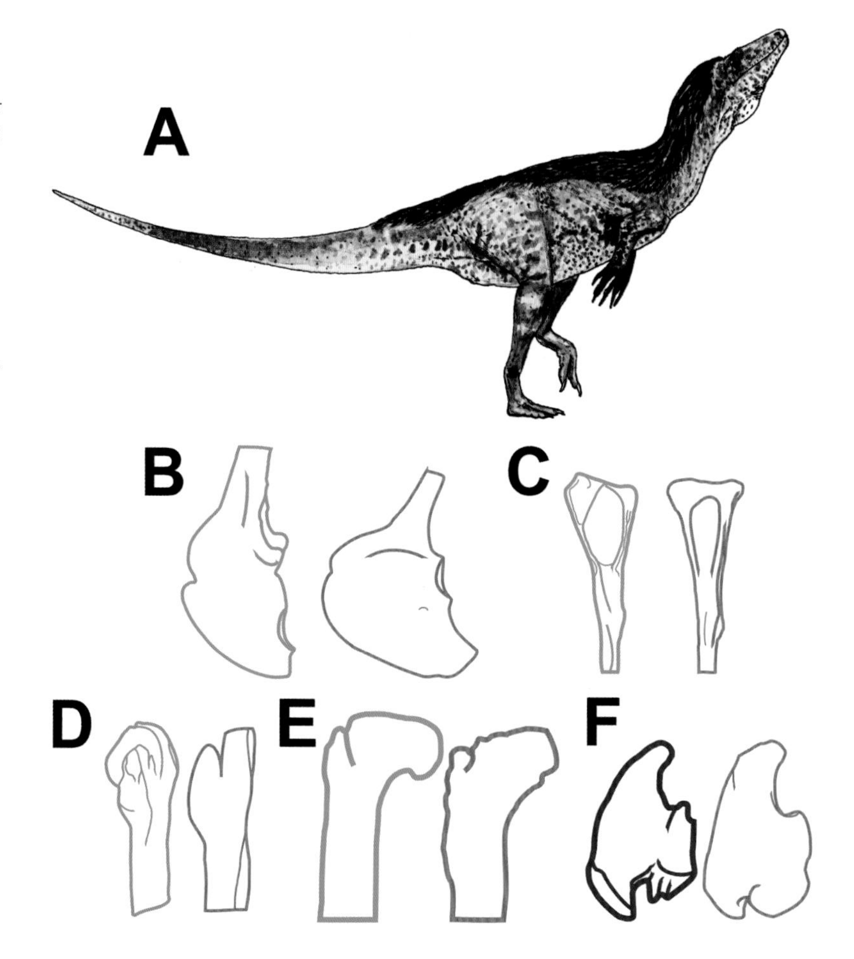

Fig. 5.34. *A*, a reconstruction of *Bahariasaurus ingens*; and a selection of *Bahariasaurus* and *Deltadromeus* material for comparison: *B*, pectoral girdle; *C*, proximal fibulae, medial view; *D*, proximal femora, lateral view; *E*, femora, anterior (green) and posterior (red) views; and *F*, tibiae, proximal view. Green, *Bahariasaurus*; red, *Deltadromeus*; blue, aff. "*Erectopus*" from Bahariya. All specimens scaled to match. Reconstruction by Joschua Knüppe, artwork adapted from Mortimer (2014), based on Stromer (1934), Sereno et al. (1996), and Chiarenza and Cau (2016).

However, I think an argument can still be made for the validity of *B. ingens* and the referral of most of that material to that taxon. At the very least the coracoid, vertebra, and the pectoral girdle belong to *B. ingens* as they were part of a partial skeleton, and they clearly don't belong to any other North African theropod. And any one of these specimens is enough to anchor *B. ingens* as a valid taxon (contra Ibrahim et al. 2020a). Mortimer (2014) also thinks that that Stromer might have been correct in referring the femur to *B. ingens*, and I agree.

The exact size of *B. ingens* is unknown, given how poorly known this genus is, but some estimates place it at around 12.2 meters (40 feet) in length (Sereno et al. 1996).

More *Bahariasaurus* material—six caudal vertebrae—may have been found in Niger (de Lapparent 1960). But no explanation for this referral was given, which makes me suspicious. For that reason, I've left the Algerian material as Theropod *incertae sedis*.

For a while, the matter rested there, but Sereno and colleagues (1996) would later refer some of the *B. ingens* holotype to their new taxon *Deltadromeus*: the femur, coracoid, tibia, fibula, and pubes (apparently the second set of pubes, which Stromer only cautiously referred to *Bahariasaurus*).

However, there is doubt as to which bones Sereno and his team were referring to. Their specimen numbers don't correspond with the specimen numbers given by Stromer. Presumably, they meant IPHG 1912 VIII 60, IPHG 1912 VII 69, and IPHG 1912 VIII 70 (Mortimer 2014). But whatever their intentions, they began a saga that has poisoned our understanding of both taxa to this day, with many subsequently assuming that the two taxa are synonyms.

Sereno and colleagues (1996) claimed that the femur and pectoral girdle of the two taxa are almost identical, but even at a cursory glance, the two pectoral girdles are not remotely similar to each other. For instance, the scapular shaft in *B. ingens* is broader and less angled, and the acromion and glenoid are much smaller than those of *Deltadromeus* (Mortimer 2014).

The tibiae and fibulae may be referable to *Deltadromeus*, although there are differences as well as similarities and no shared apomorphies were identified; however, since Stromer never referred the tibia to *B. ingens* (contra Sereno et al. 1996) but actually identified it as aff. *Erectopus*, it has no bearing on *B. ingens* taxonomy either way (Mortimer 2014).

Further, it's clear that the two femurs also differ in the morphology of the femoral head amongst other features (see fig. 5.34). *Deltadromeus* has a medially oriented head, and the anterior trochanter extends all the way to the fourth trochanter—traits that are absent in *B. ingens* (Mortimer 2014). It's also noteworthy that the neck of the *B. ingens* femur lacks the incline seen in *Deltadromeus* and that the lateral accessory crest on the distal end of the femur is also shorter (Chiarenza and Cau 2016).

The pubes also show marked differences between the two taxa. The lateral flaring is more proximally placed than in *Deltadromeus*, and the position of the interpubic foramen is also different, 80% down the shaft versus 71%. Also, the pubes described by Stromer (1934a)—BSP 1912 VIII 81 and BSP 1922 X 47—differ from each other (Brea 2020), so one of which is likely not referable to *B. ingens* anyway. Carrano & Sampson (2008) also believe that the distal pubes of *Deltadromeus* might actually be a misidentified ischial boot rendering the subject moot.

Ibrahim and colleagues (2020a) attempt to refute this, and while they provide some much-needed anatomical data, they're forced to settle for noting similarities for the most part; the criticisms raised, such as the differences in the pectoral girdle, went unanswered. More importantly, no photographs of the *Deltadromeus* specimens in question, particularly the femur, were provided, preventing us from reviewing the data for ourselves and settling this issue once and for all. Given that the *Deltadromeus* holotype is now fully prepared, this failure is inexcusable, and frankly rather telling. Nor was any explanation offered for the huge discrepancy in size between the almost fully grown *Deltadromeus* holotype and *B. ingens*, with the latter having a femur that is 165% bigger.

Clearly they are two distinct taxa (contra Sereno et al. 1996 and Ibrahim et al. 2020a), with Brea (2020) offering the closest thing to a redescription *B. ingens* is ever likely to get. Huene (1948) gave *B. ingens* its own family, while others have labeled it a megalosaurid (de Lapparent 1960) or a carcharodontosaurid (Rauhut 1995). Chiarenza (2016b) suggested a noasaurid identification, which would result in the Noasauridae being renamed, given that Bahariasauridae has priority (Huene 1948). Current consensus places this taxon among either the ceratosaurs (Carrano and Sampson 2008; Pol and Rauhut 2012), tyrannosaurids (Motta et al. 2016; Chure 2000; Brea 2020), or megaraptorians (Motta et al. 2016), but who knows how long it will stay in any of them?

Deltadromeus

This brings us to *Deltadromeus agilis* itself (fig. 5.35), although one might wonder what's left after the discussion above. While much of the material assigned to this taxon has gone (Mortimer 2014), there is still more than enough to warrant the retention of *D. agilis* as valid.

The holotype (see fig. 5.36) includes cervical and dorsal ribs, incomplete ilium and ischia, gastralia, caudal vertebrae, scapula, and coracoid,

Fig. 5.35. A reconstruction of *Deltadromeus agilis*. Artwork by Joschua Knüppe.

and the limbs consist of a femur and incomplete humerus, radius, ulna, fibula, tibia, and metatarsals (Sereno et al. 1996; Mortimer 2014). One thing to note is the length of the hind limbs, which are more like those of an ornithomimid than a traditional theropod (Sereno et al. 1996).

The pubis of *D. agilis* is in fact a misidentified ischium (Carrano and Sampson 2008; Mortimer 2014). Interestingly, the morphology of the distal boot is very similar to a pair of ischia left as Theropoda indet. by Stromer (1934a). So it's possible that IPHG 1912 VIII 82 is referable to this taxon (Mortimer 2014).

Because material from *B. ingens* has been added to it, *D. agilis* is usually assumed to be a giant predator. Without that material, the suggestion that *D. agilis* was a large species is effectively debunked, especially as the holotype has a fused ischial boot, showing that it was almost fully grown when it died (Mortimer 2014). In reality, *D. agilis* was an average-sized theropod with an estimated length of 8 meters (26 feet).

While Sereno et al. (1996) originally believed that *D. agilis* was a coelurosaur, most researchers have gone over to the view that *D. agilis* was a ceratosaurid. Some suggest that it belongs with the Noasauridae (Wilson et al. 2003), occupying an unranked basal position in that family, along with *Elaphrosaurus* and *Limusaurus* (Carrano and Sampson 2008; Xu et al. 2009). Rauhut and Carrano (2016) and Wang and colleagues (2017) also found *D. agilis* to be a noasaurid, although *Elaphrosaurus* and *Limusaurus* were found to occupy a more derived position within this family. However, the position of *D. agilis* was unstable, as it could

Fig. 5.36. A speculative reconstruction of *Deltadromeus*. Credit Ryan Somma / CC-BY-SA-2.0. See Wikimedia Commons.

be either a member at the base of the Noasauridae or a derived member of the Noasaurinae (Wang et al. 2017).

The fact that *Elaphrosaurus* and *Deltadromeus* keep dancing around each other is interesting, as Stromer (1931) reported *Elaphrosaurus* material from Bahariya. Could that material pertain to *D. agilis*? Intriguing speculation, but that's all it remains, given that the Egyptian material is lost to us. And if Rauhut and Carrano (2016) and Wang and colleagues (2017) are correct, then *Deltadromeus* and *Elaphrosaurus* are not closely related, which makes the above suggestion unlikely.

As if to complicate the matter further, it's suggested that *Bahariasaurus* and *Deltadromeus* might be members of the Tyrannosauroidea. Motta and colleagues (2016) suggest that both of these taxa belong to the Megaraptora, a group of theropods traditionally placed within the carcharodontosaurids.

Apesteguía and colleagues (2016) also agree with the megaraptorid identification, placing *D. agilis* as a sister taxon to *Gualicho*/*Aoniraptor*. However, they place the Megaraptora within Neovenatoridae or leave them as basal coelurosaurians. While I suspect *Bahariasaurus* is indeed a megaraptorid, with the Megaraptora actually belonging to the Tyrannosauroidea, I don't think *D. agilis* groups with either (contra Apesteguía et al. 2016 and Motta et al. 2016) as it is clearly a ceratosaurian.

Rugops

North Africa was also home to the Abelisauridae, a family of theropods famed for their truncated forelimbs and elaborate cranial ornamentation (Senter 2010; Burch 2013; Tykoski and Rowe 2004). This group is known only from fragmentary remains in this region (Russell 1996; Mahler 2005; Carrano and Sampson 2008; Porchetti et al. 2011; Richter et al. 2012; Evans et al. 2014).

Some of the best material includes two partial maxillary bones, MPUR NS 153/02 and 517 MPUR NS 153/01 (Porchetti et al. 2011). Phylogenetic analysis by Chiarenza (2016a) suggests that both maxillae belong to a genus similar to *Rugops primus* from Niger. This raises the question of whether all these isolated specimens belong to the same species. The

Fig. 5.37. A reconstruction of *Rugops* sp. Credit Liam Elward / CC-BY-SA-4.0. See Wikimedia Commons.

two maxillae code together, but they could represent two closely related species as well as one single species.

Here Chiarenza and I disagreed, as Chiarenza suspected that multiple species of abelisaurid were present (Chiarenza 2016b) while I lumped them together. My hypothesis has later proved to be accurate, at least in some regards, as apparently some maxillary bones do indeed belong to one unnamed species of *Rugops* (see fig. 5.37).

The only real anatomical trait separating MPUR NS 153/02 from *Rugops* is the lack of ridge on the margin of the antorbital fossa (Smyth, Ibrahim, et al. 2020). Another maxilla, UCPC 10, can also be assigned to *Rugops*, as many of the features separating it, such as fewer striations and more pitting of the surface, are also of dubious taxonomic value (Smyth, Ibrahim, et al. 2020).

But how much of the other abelisaurid material from Morocco belongs to *Rugops* sp. is open to doubt. MPUR NS 153/02 and MPUR NS 153/01 do show differences from each other, such as the amount of ornamentation, the height of the maxilla, the absence of alveolar bulges and interalveolar depressions, and the various angles at the maxillary-jugal contact and the maxillary margins (Smyth, Ibrahim, et al. 2020).

While this Smyth, Ibrahim, and colleagues (2020) explain this with ontogenetic changes and individual variation, I prefer to keep MPUR NS 153/01 as Abelisauridae *incertae sedis* for now, since Chiarenza was also right in some regards, as there is clearly one other species of abelisaurid in this region besides *Rugops* (Cau and Maganuco 2009; Chiarenza and Cau 2016; Chiarenza 2016a).

Ultimately, with only fragmentary specimens to work with, no one will ever be able to answer the questions of abelisaurid taxonomy in Morocco for sure until we find a partial skeleton to compare these isolated specimens against.

Rebbachisaurus

Rebbachisaurus garasbae (fig. 5.38) is the only named sauropod from the Cenomanian of Morocco (Lavocat 1954b, 1955a, 1955b), and given this importance, it's surprising that it took decades for it to get a full description. The holotype of *R. garasbae* is composed of the sacral and caudal vertebra, ischium, a scapula, partial ribs, and the right humerus (Wilson and Allain 2015).

Fig. 5.38. A reconstruction of *Rebbachisaurus garasbae*. Artwork by Joschua Knüppe.

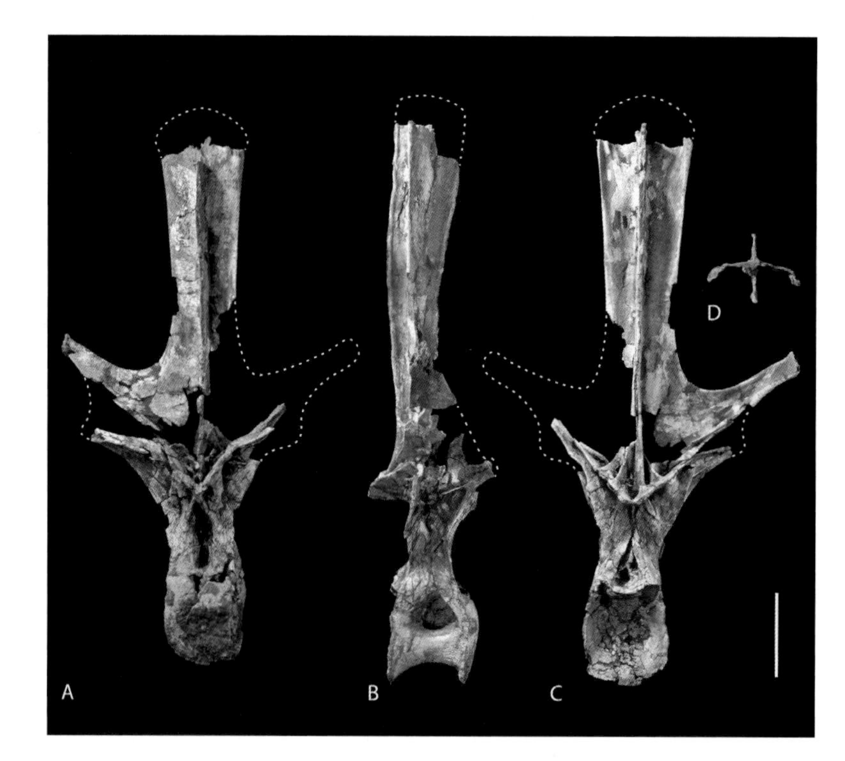

Fig. 5.39. The vertebra of *Rebbachisaurus*: *A*, anterior; *B*, right lateral; *C*, posterior; and *D*, dorsal cross-sectional views. Copyright 2015. From Wilson and Allain (2015, fig. 3). Reproduced by permission of the Society of Vertebrate Paleontology, http://www.vertpaleo.org. Scale bar equals 200 mm (20 cm).

The most iconic specimen of *R. garasbae* is the vertebra (fig. 5.39). The obvious thing to note is the incredible height, 57.0 inches tall. These proportions become even stranger when one considers that the centrum is small (Lavocat 1954b; Wilson and Allain 2015). It's theorized that the dorsal vertebrae of rebbachisaurids evolved to allow torsional movements (Apesteguía et al. 2010).

Wilson and Allain (2015), however, claim that some of the articulations in the vertebrae would have had the opposite effect, limiting movement. They suggest that this morphology evolved to provide resistance to forces pressing down on the spine. One possible reason could be the presence of an enlarged gut, which would have increased the weight on the ribs. Wilson and Allain also provide other suggestions, such as efforts to graze on high vegetation or even brief periods of bipedal locomotion. Wilson and Allain find the latter suggestions unlikely, as do I.

Many are under the impression that *R. garasbae* was a giant. Some estimates even have it at 20 meters (66 feet) in length (Holtz and Rey 2011). Unfortunately, this is an illusion created by those elongated vertebrae. *Rebbachisaurus garasbae* would have been small for a sauropod; the whole animal was probably no bigger than 8 to 13 tons in weight (Wilson and Allain 2015).

Rebbachisaurus garasbae would have been a highly specialized, ground-feeder. Its relatives have neurovascular grooves along the front of the jaws that indicate the presence of a beaklike sheath for cropping plant material (Sereno et al. 1999; Sereno et al. 2007).

Aegyptosaurus

Aegyptosaurus baharijensis (fig. 5.40), the first sauropod to come out of Egypt, remains a much sought-after prize, as the holotype has been destroyed and no other confirmed specimens have been found since. *Aegyptosaurus baharijensis* was represented by most of the limbs and caudal vertebra (Stromer 1934a). For a large animal, *A. baharijensis* had surprisingly gracile limbs (Knüppe 2014).

Fig. 5.40. A reconstruction of *Aegyptosaurus baharijensis*. Artwork by Joschua Knüppe.

Other supposed *Aegyptosaurus* material was uncovered by de Lapparent (1960). This partial specimen is composed of caudal vertebrae, a rib, and two metatarsals. De Lapparent gives no reason for the referral of this material to *A. baharijensis*, so I'm going to err on the side of caution and leave the Algerian material as Sauropod *incertae sedis*.

The family to which *A. baharijensis* belongs is uncertain, although Curry Rogers (2006) suggests that it belongs somewhere near the Saltasauridae. Ecologically, *A. baharijensis* would have been a midsize herbivore. However, it's possible that the holotype specimen was not fully grown and that this species was larger than currently assumed (Ibrahim et al. 2016).

Paralititan

While some taxa have been cursed with names that are barely pronounceable, *Paralititan stromeri* (fig. 5.41) positively rolls off the tongue. The aptly named "tidal titan" is known from a partial skeleton (plate 7) that includes a scapula, ischium, metacarpals, humeri, and caudal vertebrae (Smith et al. 2001b; Lamanna and Hasegawa 2014). A dorsal vertebra found by Stromer (1932), BSP 1912 VIII 64, was also assigned to *Paralititan* (Smith et al. 2001b).

The humerus is 1.69 meters (5.54 feet) long, the second longest known in a Cretaceous sauropod (Smith et al. 2001b). The morphology of the metacarpal bones (the bones between the fingers and the wrist)

Fig. 5.41. A reconstruction of *Paralititan stromeri*. Artwork by Joschua Knüppe.

Plate 7. The *Paralititan* holotype. *A*, a partial scapular; and *B*, the radius, ulna, and sternum. *C*, a caudal vertebra; and *D*, a humerus. Image *A* provided by Joshua B. Smith/ University of Pennsylvania. Images *B*, *C*, and *D* by Hatem Moushir / CC-BY-SA-2.0. See Wikimedia Commons.

is remarkably flat, suggesting that the phalanges (finger bones) were reduced in size or had been lost altogether (Smith et al. 2001b). That would not be unusual, as ungual and phalanges reduction is well known among sauropods (Apesteguía 2005).

Some estimates suggest that *P. stromeri*, a giant herbivore, was 26 meters (85 feet) long and 65 tons in weight (Burness and Flannery 2001; Carpenter 2006). Interestingly, *P. stromeri* is considered to be a member of the Argyrosauridae (Curry Rogers 2006), which suggests that it was more closely related to the South American sauropods than it was to other African sauropods.

Dinosauria *incertae sedis*

SPECIMEN FSAC-KK 01

FSAC-KK 01 is a partial sauropod humerus. While it is incomplete, some features suggest that FSAC-KK 01 is a saltasaurid. However, the specimen also shows similarities to *Paralititan*. It may belong to the same species as FSAC-KK 7000 (Ibrahim et al. 2016).

SPECIMEN FSAC-KK 7000

FSAC-KK 7000 is a caudal vertebra from Morocco belonging to a giant Lithostrotian titanosaurid (Ibrahim et al. 2016). It's mostly complete, but some features such as the neural spine are not preserved. The centrum also shows signs of damage to the cortical bone layer (Ibrahim et al. 2016).

SPECIMEN GMNH-PV 2399

Another sauropod from Morocco is represented by GMNH-PV 2399, an anterior dorsal vertebra (Lamanna and Hasegawa 2014). Lamanna and Hasegawa considered the possibility that this specimen belongs to *Aegyptosaurus* but concluded that *Aegyptosaurus* is a more derived titanosaurid. *Paralititan*, while sharing a similar phylogenetic position, lacks overlapping material for comparison and differs in the shape of the centrum (Lamanna and Hasegawa 2014). So GMNH-PV 2399 is best left as Somphospondyli *incertae sedis*.

SPECIMEN GMNH-PV 2314

GMNH-PV 2314 is a right ischium that may belong to the same species as GMNH-PV 2399, or maybe even to the same individual animal, although that is impossible to prove without information on its provenance (Lamanna and Hasegawa 2014). The fact that it's broken allows us to view the interior, which is composed of huge, pneumatic chambers that are roughly polygonal in shape.

SPECIMEN CR376 JP

Cr376 JP, is a dorsal vertebra assigned to the Diplodocimorpha (see fig. 5.42) that has yet to be formally described in the accredited literature (Singer 2015). While I admit that the presence of a late-surviving Diplodocidae is possible (Upchurch and Mannion 2009; Gallina et al. 2014), I will need an analysis demonstrating a lack of rebbachisaurid synapomorphies before I can fully support that referral. With any luck, a formal description will be forthcoming.

SPECIMEN NHMUK R36636

NHMUK R36636 is a caudal vertebra with the base of the neural arch and part of the caudal ribs. While it belongs to a rebbachisaurid, Mannion and

Fig. 5.42. Cr376 JP, the holotype of "*Megapleurocoelus.*" Hopefully, this specimen will be officially named in the future. Image provided by JuraPark.

Barrett (2013) refrain from referring NHMUK R36636 to *Rebbachisaurus*. NHMUK R36636 has a suite of features that suggest that it represents a new genus. Because it is plausible that there was greater rebbachisaurid diversity in the region than we have currently seen, I agree that NHMUK R36636 is best left as Rebbachisauridae *incertae sedis*.

SPECIMEN JP CR340

JP Cr340 is an incomplete neural arch of a dorsal vertebra (see fig. 5.43). Singer (2015) considers JP Cr340 to be a carnosaurid, as does Mortimer (2015), who notes similarities to the dorsal vertebra of *Mapusaurus*. Mortimer also suggests that JP Cr340 may belong to either *Carcharodontosaurus* or *Saurionops*, but that cannot be proved at this time.

This specimen is unique in possessing large hollow chambers, which are unprecedented in this degree in other North African theropods (Singer 2015). These gaps in the vertebra are the result of pneumatization.

Fig. 5.43. JP Cr340, the holotype of "*Osteoporosia*." Hopefully, this specimen will be officially named in the future. Image provided by JuraPark.

They are used for respiration and, in huge species, to help lighten the load (Paladino et al. 1997).

SPECIMEN CMN 50852

The presence of dromaeosaurids (see fig. 5.44) in North Africa has been known for some time (Rauhut and Werner 1995; Amiot et al. 2004; Riff et al. 2004; Richter et al. 2012). Teeth and claws assigned to this group are common in most Moroccan fossil shops. The idea of *Velociraptor*-like creatures stalking the ancient mangroves fires the imagination. Unfortunately, with such limited material, it's impossible to identify these remains on a family level, let alone a species level.

The only material to get a proper description is a vertebra—CMN 50852—initially thought to belong to a bird but now assigned to the Unenlagiidae (Riff et al. 2004). This specimen is similar to *Rahonavis* from Madagascar, although it lacks some features seen in that genus. So, do all those isolated teeth and claws belong to the same genus as CMN 50852? The fact that we have no basis for comparison between any two specimens makes assigning such isolated remains impossible.

It's also not certain that *Rahonavis* is a member of the Unenlagiidae. The prior consensus favored that identification (Makovicky et al. 2005; Norell et al. 2006; Turner et al. 2007), but some disagreed (Agnolín and

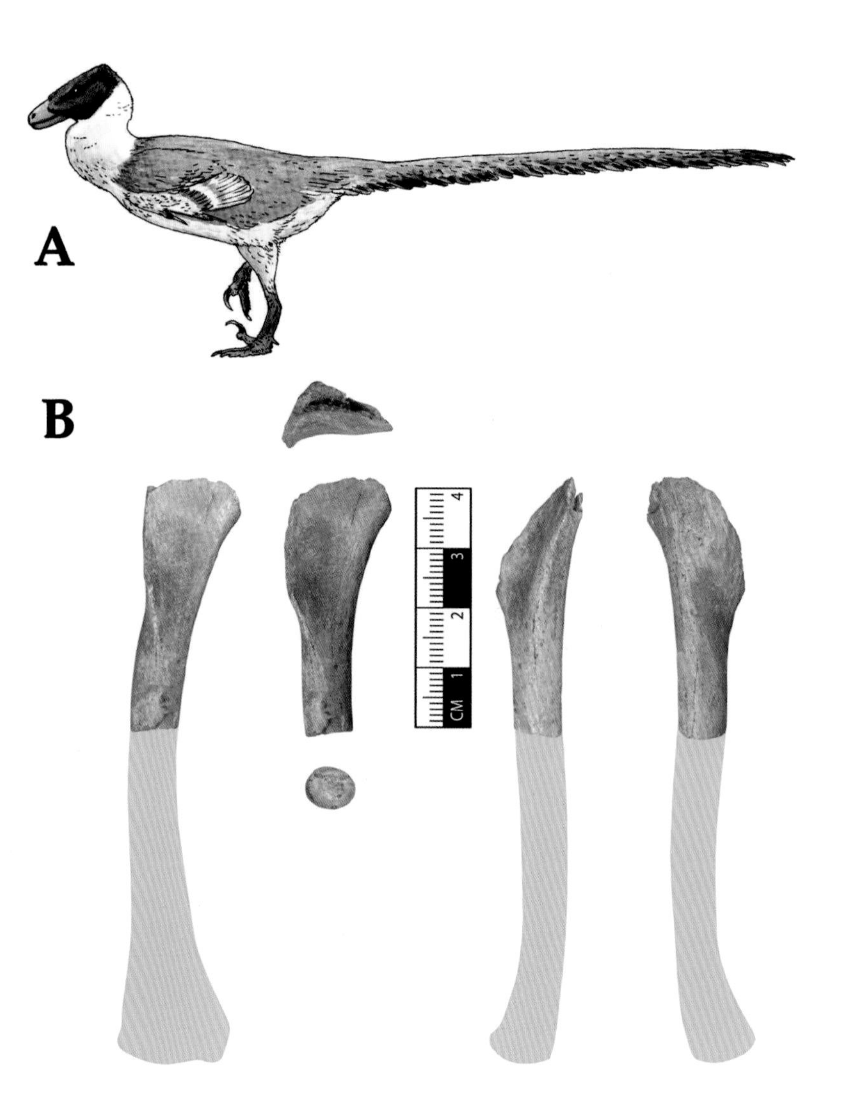

Fig. 5.44. *A*, a speculative reconstruction of the African *dromaeosaur*; and *B*, a dromaeosaurid humerus from Morocco, currently in a private collection. Image provided by Olof Moleman and artwork by Joschua Knüppe. See image *B* at http://www.thefossilforum.com/index.php?/topic/75767-kemkem-raptor-humerus/.

Novas 2013; Cau 2018), only to be later challenged themselves and the former identification defended (Pei et al. 2020).

If *Rahonavis* is not a member of the Unenlagiidae, then it's unlikely that CMN 50852 is either. This is made more complicated by the fact that the Unenlagiidae weren't the only dromaeosaurids present in Africa during this age, as velociraptorine dromaeosaurids are known from Sudan (Rauhut and Werner 1995). However, Holtz (2016) cautions that the velociraptorine identification of the Sudan specimens was made before anyone knew that the Unenlagiinae existed.

Adding insult to taxonomic injury, Chiarenza and Cau (2016) noted that CMN 50852 lacks any clear-cut Unenlagiidae synapomorphies and so could belong to any number of theropod groups. It may even be a crocodyliform. The issue of North African dromaeosaurid taxonomy will never be solved until we have a partially complete specimen, which we can then use as a Rosetta stone to compare all these fragmentary specimens with.

SPECIMEN FSAC-KK-5015

FSAC-KK-5015 is a damaged axis vertebra from a basal abelisaurid, and the ramshackle repair work (likely by its discoverer) actually damaged it further by obscuring some features. Hopefully, this can be undone, but so far, no attempts have succeeded (Smyth, Ibrahim, et al. 2020).

Some features appear contradictory, such as the broad anterior neural spine reminiscent of *Majungasaurus*, yet the top of the neural spine is wide and blunt, more like that of *Masiakasaurus* or *Ceratosaurus*. These characteristics are probably plesiomorphic (Smyth, Ibrahim, et al. 2020). It's possible that FSAC-KK-5015 belongs to a juvenile *Rugops*, but given the lack of overlapping material for comparison, I prefer to keep it separate for now.

SPECIMENS: OLPH 025

Another abelisaurid find was of a partial femur from Morocco (Chiarenza and Cau 2016). This specimen warrants its own section because of its size; the animal to which it once belonged was an estimated 5.5 meters (18.05

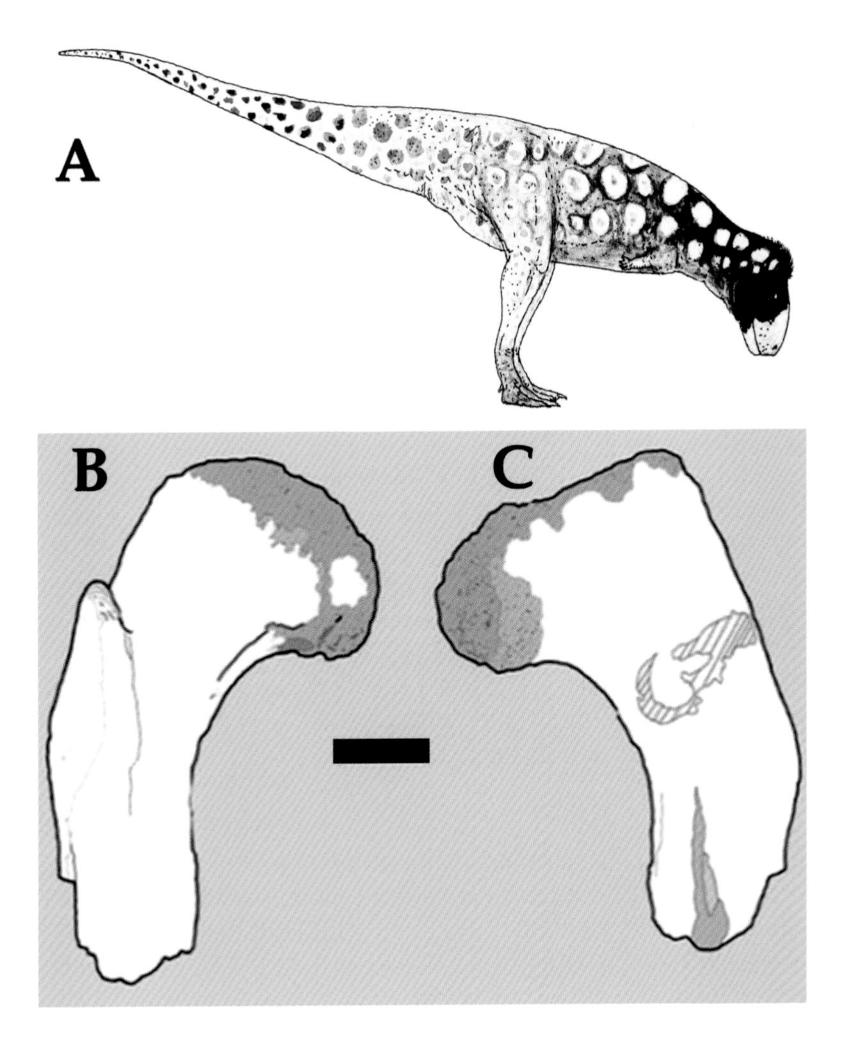

Fig. 5.45. *A*, a highly speculative reconstruction of OLPH 025; *B*, OLPH 025 in posterior view; and *C*, OLPH 025 in distal view. Scale bar is 1.9 inch (5 cm). Artwork by Joschua Knüppe. Fossil from Chiarenza and Cau (2016, fig. 3B; see also figs. 1B, 1D for photos of these same two views).

feet) long, one of the largest abelisaurids known (Cau and Maganuco 2009; Chiarenza and Cau 2016).

Far more interesting is the fact that OLPH 025 seems to belong to a different genus than the *Rugops* specimens described earlier (see fig. 5.45). Phylogenetic analysis by Chiarenza (2016a) suggests that the femur belongs to a more basal species than *Rugops*.

SPECIMENS NMC 41589 AND NMC 41861

Two pieces of a dentary bone labeled NMC 41589 and NMC 41861 were found by Russell (1996) and later assigned to the abelisaurids (Carrano and Sampson 2008; Cau and Maganuco 2009). Judging by comparison with other African abelisaurid taxa, NMC 41589 is estimated to be from a huge individual 7 to 9 meters (22 to 29 feet) long (Cau and Maganuco 2009).

Given the huge size, I suspect that the bones may belong to the same species as OLPH 025, but without any overlapping material, such a referral cannot be supported at this time. So these are best left as Abelisauridae *incertae sedis* until more complete specimens come to light.

SPECIMEN MUVP 477

MUVP 477 is an abelisaurid cervical vertebra and the first recorded from Egypt. Coming from a "medium-sized" species, its taxonomic position is unclear but it may belong to the Brachyrostra (Salem et al. 2021). Hopefully a full description will be forthcoming in the near future.

SPECIMEN ROM 65779

ROM 65779 also deserves special mention, even though it's just a single femur, as it represents a small theropod in a region where the ecology was remarkably unbalanced in favor of giants (Evans et al. 2014). Histological sections also show LAGs, brief periods when the animal stopped growing, in the outer layer. This shows that the specimen belonged to an adult or a late adolescent. The individual was an estimated five years old when it died, which matches life-span estimates for other small theropods (Evans et al. 2014; Varricchio 1993).

The affinities of this species are uncertain, and it's currently left as Averostra *incertae sedis* (Evans et al. 2014). It's possible that ROM 65779 belongs to the same species as FSAC-KK-5016 (Smyth, Ibrahim, et al. 2020), but the lack of overlapping material prevents a definitive identification. Regardless, the fact that ROM 65779 was fully grown means that it was a very small animal, filling in an important gap in the food chain.

SPECIMENS: ROM 64666

ROM 64666 is another femur, but it's distinct from ROM 65779. Unlike ROM 65779, this one can be confidently assigned to the Noasauridae (Evans et al. 2014).

Evans and colleagues (2014) suggest that the specimen could belong to *Deltadromeus*, noting general similarities in shape. While it's a possibility, I personally don't favor this hypothesis (Carrano and Sampson 2008; Evans et al. 2014).

Regardless, ROM 64666 adds important new data for reconstructing North African ecology, as some noasaurids had derived diets and hunting strategies, specializing in hunting burrowing mammals (Carrano et al. 2002; Nash 2015b).

SPECIMEN FSAC-KK-5016

FSAC-KK-5016 is a cervical vertebra from Morocco. It shows similarities to *Masiakasaurus* but also has some differences, such as the curved surface of the centrum and the presence of a small keel, both features that *Masiakasaurus* lacks (Smyth, Ibrahim, et al. 2020).

The specimen appears to be from an animal closely related to *Deltadromeus* and may even be synonymous with that taxon. However, the neural arch and centrum had begun to fuse, suggesting it was near adulthood when it died (Smyth, Ibrahim, et al. 2020), which would make it much smaller than *Deltadromeus*, even with all the *Bahariasaurus* material removed. The lack of any overlapping material for comparison between the two also makes me hesitant about any referrals. As a result, I prefer to leave this specimen as Noasauridae indet. for now.

SPECIMEN MHNM KK 04

MHNM KK 04 is a damaged and incomplete right ilium (one of the bones of the pelvis), purchased from a local collector. Given a very brief description, the specimen is described as possessing features Zitouni and his team (2019) consider to be characteristics of the Abelisauria, such as a brevis fossa (a grove on the underside of the pelvis) that widens posteriorly and the general shape of the ilium's dorsal margin.

However, Cau (2019) points out that some of the characteristics they noted are not exclusively found in abelisaurids. MHNM KK 04 also has features not seen in the abelisaurids, such as a small ischial peduncle (a bony extension where the ilium and ischium bones of the pelvis join) that is wider than it is long. For that reason, Cau doubts the referral of MHNM KK 04 to the Abelisauria and left the specimen as Tetanurae *incertae sedis*.

I agree with Cau, given the similarities I see when comparing the iliac blade (part of the ilium bone) of MHNM KK 04 to that of the spinosaurid *Ichthyovenator* from Laos—similarities Zitouni et al. (2019) also briefly acknowledged. Nothing else can be said about this specimen until it is fully described.

SPECIMEN JP CR678

Although large ornithopods were well represented in the Aptian age of Africa (Taquet 1976; Rauhut and Lopez-Arbarello 2009), by the Cenomanian, they had become remarkably rare (Lamanna et al. 2004). Singer (2015) reports the discovery of a left ischium (JP Cr678) and some caudal vertebrae, currently undescribed, which are tentatively referred to the iguanodontids.

Fig. 5.46. A specimen tentatively identified as a partial iguandont ischium. Image provided by JuraPark.

Other ornithopod material includes teeth from Algeria and Morocco (de Lapparent 1960; Ibrahim et al. 2020a). Dinosaur teeth are not usually diagnostic on a species level, so it's impossible to identify it with certainty, and the Moroccan tooth may not actually belong to an ornithopod at all; it may be from a crocodyliform.

Dinosaur Trace Fossils

"*KOREANAORNIS*"

Koreanaornis is an ichnogenus that deserves its own section, as it represents the sole avian known from North Africa (Contessi and Fanti 2012a). First discovered in Korea (Kim 1969), *Koreanaornis* is also found in Africa and North America. This ichnogenus is poorly defined (Anfinson et al. 2009), and Contessi and Fanti (2012a) opted to leave these specimens as *Koreanaornis* sp.

The toes are slender, with the longest being 12.5 mm in length, and they show evidence of claws or webbing (see fig. 5.47). The African tracks are differentiated from other *Koreanaornis* specimens by their small size (Contessi and Fanti 2012a). Contessi and Fanti considered whether this is a result of ontogeny but point out that track size in extant shorebirds shows little change as the chick reaches adulthood (Tjørve et al. 2009; Tjørve and Tjørve 2010). Interestingly, the ichnogenus *Koreanaornis* has tracks comparable to those of modern plovers and sandpipers (Contessi and Fanti 2012a), which inhabit similar habitats today (O'Brien et al. 2006).

An episode of the American documentary series *NOVA*, "Bigger Than *T. rex*" (Cohen 2014), also mentions the presence of carnivorous birds in this region. But from whence these birds came is a mystery, as I can find no reference to them anywhere else.

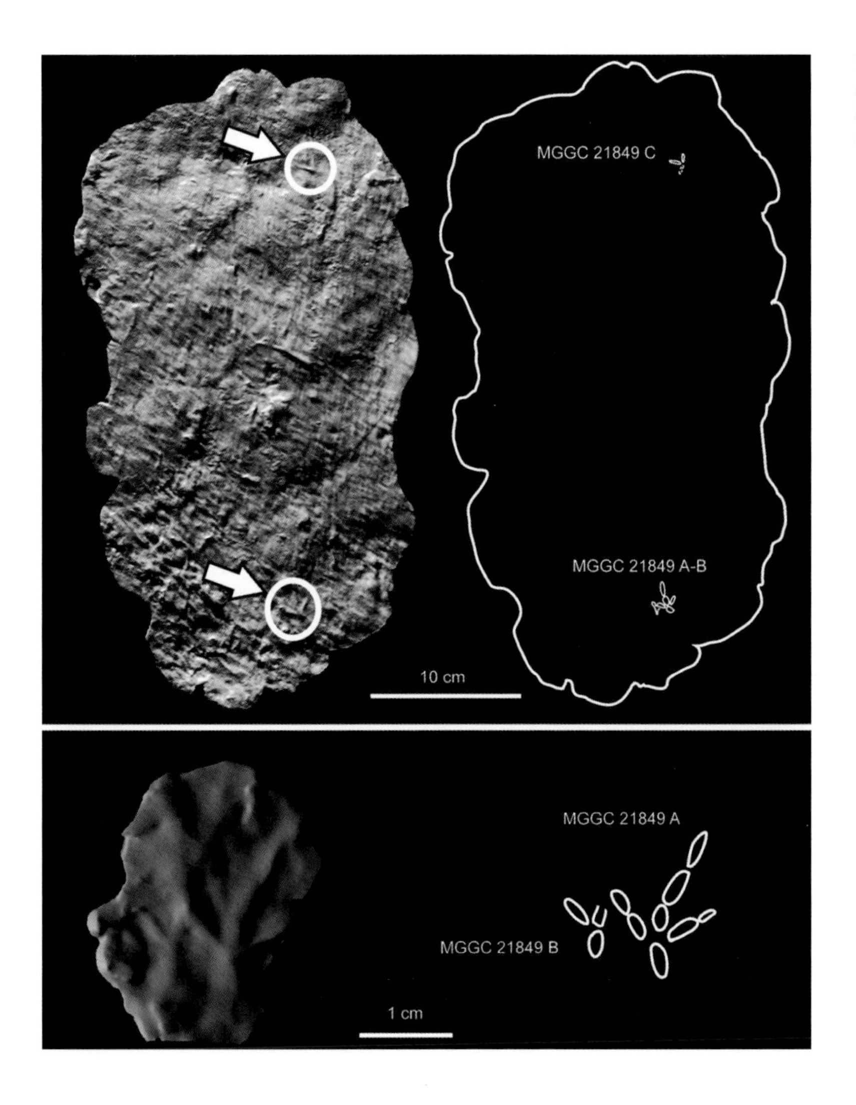

Fig. 5.47. The footprints of *Koreanaornis*. Image provided by Michele Tomlinson from SEPM Publications.

Other Footprints

Dinosaur tracks are exceptionally well known from Morocco. Theropod footprints are common, with most belonging to *Megalosauripus* (Belvedere et al. 2013; Ibrahim et al. 2014a). Sauropods are represented by two footprints (Ibrahim et al. 2014a). The shape of the first print suggests an animal resembling *Deltapodus*. Ornithopod footprints are known from one track, easily equal to that of *Iguanodon* in size (Ibrahim et al. 2014a).

An interesting preservational paradox is that the theropod fossil biota is almost completely dominated by large species, yet the footprints are mostly from small species (Belvedere et al. 2013; Ibrahim et al. 2014a). Another paradox is found at the Tarda oasis (Smith 2020). Unlike other locations, this site has many footprints assigned to the ornithopods. Why so many ornithopod trace fossils are found here, when they are so rare elsewhere, is unclear. Hopefully, this will all become clearer when the specimens and the sedimentological aspects of this location are properly described.

The most exciting scientific findings of the past half-century has been the discovery of widespread trophic cascades . . . as soon as the wolves arrived . . . the number of songbirds, of migratory birds, started to increase . . . the number of beavers started to increase . . . They create niches for other species . . . otters and muskrats and ducks and fish and reptiles and amphibians.

—**GEORGE MONBIOT**, 2013

North African Ecology

6

As you can see from the prior quotation, even something minor can have a profound impact on an ecosystem. Ecology is a complex web of interactions among species, much of which is still not fully understood for the modern world. Reconstructing extinct ecosystems is even more difficult.

North Africa must represent a paleoecologist's dream. When all the various formations are correctly aged and combined, North Africa represents an ecosystem in its entirety: mountains leading down to river floodplains, then mangrove swamps, and finally coastal reefs, with each one blending into and influencing another.

How to Build a Cenomanian Reef

Starting in the epicontinental seas, we find ourselves in shallow reefs off the African coast. These shallow reefs would have had low levels of salinity and ranged in depth from 50 meters (164 feet) to less than 20 meters (65 feet) (Gertsch et al. 2010a).

Algeria gives the best example of a typical reef structure for this region (see fig. 6.1). The intertidal zone, the part of the shoreline periodically exposed at low tide, was covered in a narrow band of stromatolites. The saline waters produced by the surrounding sabkhas and salt flats were perfect for stromatolite growth. Beyond the stromatolites was a huge expanse of bivalves and rudists. This finally gave way to sponges in the deep waters of the open ocean (Benyoucef et al. 2019).

While the composition of the African reef system changed over time, a general pattern can be deduced. In Eastern North Africa, the reefs were mostly composed of rudists and oysters, with the Exogyrinae and Liostreinae being dominant (Ayoub-Hannaa 2011). Some of these reefs are 2.4 miles (4 km) in length (Gertsch et al. 2010a). Judging by the work of Ayoub-Hannaa (2011), the Galala Sea went through at least two major ecological changes over time. Regardless, the faunal composition in Eastern Gondwana seems to have remained stable, usually with only the relative abundances of genera seeming to change (see table 6.1). Only at the Cenomanian/Turonian boundary do many families become extinct.

Table 6.1. The general faunal composition of the Galala Sea. Adapted from Ayoub-Hannaa (2011)

Fauna	Order	Family
Bivalves	13	28
Gastropods	7	19
Echinoids	8	14
Corals	1	3
Coraline Sponges	1	1
Cephalopods	2	6

Anyone free diving in eastern North Africa during the early to mid-Cenomanian would have been impressed by the high species diversity but surprised by the nature of these reefs, as they would have appeared strangely flattened compared to extant ones. Instead of multistory coral layers, this reef was composed of rows and rows oysters, such as *Ilymatogyra africana*, *Ceratostreon flabellatum*, and *Rhynchostreon suborbiculatum*. These reefs would be occasionally punctuated by marine plants,

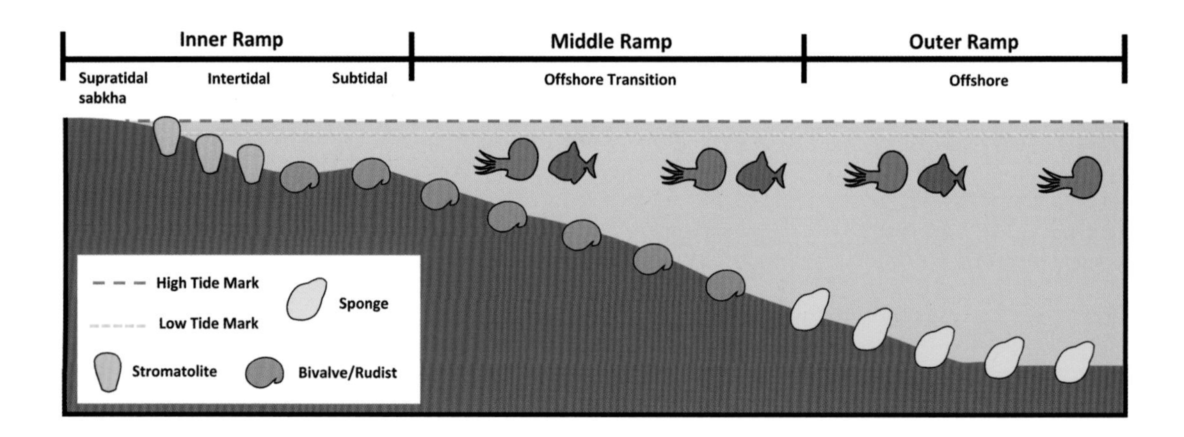

Fig. 6.1. A reconstruction of an Algerian reefal system. No marine reptiles are known from this area, and fish are not found in the distal part of the outer ramp region, but this is probably due to preservation bias. Images represent groups and not species. Images not to scale. Adapted from Benyoucef et al. (2019).

large domed corals, branching sponges, and the small tubes of serpulids (Ayoub-Hannaa 2011).

Large numbers of snails and echinoids patrolled the surface of the reef. Below the surface, the sediments were alive with burrowing bivalves and other subterranean taxa. Above them all, the surface waters were patrolled by ammonites and presumably by a carnival of fish and marine reptiles.

The waters in this region would have been subtropical, well oxygenated, and of average salinity. The reef itself would have ranged from 25 to 50 meters (82 to 164 feet) in depth. The waves were of moderate energy, although there is evidence of periodic hurricanes that would have dislodged and buried parts of this reef (Ayoub-Hannaa 2011).

Epifaunal species (those that live on the surface of the seabed) were dominant, constituting 77.15% of all the documented species at Wadi Quseib, 73.30% of those found in the East Themed area, and 56.10% at Gebel Areif El-Naqa. Burrowing species (infaunal) compose only 22.85% of the species found at Wadi Quseib and 26.69% in the East Themed area, although they did better in the deeper-water regions recorded at Gebel Areif El-Naqa, where 43.9% all of the species found are infaunal (Ayoub-Hannaa and Fürsich 2012b).

The vast majority were also suspension feeders. These make up 65% of all the fossils found in the Halal/Galala Formation. The deposit feeders were the next most abundant, at 14.65%. Others such as carnivores, omnivores, and herbivores are rare (Ayoub-Hannaa and Fürsich 2012b).

Both Stromer (1936) and Ayoub-Hannaa (2011) note the large numbers of echinoderm spines. Echinoids are the third largest benthic group on these reefs in terms of both species count and individual specimens (Ayoub-Hannaa 2011). On a global scale, the echinoderms produce 0.1 gigatonnes of calcium carbonate a year, making them a vital part of the global carbon cycle. They also feed on macroalgae, preventing it from overgrowing and suffocating the reef. Some also burrow into the sediments, releasing nutrients back into the ocean while oxygenating the substrate (Uthicke et al. 2009; Kaplan 2010).

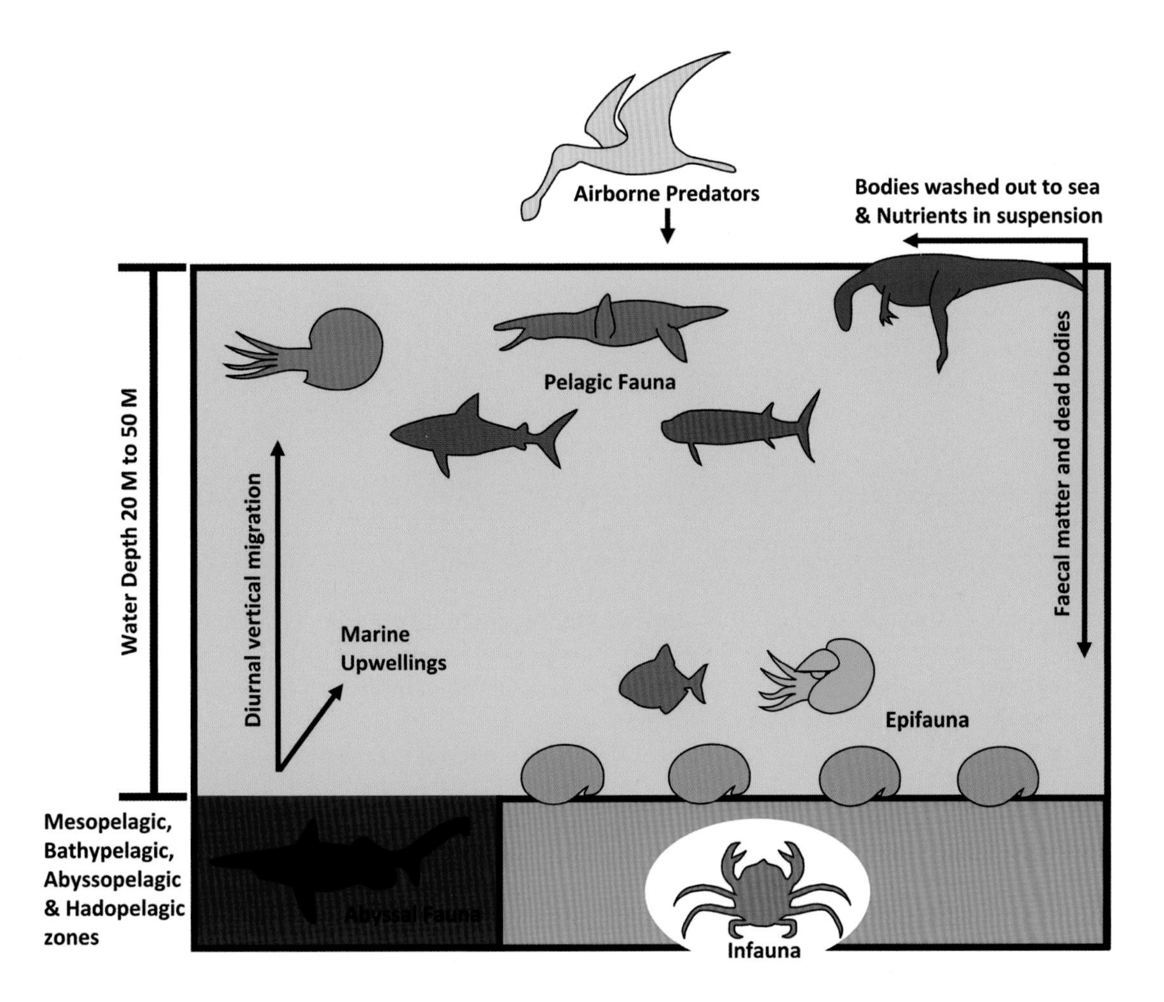

Fig. 6.2. A reconstruction of pelagic ecosystem. Based on Martill et al. (2011) and Khalloufi et al. (2010). Images represent groups and not species. Images not to scale. Adapted from Witton (2013b).

Gastropods are also common. Their remains are found in all environments from intertidal mudflats to depths of 200 meters (660 feet) (Gertsch et al. 2010a). It is surprising that there are no crustaceans mentioned in this habitat, especially as they are found in terrestrial and coastal habitats of this time. This could be a case of nonrecovery—in which their remains were not preserved as fossils, perhaps because their exoskeletons were made of chitin instead of calcium carbonates, as we'll discuss below—rather than of their absence from the ecosystem. It may also be that they were outcompeted by the gastropods, a concept we'll be discussing later.

The free-swimming members of this reefal community (see fig. 6.2) are less well known. Ammonites make up 12.28% of all the fossils found in the East Themed area and 2.95% at Wadi Quseib (Ayoub-Hannaa and Fürsich 2012b). This is probably due to most of the genera present preferring depths of up to 300 meters (984 feet) rather than shallow waters (Gertsch et al. 2010a). One exception, however, *Neolobites vibrayeanus*, is so common that it's used as an index fossil. The *Neolobites* biozone is found throughout Morocco, Egypt, Algeria, Tunisia, Niger, Europe, and the Middle East (Ayoub-Hannaa 2011).

Nothing is known of the fish and marine reptiles that inhabited these seas. This is due to the harsh depositional environment in this region, which favored species with calcitic shells (Ayoub-Hannaa and Fürsich 2012b). Variations in the plesiosaur fossils from Bahariya hint that multiple taxa inhabited these coastal waters, but these specimens are now lost (Stromer 1936).

This gap in our knowledge is especially galling as fish and marine reptiles would have had a significant impact on the ecology of the reef in ways that will, in all likelihood, never be known. Coccolithophorids are even more critical to the global carbon cycle than echinoids (Marsh 2003), yet their remains are not documented, so their ecological impact in this region is undetermined. The presence of herbivorous echinoid species also shows that there were marine plants on these reefs that have not been preserved (Ayoub-Hannaa and Fürsich 2012b).

However, some general comments can be made. *Ichthyotringa* inhabited surface waters, while *Clupavus*, which appears to have formed huge shoals like modem sardines, inhabited the cooler waters of the middle and lower water column. Also, like sardines, the *Clupavus* would have been at the base of the food chain. *Tselfatia* and *Elopopsis* also preferred the middle of the water column (Diedrich 2012).

By the late Cenomanian, the reefs were dominated by the sea snail *Pchelinsevia coquandiana*, which inhabited a reef system composed of rudist bivalves and corals. The surface of the reef would have had large cone-shaped bivalves emerging from the sediments, interspaced with flattened disk-shaped bivalves, while the rarer corals and sponges anchored to whatever hard surface they could find, slowly crawling amid all these growths were snails and echinoids (Ayoub-Hannaa 2011).

The salinity levels were average, and the substrate was soft but reasonably stable. This reef would have had reasonably high wave energies, and many of the shell fragments have been carried here from elsewhere down the coast. Below the surface, the sediments were barren of deep-infaunal life. The low levels of species diversity and the dominance of nerineoid snails indicate that the ecosystem was in decline.

Things were different on the western side of the Trans-Saharan Sea. In western North Africa, the bivalve reefs only extend as far as southern Algeria. Beyond that, the reefs are often composed of corals and coralline sponges, although oyster beds are known (Martill et al. 2011). Modern sponges are known to build reefs (Lang et al. 1975), but the fact that they were doing so in the Cenomanian of North Africa is entirely unexpected given the lack of attention they've thus far received. Also, unlike in Egypt, we have evidence of abundant crab burrows (Martill et al. 2011).

Interestingly, Martill and colleagues (2011) also make no mention of fish in these early Cenomanian reefs. It's not until the late Cenomanian that fish fossils are found, although by this point, as if just to be contrary, there is no evidence of reefs. Lézin and colleagues (2012) also note the disappearance of reefal species by this point in time. I suspect that the

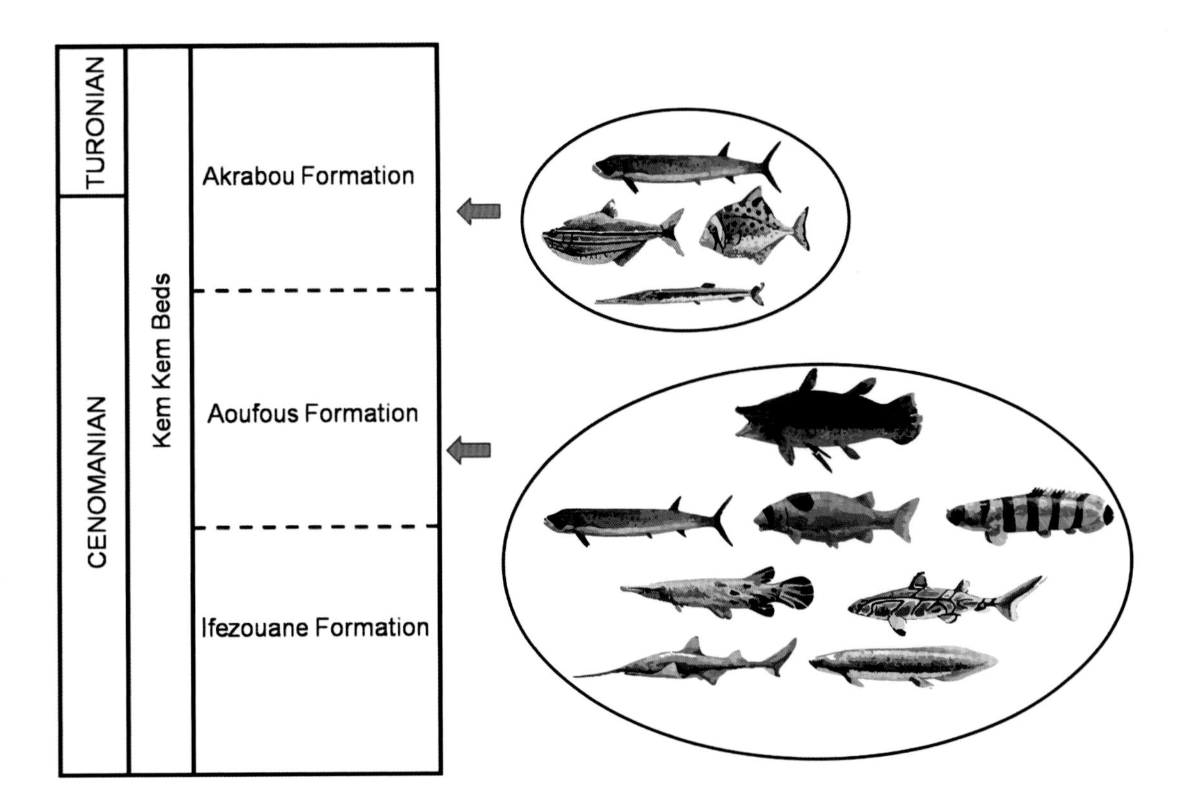

Fig. 6.3. The differing marine biota of the Kem Kem Formation (Martill et al. 2011). Images represent groups and not species. Images not to scale.

disappearance was due to environmental factors, as the waters were becoming increasingly deeper and anoxic. Cavin and colleagues (2010) also note the low species diversity of plankton and the rarity of foraminifera in the late Cenomanian, which concurs with that environmental analysis.

The aquatic fauna of the Gara Sbaa member of the Kem Kem Formation also differs from the fauna seen lower in the formation (see fig. 6.3). The underlying beds have a far more diverse fish community: polypteriformes, gars, coelacanthiformes, elasmobranchs, ichthyodectids, and lungfish. Gara Sbaa only has ichthyodectids, needlefish, pycnodontiforms, clupeids, and macrosemiiformes (Martill et al. 2011).

This opens the possibility that these taxa weren't coeval in this ecosystem. The Kem Kem beds record a period that lasted for 15 million years (Martill et al. 2011), so it's possible that *Agoultichthys*, for instance, only evolved after *Mawsonia* was already extinct. Solving this issue is vitally important to our understanding of Cenomanian North Africa because it could mean that all the preceding taxa did not even live at the same time, let alone in the same ecosystem. In that case, we are dealing with two separate biotas instead of one.

The fact that the fauna of the lower formation is completely replaced with a new fauna in the upper formation would suggest that possibility. However, the preceding layers were all freshwater/delta environments, and Gara Sbaa was fully marine, a result of rising sea levels that flooded the land. So it's possible that this is a single biota and that we are seeing the migration of freshwater species farther inland and the arrival of marine species into this region as a result of environmental change.

So which is it, extinction and replacement or extirpation and immigration? Although I suspect the latter personally, it's difficult to say given the rapid environmental changes that mask the ecological signals. We will need freshwater sediments that are coeval with the Gara Sbaa for comparison to see whether the preexisting fauna really is gone.

Another interesting fact is that the western reefs seem to have a large number of abyssal species preserved in their shallow coastal waters. Joschua Knüppe (2016) suggests that they lived at depth and migrated to the surface at night to feed. Such diurnal vertical migration is well documented among extant marine fauna (Neilson and Perry 1990), and some circumstantial evidence indicates that this might apply to *Protostomias*, as almost all the specimens found had fed shortly before death (Diedrich 2012).

Also, the sedimentation in some regions of Jbel Tselfat shows that there were areas of very deep water near the coastline (Khalloufi et al. 2010). This suggests that the continental shelf was very narrow in some areas, bringing the abyssal plain right up to the shoreline, potentially allowing abyssal species access to shallow water (Tlïg et al. 2008; Diedrich 2012).

How Does Your Garden Grow?

That's fine for the marine side of this ecosystem, but what was happening on land? At the moment only two formations, the Kem Kem beds of Morocco and the Bahariya Formation of Egypt, are known well enough for a detailed analysis. Fortunately, when taken together, they provide a clear picture: the first formation represents the freshwater part of the ecosystem, and the other formation represents the coastal part. Most of what follows are based directly on these two paleoenvironments.

Primary Production

From above, the Bahariya Bight would have shimmered like a blue mirror, covered with curving, thick green lines where plant life grew in the shallower waters, creating a labyrinth of waterways, isolated pools, and oxbow lakes. Indeed, it would have been difficult to tell whether it was a forest with an extremely large number of rivers or a vast bay with tangled rows of plant life growing over it in all directions.

Only on the ground level would the truth be revealed, with rows and rows of *Weichselia* dominating the bight for as far as the eye could see. The actual coastline is farther inland, as most of these tree ferns are actually growing directly out of the ocean, with their stilt-like prop roots elevating them above the water.

Below the waterline, the snarl of prop roots trapped sediments and stabilized them, leading to the gradual development of mud bars and fine-grained sandbanks around the *Weichselia* mangals. By filling the bay with

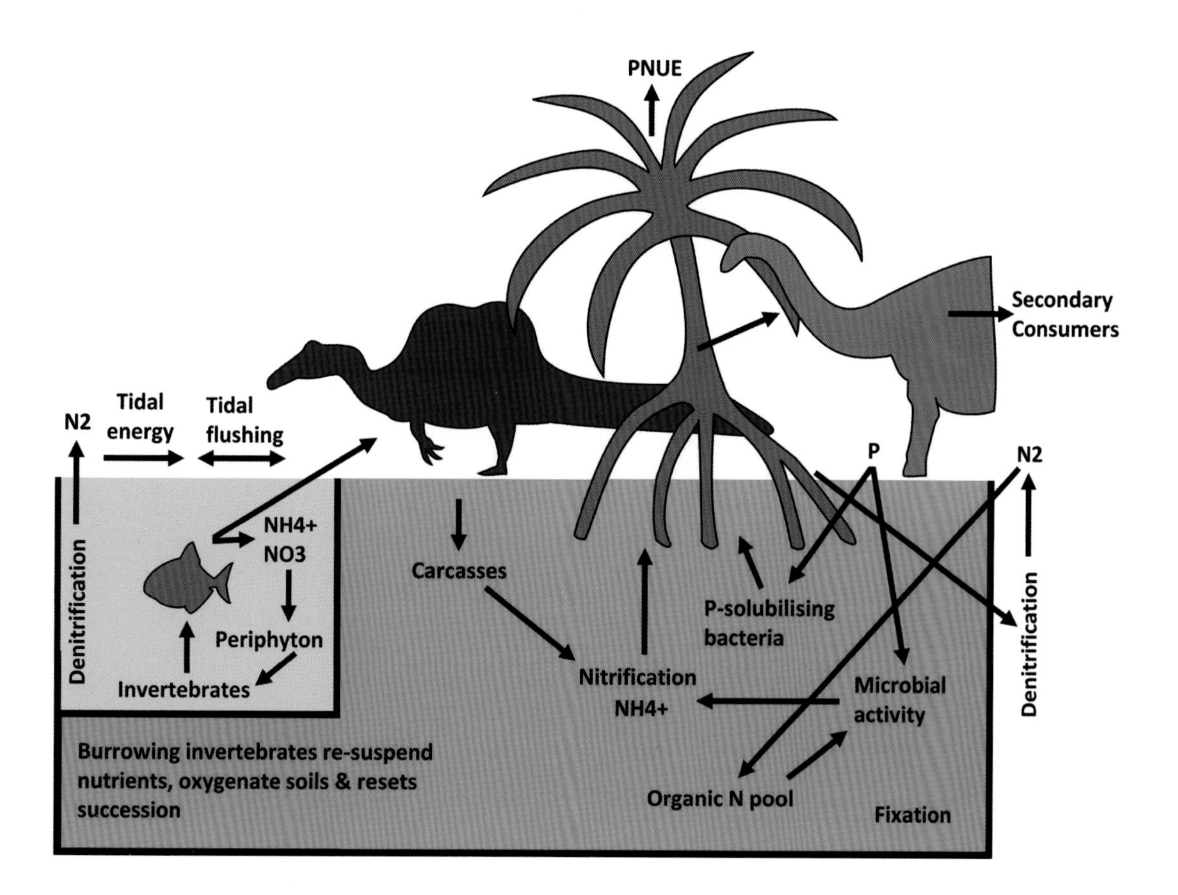

Fig. 6.4. A reconstruction of the Egyptian mangrove system. Abbreviations: N, nitrogen; P, phosphorus; NH4+, ammonium; N_2, nitrogen gas; and PNUE, photosynthetic nitrogen use efficiency. Images represent groups and not species. Images not to scale. Adapted from Hossain and Nuruddin (2016).

sediments, the *Weichselia* proved advantageous for other flora, such as *Cladophlebis* and a host of unnamed flowers, which grow amid the ferns. The surface of the shallower river channels is covered in *Nelumbites*, and even the gently lapping, green-tinged water itself is alive with plankton.

At the back of the mangroves, toward the actual shoreline, the vegetation becomes more diverse. The marine influences are waning, and these older mangrove sediments are compacted enough to support actual trees, such as *Terminalioxylon* and *Magnolia*, that grow around huge deltas and large but shallow lagoons.

So it's no wonder that the Bahariya Bight is often claimed to be an incredibly productive biome (Nothdurft et al. 2002). Comparisons with extant mangroves can give a general picture of productivity at Bahariya (Ijouiher 2016a). Lacovara and colleagues (2003) have noted that the depositional environment of Florida is the closest extant analogue to Bahariyan sedimentation, so Bahariya is not without precedent. And the flora was clearly behaving *like* an extant mangrove community (see fig. 6.4) and would have had to survive under identical conditions as any "true" mangrove swamp.

A habitat's overall net primary production depends on its total area (Pauly and Ingles 1986). With this one mangrove bight alone extending for an estimated 300 km along the coast, Bahariya had sufficient landmass for a sizable trophic base. Schweitzer and colleagues (2003) record that the mud and sandstones were also rich in organic matter. The abundance of siderite is also evidence of organic-rich sediments (Tanner and Khalifa 2009). And this would have been just one of many such bights in North Africa at the time.

However, productivity is based on more than just habitat. It's also based on the biology of the plants themselves. So how do the Cenomanian plants hold up when compared to extant species? Surprisingly well, actually. Studies have shown that the energy content of some gymnosperms (conifer, ginkgo, and cycad trees) is only slightly lower than that of angiosperms (Hummel et al. 2008; Hummel and Clauss 2011). Some Mesozoic trees like *Agathis* sp. are also productive regarding biomass (Gee 2011). The exception to this rule appears to be the tree fern *Dicksonia*, which is horrendously poor in terms of energy (Hummel et al. 2008).

This opens up an interesting question about *Weichselia*. Most ferns are practically equal to angiosperms in terms of the energy herbivores gain by digesting their foliage. However, if we have one exception with regard to harvestable energy content in *Dicksonia*, could *Weichselia* also be an exception? The Matoniaceae have yet to be included in any in vitro digestibility studies. While the current assumption would be that *Weichselia* was energy rich, given that most ferns are, this conclusion could change in the future. This would have a significant impact on Bahariyan productivity estimates, as *Weichselia* was a keystone species.

High tidal energy along the coastline and changes in the water circulation in the Tethys Sea would also have resulted in increased oceanic nutrient flows, importing large quantities of nutrients into Bahariya (Lacovara et al. 2002; Leckie et al. 2002). And we need to remember that, regarding primary production, the ocean is the second most productive region on earth. It produces 2000 $g/m^2/yr$ on average (MarineBio Conservation Society 2016).

Yet the nutrient content of such tidal influxes can be ten to twenty times lower than that of riparian sources (Lugo et al. 1976), and data suggest that terrestrial runoff is more important than tidal influx in delivering nutrients to modern mangroves (Hogarth 2007). So while Bahariya would have increasingly stored nutrients rather than exporting them, the mangroves at the mouth of the extensive river networks in the south would have been more productive than those in the bight itself.

While river networks for Bahariya are known (Werner 1989; Lyon et al. 2001), the much better known Kem Kem Formation now takes point as the best example we have of a northern Gondwanan river system.

While it can be challenging to wrap our minds around the concept, these river networks were continental in extent (Nothdurft et al. 2002). For comparison, the Amazon can shift 1200×106 tons of sediment a year, 1% of which is organic. This puts the carbon load for the Amazon at an annual 13×106 t/y, with dissolved carbon being double that amount (Luiz et al. 1989).

While the exact nutrient and sediment load for the Kem Kem delta will probably never be known in such detail, I would not be surprised if the North African rivers could equal that seen in the Amazon, given their size and extent. Models of hydrologic cycles by Floegel and Hay (2004) also show that the tropical climate would enhance the rates of groundwater flow, further increasing the nutrient load entering the ecosystem.

The source of the Amazon is the Andes Mountains (Luiz et al. 1989; McClain and Naiman 2008). North Africa also had a series of mountain ranges (Lüning et al. 2004; Moustafa and Lashin, 2012). Certainly, the levels of mineral nutrients provided by these mountains were high; Martill and colleagues (2011) have documented the presence of extensive iron oxide and iron staining, along with what appears to be vivianite. Indeed, the West African Craton could today still produce 1 million tons of iron ore annually (Wright 1985).

What's shaping up to be the best freshwater deposit in the Kem Kem Formation has been found at Jbel Oum Tkout. This was likely an oxbow lake surrounded by a gallery forest dominated by araucarian conifers, interspersed with angiosperms, gymnosperms, and spermatopsids. This waterway was home to *Serenoichthys*, *Spinocaudichthys*, *Diplospondichthys*, and roughly thirteen new unnamed taxa. Shark and ray teeth are also known from this site. The stagnant waters and fine-grained mud resulted in Lagerstätte-quality preservation (Khalloufi et al. 2018; Ibrahim et al. 2020a).

Floral Population Dynamics

Such nutrient-rich waters may have been the source of the coastline's rich diversity of plant life compared to the river networks that fed it, although it must also be noted that megaflora can also flourish in depleted soils (Aragão et al. 2009). So nutrient loading is not necessarily a factor in the development of an extensive forest.

Regardless, these coastal forests were so diverse that the oil deposits in Egypt's Western Desert are composed of their remains, with small amounts of marine algae included (Younes 2012). It's incredibly uncommon for oil deposits to be composed primarily of land plants, so there must have been abundant plant life along the coastline. Because many botanical specimens are undescribed, the current count is based solely on Miospore taxa.

Bahariya went through four ecosystems (Tahoun et al. 2013), each named after its dominant miospore taxa: zone 1 (*Afropollis jardinus*—an angiosperm); zone 2 (*Elaterosporites klaszii*—a gymnosperm); zone 3 (*Cretacaeiporites densimurus*—a magnoliid); and zone 4 (*Trilobosporites laevigatus*—a type of "pteridophyte"). The changes in flora correlate to changes in sea level and climate.

We can study floral diversity on a localized level as well. Baioumi and colleagues (2012) note that the sediments at the Horus-1 oil well, to the north of Bahariya, are slightly older than those preserved at Bahariya and represent a freshwater river as opposed to the brackish, marine conditions that later became predominant. It also demonstrates localized humid conditions in a region that was predominantly arid.

Afropollis jardinus is abundant throughout the Cenomanian at Horus-1 (20% or higher in the sampled sediments). At only three points in time does it become rare (less than 10%). *Crybelosprites pannuceus*, *Cyathidites* sp., and *Dictyophyllidites harrisii* were also common, with brief periods of decline and recovery. The amounts of *Cicatricosisporites minutistriatus*, *Elaterosporites klaszi*, *Classopollis* sp., and *Ephedripites* sp., however, remain consistently low except for a brief period when abundances increase (Baioumi et al. 2012).

Horus-1 would have been an open gallery forest along a riverbank, rich in angiosperms and gymnosperms of various types. Amid all this would be the ground-level leptosporangiate ferns and the taller matoniaceae ferns. Interspaced between were the much rarer schizaealid ferns and conifer and ginkgo trees.

The section of the formation found at the El-Waha-1 oil well also has a unique palynomorph assemblage (Moustafa and Lashin 2012), initially containing *Classopollis brasiliensis*, *Cretaceioporites scabratus*, *Ephedripites* sp., and *Araucaiacites* sp. This ecosystem later shifted to an assemblage dominated by *Afropollis jardinus*, *Classopollis brasilienses*, *Retimonocolpites* sp., *Matonisporites simplex*, *Crybelosporites pannuceus*, *Cretaceioporites scabratus*, and *Retitricolpites* sp. in the later Cenomanian.

Growing on a mountainside, El-Waha-1 would have been a dense conifer and araucarian forest, with smaller gymnosperm shrubs growing amid the towering trees. Later, as conditions became more humid, this forest would be dominated by angiosperms and semiaquatic ferns. Leptosporangiate ferns replaced the gymnosperm shrubs, and araucarian trees became rarer, although conifers would remain plentiful, joined by large numbers of ginkgo trees.

However, the presence of some of the other taxa recorded in Bahariya is strange. *Agathis* can only survive short periods of waterlogging and mild salt spray (Tomlinson 1986; Thomson 2006). Mahmoud and Moawad (2002) and Moustafa and Lashin (2012) show that such coniferous forests were found only in the uplands during the Aptian, and this may well have been the case in the Cenomanian, as Baioumi and colleagues (2012) note the decline of araucariacean pollen as marine influences increased. This suggests to me that any *Agathis* remains found in Bahariya were carried downstream from farther inland.

Both the rivers and coastal waters were also rich in phytoplankton (Stromer 1936). *Mudrongia*, *Cyclonephelium*, *Xiphophoridium*, *Florentinia*, *Cribroperidinium*, *Dinopterygium*, *Exochosphaeridium*, *Peseudoceratium*, and *Subtilisphaera* were the dominant dinoflagellate taxa, composing 40%–90% of the total plankton biomass in some areas (Ibrahim et al. 2009; Tahoun et al. 2013). Multiple algal blooms also saw *Pediastrum* and *Botryococcus* periodically increase in abundance (Ibrahim et al. 2009).

In Morocco, the situation is vastly different because of the paucity of plant life (Läng et al. 2013; Läng 2014). Regardless, species diversity seems

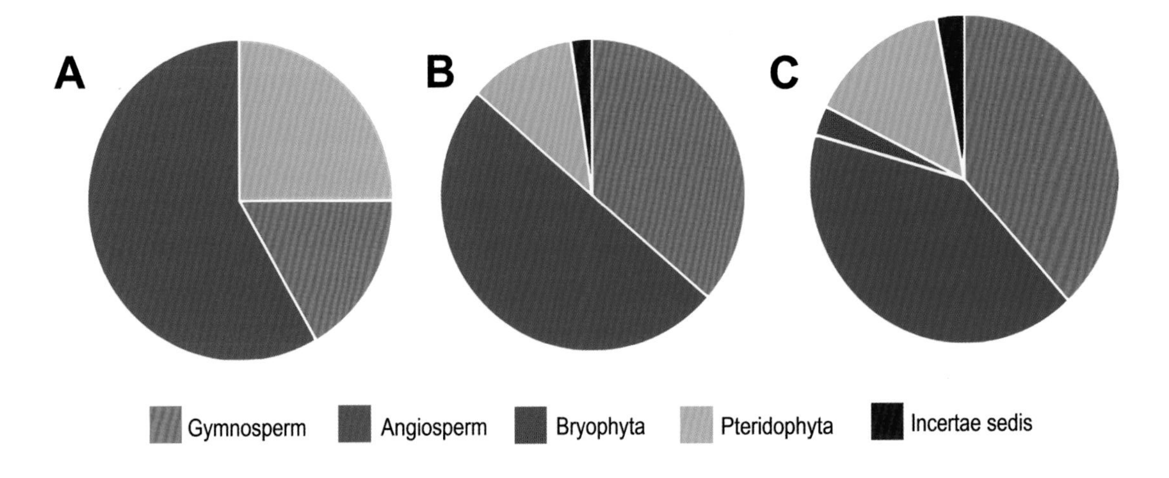

Fig. 6.5. Floral diversity in North Africa. Results show species-level diversity. *A*, Morocco; *B*, Egypt; and *C*, species diversity for Egypt based on pollen. Similar palynological information for Morocco is unavailable.

similar (see fig. 6.5). Of course, this implies that the two ecosystems had the same plants, not that they had the same amount of plant cover.

Interestingly, the one area of Bahariya that does correspond ecologically to the inland Kem Kem river network—the Southern Meander System—also seems to lack plant life (El-Sisi et al. 2002). The same seems true for Libya, Tunisia, and Algeria. Plant life in North Africa seems to have been concentrated at the coast and in the uplands, with a belt of mostly barren floodplain between them.

This is an odd pattern; given the abundance of huge rivers, you would typically expect diverse gallery forests along the waterways. Clearly, there were ecological factors at play here, besides water availability, which we don't fully understand, especially as isolated inland forests and groves are known in a few areas (de Lapparent 1960; Engel et al. 2012; Baioumi et al. 2012; Murray et al. 2013). But why they grew in these places, and not anywhere else on the floodplain, is a mystery.

Imports and Exports

Coastal regions import not only mineral nutrients but also particulate and dissolved organic matter (Hogarth 2007). Other imports come in the form of adult crustaceans returning after spending their larval stage at sea, and fish using mangroves as nurseries for their young (Morrissey and Gruber 1993a, 1993b; Cerutti-Pereyra et al. 2014). This has great implications for North Africa, as not only were the fish populations extensive but they contained numerous large species (Cavin et al. 2015).

It's possible that some Moroccan pycnodonts, an order usually reconstructed as being marine, periodically migrated upstream to spawn or to feed. This has been inferred for some European species (Cavin et al. 2020), and it would explain why pycnodonts are so common here while they are rare in other freshwater deposits. But the Moroccan taxa are too poorly known to say for sure (Cooper and Martill 2020).

So while such fish nurseries can be inferred for some species (Yabumoto and Brito 2013; Cooper and Martill 2020), the only confirmed case is in *Enchodus*. While *Enchodus* seems to have gathered in certain areas to spawn, it's unlikely that these groups reached anywhere near the size of the sardine or salmon runs (whose shoals can be measured in miles and number in the billions). Sardines feed on zooplankton and phytoplankton, while *Enchodus* was carnivorous. It's unlikely that an *Enchodus* shoal even half the size of a salmon or sardine shoal could have supported itself (except for the sockeye salmon, salmon are also carnivorous, but they stop eating when they begin migrating upriver, making the question of shoal sustainability moot).

As far as I'm aware, no one has ever done the calculations to show how large a predatory shoal can get. However, even the arrival of a moderately large shoal would have brought an important influx of protein into the region. The fact that *Goulmimichthys* were found in the area suggests that they were attracted to this region by the presence of spawning *Enchodus* sp. (Cavin 1999b).

However, wetlands also lose productivity. Lyon (2001) and Krassilov and Bacchia (2013) report extensive insect damage to the plants in both the freshwater and mangrove communities. But extant swarming insects only account for an inconsequential 2%–5% of forest productivity (Hogarth 2007), and there is no reason to assume that these Mesozoic forests were different.

Of course, while insects drain plant productivity, they also contribute nutrients to the food chain, as they are prey for a host of other animals. This is especially true for emergent insects, like mayflies and mosquitoes, which can form huge swarms as they as emerge from the water as adults. Deducing how much they actually contribute is difficult even in modern ecosystems, given the number of variables involved (species, climate, terrain, etc.), and nearly impossible in prehistoric ones.

However, there are some general trends. For instance, the larger the body of water, the greater the insect population. Also, the influx of aquatic insects into a terrestrial ecosystem can sometimes equal the net secondary production of that entire ecosystem, particularly in ecosystems with low productivity, such as desert or arctic tundra (Gratton and Vander Zanden 2009).

Likewise, the microfauna is also a drain on mangrove productivity, although it plays an important role in leaf-litter turnover. The carbon they release gets recycled within the microbial community alone, with little escaping into the wider ecosystem. Roughly 10% of production is lost this way (Ong 1993).

Microbial communities are poorly known for North Africa. Fortunately, such communities tend to form distinct food chains, with only limited, and inconsequential, interactions with the mangrove ecosystem as a whole (Hogarth 2007). Thus, our lack of understanding of this fauna does not hinder our understanding of the local ecology as a whole.

Lacovara doubts that the dinosaur population lived permanently in the mangroves (in Nothdurft et al. 2002). This is in accordance with fauna populations in extant mangroves. In Australia, out of the two hundred bird genera recorded in this habitat, only fourteen are found solely in

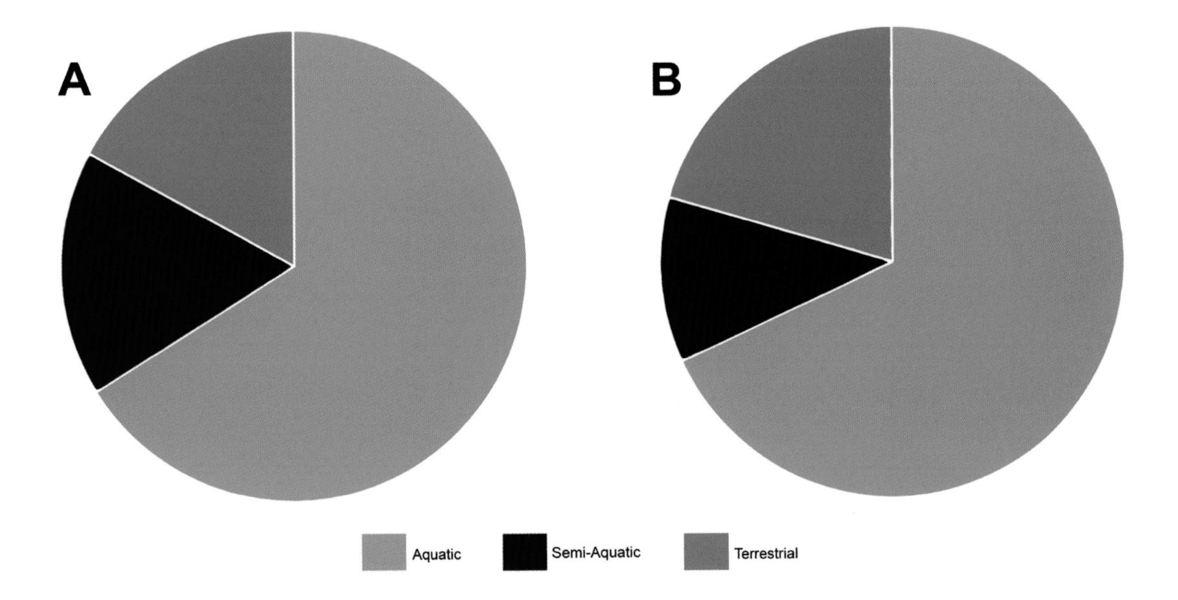

Fig. 6.6. Vertebrate ecological diversity for North Africa. *A*, Morocco; *B*, Egypt. Results show genus-level diversity.

mangroves. The rest commute daily or annually (Noske 1996). Even fewer mangrove-adapted birds are known from elsewhere in the world, and the number of mangrove-adapted mammals is smaller still (Hogarth 2007).

It must also be noted that no exclusively mangrove-adapted herbivore over a ton in weight ever appears to have evolved, although in the recent past elephants, buffalo, and rhinoceroses frequented the Sundarban mangroves to feed (Hogarth 2007). Indeed, the bulk of the vertebrate fauna from Bahariya comprises aquatic or semiaquatic forms, with terrestrial species in the minority (see fig. 6.6). The titanosaurids, the dominant herbivores in Bahariya, also appear to have been increasingly adapted for inland habitats (Brusatte 2012). The exception to this lack of specialization among large vertebrates appears to be the spinosaurids, which were specifically adapted for coastal environments (Fanti et al. 2014).

This is because the basic structure of the mangrove ecosystem prevents specialization (Nagelkerken et al. 2008). There aren't enough ecological niches to occupy, and those fauna that do become specialized often have to compete with the generalist fauna for the same resource. So rather than a host of mangrove-adapted taxa, Bahariya probably had a generalist population entering the mangroves from the inland floodplains to take advantage of the abundant food supply. We can see this with at least one species of Cenomanian bird, which frequented the mudflats of Tunisia at low tide to beachcomb for food (Contessi and Fanti 2012a), mirroring the behavior of extant birds (Frey and Pemberton 1987).

This means that the Bahariya Bight was losing productivity, as the fauna fed and then took that carbon outside the mangrove biome. However, this loss may be overstated, as little carbon is exported to surrounding habitats in modern mangroves (Nagelkerken et al. 2008; Bouillon et al. 2008), although it must be noted that these studies were all conducted after most modern mangroves had had their large herbivore visitors extirpated (Hogarth 2007). The results for the preindustrial mangrove communities might have been different from what we see now.

It would also be offset by the fact that commuter species also import carbon, as fecal matter and as carcasses (Food and Agriculture Organization of the United Nations [FAO] 1994; Stafford-Deitsch 1996). This would be especially significant with dinosaurs, which would become what Russell called "cadaver decomposition islands" upon death (Carter et al. 2007; Russell 2009). We can see this with the holotype of *Paralititan*, which was scavenged after death (Nothdurft et al. 2002).

Carrano and Velez-Juarbe (2006) also suggest that the relative immunity of adult sauropods to predation would have created a distinct food chain of their own: plants-sauropods-decomposers. At most, all that carbon would be recycled through one extra trophic level (giant theropods). With fewer levels in the food chain to lose energy, most of the nutrients would be recycled straight back into the environment directly, resulting in greater productivity and thus a greater diversity of large predators.

Nutrient Recycling

The presence of such a loop would give increased importance to decomposers, which brings the macroinvertebrates into the picture. By comparing Morocco and Egypt, we can gain a complete picture of recycling. In the freshwater river networks, insects were the chief invertebrate scavengers (Ibrahim et al. 2014a).

If termites were present in this region, they would have had a considerable role in nutrient recycling (DeSouza and Cancello 2011; Ali et al. 2013). As detritivores, termites increase nutrient recycling rates and enrich the soil by breaking down plant material faster by predigesting it. They have also been recorded feeding on herbivore dung, with the same results (Freymann et al. 2008).

Because of the excess nutrients in the area and the moisture given off by the nest itself, termite nests are often surrounded by vegetation (Bonachela et al. 2015). This forms a food supply for herbivores by increasing plant yield by 36% (Evans et al. 2011). It also forms a barrier against desertification (Bonachela et al. 2015). In the Cenomanian, because of global warming, the spread of the Central Gondwanan Desert (Jacobs et al. 2009; Mateus et al. 2011) would have been a significant issue, much like the expansion of the Sahara Desert into central Africa today (Thomas and Nigama 2018).

In the coastal mangroves, snails and crabs seem to have taken over as the main invertebrate scavengers. The chief detritivores in extant mangroves are often crabs (Tan and Ng 1994; Gillikin and Schubart 2004; Hogarth 2007). Crab burrows oxygenate the sediments and reduce salinity (Micheli et al. 1991), while their leaf-litter turnover rates are 75 times faster than microbial decay (Robertson and Daniel 1989). Decapod digestion also enhances the nutritional value of leaves excreted, as they digest the carbon but expel the nitrogen in their fecal matter, which itself is much easier than rotting leaves for microbes to break down (Lee 1997).

While Bahariyan crabs were undoubtedly scavengers of vegetation and animal carcasses (Schweitzer et al. 2003; Ibrahim et al. 2014a), Nothdurft and colleagues (2002) also recorded large numbers of snails. This is not without precedent, as extant species can achieve high population densities in mangrove habitats (Sasekumar 1974; Nagelkerken et al. 2008). Snails sometimes even replace crabs as the primary ecosystem engineers: crab predation on seedlings in Malaysian mangroves is responsible for 95%–100% of losses, while in the Floridian mangroves—the best modern analogy for Bahariya—crabs only accounts for 6% of seedling losses, with 73% caused by gastropods (Hogarth 2007).

Personally, I suspect that Schweitzer and colleagues (2003) were wrong and that gastropods were the real ecosystem engineers in Bahariya. Little has been published on Bahariyan snails, not even the families to which they belong, so their role cannot be studied in detail.

Spinosaurus and *Sigilmassasaurus* might have also played a role in nutrient recycling (see fig. 6.7). The TV series *Planet Dinosaur* received criticism for its suggestion that "Spinosaurus can afford to be wasteful" (Paterson 2011). Yet there is some basis for that claim, assuming that spinosaurids were like extant piscivores. During the salmon run, bears leave half of their catch uneaten, resulting in an influx of nitrogen, an increase of 24% in some cases, into the surrounding woodland (Reimchen 2001; Helfield and Naiman 2006).

This could have potentially made the spinosaurids keystone species, as any carcasses they carried away from the water would have provided extra nitrogen for the plants (Nash 2012). They would also provide food for

Fig. 6.7. A reconstruction of the Moroccan riparian system. This also shows the actual dynamics of how the spinosaurids would have helped reinforce their ecosystem through their contribution to the nitrogen cycle. Abbreviations: N, nitrogen; NH4+, ammonium; NO3, nitrate; OM, organic matter; and CWD, coarse woody debris. Images represent groups and not species. Images not to scale. Adapted from US Environmental Protection Agency 2011.

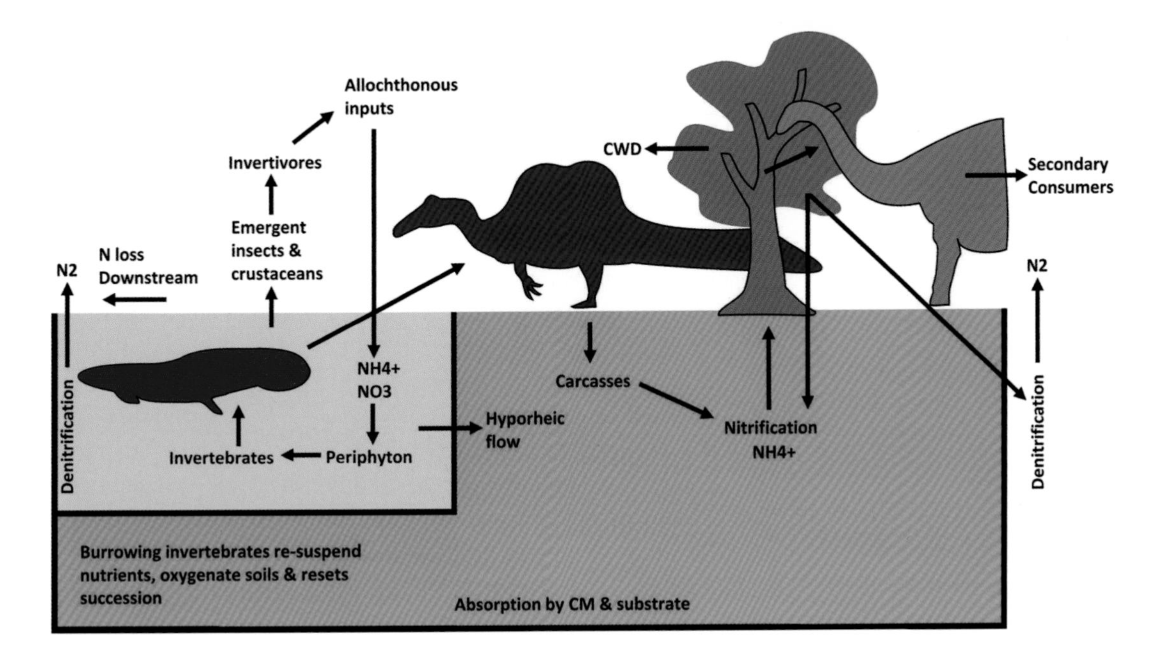

smaller theropods, terrestrial crocodilians, pterosaurs, and invertebrates, which would scavenge any leftover carcasses. Even species that don't feed on the fish directly, such as insectivores, would benefit indirectly as the fish-derived nutrients made their way up the food chain (Field and Reynolds 2011).

Faunal and Floral Differentiation

The Continual Ecological Province

Previously I've mentioned something called the "continual ecological province" hypothesis, which is shaping up to be the most contentious issue facing North African paleontology (see fig. 6.8). In a nutshell, some researchers suggest that North Africa was dominated by a homogeneous biome that stretched from Morocco to Egypt (Cavin et al. 2013; Ibrahim et al. 2014b; Cavin et al. 2015), while others dispute this (Evers et al. 2015). One of the biggest problems for those proposing this hypothesis is that the fauna of eastern and western North Africa do seem to be different in some key aspects.

Only one pterosaur specimen is known from Bahariya despite having proved common in the Kem Kem beds (Mader and Kellner 1999; Ibrahim et al. 2010). Other missing Bahariyan fauna include the high diversity of crocodylomorphs from the Kem Kem beds (Lavocat 1955a; Sereno and Larsson 2009). It's also interesting to note that the evidence of rebbachisaurid sauropods from Bahariya is dubious, which is consistent with the theory that the rebbachisaurids were restricted to northwestern Africa (Mannion and Barrett 2013).

Only the lack of amphibians from Egypt can be explained with certainty, as, with few exceptions, amphibians cannot tolerate brackish waters (Dicker and Elliott 1970). But how do we explain the rest? Or is the continual ecological province doomed?

I suspect that simple nonrecovery is the most likely explanation in some cases, given how poorly sampled many of these formations are. It took 120 years for ankylosaurid remains to be found in the well-sampled Morrison Formation (Foster 2007). Coastal regions can also be a harsh depositional environment for some fossils (Lacovara et al. 2002). Birds are utterly absent in most of these formations, yet isolated footprints show that they were there the whole time, just the same.

However, an interesting pattern seems to have taken place among the Stomatosuchidae and Aegyptosuchidae. Western Gondwana has *Aegisuchus* and *Laganosuchus*, while Eastern Gondwana had *Aegyptosuchus* and *Stomatosuchus*. Only one species of each family is present in each region. This could be evidence against the continual ecological province, with the Trans-Saharan Seaway forming a barrier to crocodilian dispersal. Once separated, the populations in the east evolved into *Aegyptosuchus* and *Stomatosuchus*, with the second populations evolving into *Aegisuchus* and *Laganosuchus* to fill the same niche in the west.

An alternative theory is that all the species of stomatosuchid and aegyptosuchid coexisted, but with niche partitioning similar to that between

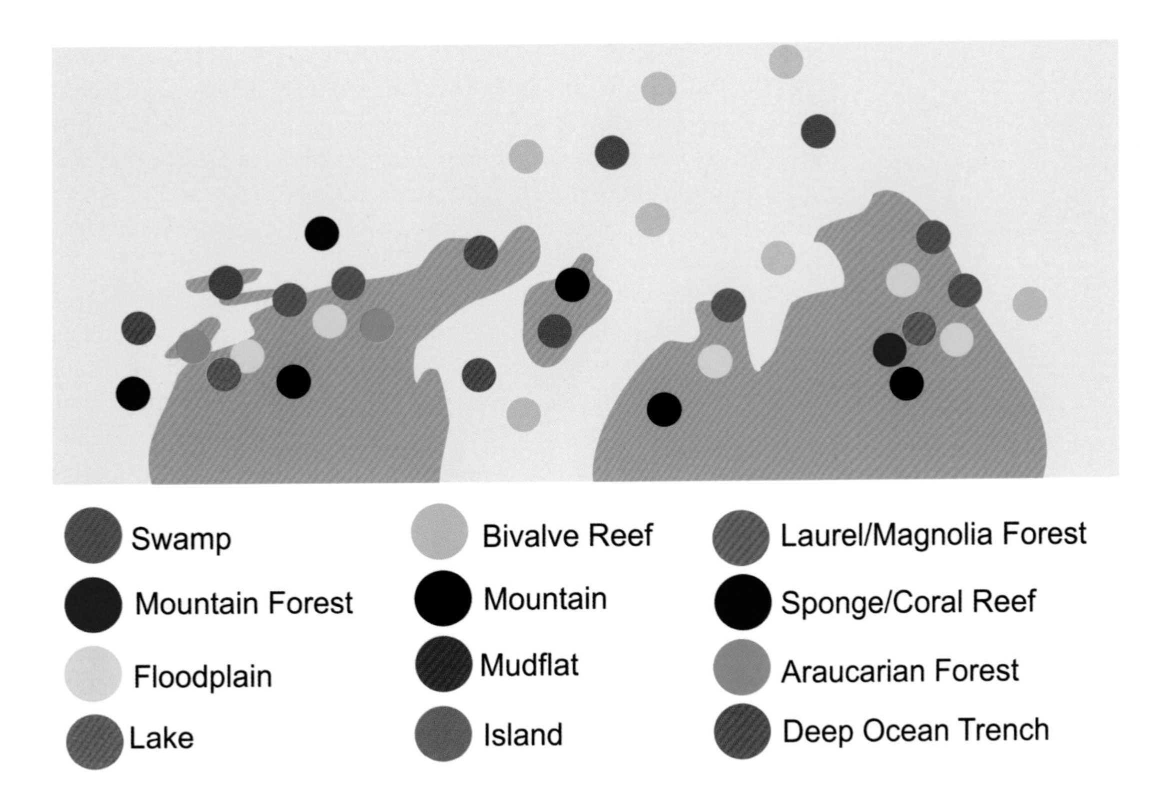

Fig. 6.8. A simplified map of North Africa, showing the various ecosystems. Contra Ibrahim et al. (2014b, 2020a), the North African ecosystem, while broadly similar, was not continuous from Morocco to Egypt. Rather than forming one monotonous ecosystem, it was broken up into a complex mosaic of habitats.

the Australian saltwater crocodile and freshwater crocodile today (Wheatley et al. 2012). *Aegisuchus* and *Laganosuchus* lived in the rivers and *Aegyptosuchus* and *Stomatosuchus* occupied the mangroves.

However, a closer look at the issue revealed that the freshwater/saltwater divide among crocodilians is more complicated than it first appears. Crocodiles—and possibly gharials—are the only extant members of this family capable of surviving in salt water, as alligators lack salt glands (Grigg and Kirshner 2015). Yet alligators, despite lacking salt glands, have still been spotted 38 miles (63 km) out at sea (Grigg and Kirshner 2015). While some might have been washed out to sea by a storm, it's clear that some undertook ocean trips of their own volition.

Given the lack of salt glands, Grigg and Kirshner suggest that alligators manage this because of their size. The larger the individual, the longer it can survive out at sea, although it will have to find land and fresh water to drink eventually. And if a large alligator can survive for prolonged periods in seawater, then a true giant like *Aegisuchus* certainly could, with or without salt glands. And since the average saltwater crocodile can manage an open ocean journey of 1,200 miles (1,931 km) (Shaw 2016), the Trans-Saharan Seaway probably wasn't a barrier for giant crocodilians.

Indeed, the barrier between the saltwater crocodile and freshwater crocodile appears to be competitive, not ecological. In areas where the saltwater crocodile has been extirpated, freshwater crocodiles expand their range down to the coast. When saltwater crocodiles recolonize these regions, the freshwater crocodiles retreat inland again (Delaney et al. 2010; Fukuda et al. 2011).

Oxygen isotope studies of crocodilian fossils could help settle this issue, but with such limited material, it's impossible to conduct any of these studies. The tests would be destructive, and we can ill afford to lose the few specimens we currently have.

Part of the problem is that the Bahariya Formation is horrendously underrepresented in terms of specimens. The Kem Kem beds have been worked for over a century. Bahariya has had only three expeditions of note, and most of the material is lost or has never been described. Therefore, it's premature to say that the absence of some taxa is the result of a fundamental difference in the prevailing ecosystems.

Another thing to note is the apparent lack of ecological overlap between the Kem Kem beds and Bahariya Formation. While they give an overview of the biome in its entirety when taken together, it's possible that the absence of particular plant and animal species is due to one formation being dominated by salt water and the other by fresh.

Chiarenza and Cau (2016) seem to have come to similar conclusions. While I agree with them in the broad strokes—environment variation being a factor in the rarity of some taxa—some researchers seem to have oversimplified this, in my opinion.

The idea that the answer to the question of how these numerous large predators coexisted is that they really didn't is contradicted by the fossil evidence in Bahariya—they were clearly frequenting the same environment—and also by another study by Sales and colleagues (2016) shows that while abelisaurids were rare in wet habitats, spinosaurids and carcharodontosaurids often coexisted. While there is a general trend of spinosaurids living along the coast and carcharodontosaurids inland, there was a reasonable degree of overlap between them. Also, Chiarenza and Cau (2016) only focus on bones and make no mention of the fossil teeth found in this region, which change the picture dramatically (Amiot et al. 2004; Porchetti et al. 2011; Richter et al. 2012).

The Bahariyan biome had at least four main ecosystems, and each one would be made up of a cross sample of the regional population. The mangroves are well documented, but there was also an open floodplain in the south (Stromer 1914; Werner 1989; Lyon et al. 2001; El-Sisi et al. 2002; Catuneanu et al. 2006; Grandstaff et al. 2012). The mountains had a series of laurel/magnolia swamps at their base, presumably formed because of the combination of humidity and upland temperatures (McCabe and Parrish 1992; Little et al. 2009). At higher altitudes, this habitat gave way to araucarian/conifer forests (Moustafa and Lashin 2012).

Given the mosaic of environments within this biome, it's possible that some species would be more common in some areas. Were abelisaurids and rebbachisaurids genuinely absent in Bahariya? Or are we dealing with a group that would have favored the drier open floodplains in the south and been uncommon in the coastal swamps? Unfortunately, the only region in Bahariya to record this environment, Gebel El Dist, is also the region least explored.

This is easy enough to verify, as we should know exactly where to look for rebbachisaurids and other missing fauna (Ijouiher 2016b). My prediction came true in part with the preliminary report of an abelisaurid from Gebel El Dist (Salem et al 2021), although no accompanying sedimentological data has been published yet, so we cannot say anything about the environment the animal was frequenting when it died.

And the two sides of Africa often do compare well, especially with the theropods, on a genus level, but not on a species level. We have two species of *Spinosaurus*, one on each side (Cau 2015a), and very possibly a different species of *Carcharodontosaurus* on each side as well (Chiarenza and Cau 2016). Indeed, *Carcharodontosaurus* shows a north/south divide as well as an east/west divide. *Carcharodontosaurus saharicus* is found in the north and another species, *C. iguidensis*, in the south.

Both sides also have their own Stomatosuchidae and Aegyptosuchidae, as discussed earlier. And while nothing can be done about the herbivorous dinosaurs, given that most are so poorly known, the Moroccan material that might indeed pertain to Egyptian sauropod taxa also shows subtle differences (Lamanna and Hasegawa 2014).

This supports the idea of "intracontinental faunal endemism," which Lehman (2001) put forward. While many of his claims later proved false, the general idea seems sound. Many dinosaur species don't seem to have continent-wide ranges like modern mammals, despite their large sizes, but were instead restricted to specific regions on a continent, possibly along latitudinal and altitudinal lines (Naish 2020).

Undoubtedly, there were some species found only in one region and nowhere else, especially among the small taxa that could not cover large distances easily. For instance, one river alone was enough to separate an ape population, which then split into two separate species, the chimpanzee and bonobo (Takemoto et al. 2015).

So while my own views are still evolving, I'm still personally in favor of the continual ecological province hypothesis, albeit with modifications. We're seeing that a fauna was broadly similar in regard to family and genus but differed on the species level, probably as a result of vicariance when the Trans-Saharan Seaway cut the continent in two. Again, it all rests on further exploration of the southern region of Bahariya, as it's the only part of that formation to be analogous to the Kem Kem Formation.

We've Got You Outnumbered

We don't have specimen counts for most North African sites to show how rare or common most taxa are. Most species are known only from their holotype specimens, with only generalized statements about abundance (Villalobos-Segura et al. 2021). However, the information does exist for site BDP 2000–19 in the Bahariya Formation, allowing for a comparison

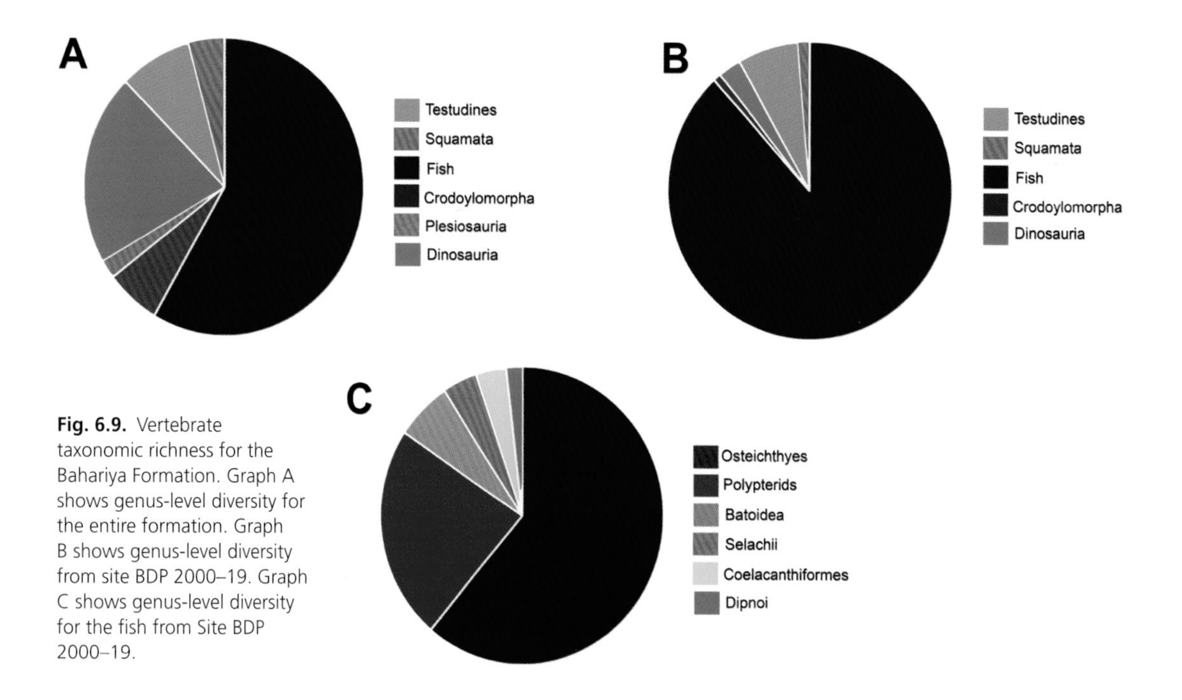

Fig. 6.9. Vertebrate taxonomic richness for the Bahariya Formation. Graph A shows genus-level diversity for the entire formation. Graph B shows genus-level diversity from site BDP 2000–19. Graph C shows genus-level diversity for the fish from Site BDP 2000–19.

between species richness and actual specimen counts at this location (see fig. 6.9).

The high diversity of fish and turtles is unsurprising, but plesiosaurids are absent, and the numbers of crocodyliforms and sea snakes are small. The number of elasmobranch and dinosaur specimens is also surprisingly small, given that their genus count makes up such a large percentage of the total vertebrate diversity. This could be further proof that Bahariya's productivity went into supporting a high diversity of terrestrial and aquatic predators rather than large populations of each (Grandstaff et al. 2012).

However, the depositional environment of site BDP 2000–19 was a rapidly infilling tidal inlet, which could have limited the faunal composition of the site. Many of the dinosaur remains also show signs of being transported and may not have inhabited this area in life (Grandstaff et al. 2012). We will need specimen counts from elsewhere in the formation before we can say with certainty whether the diversity patterns seen here are a result of localized conditions or a genuine population trend across Bahariya, let alone the entire region.

Great God! This Is an Awful Place

So says Antarctic explorer Robert F. Scott in his journal (1912). Scott and his men died of exposure during this expedition, pinned down on the ice during a blizzard. The final kick in the head was that they were less than 11 miles from a supply depot and salvation when the storm hit.

To his credit, Scott faced his imminent death with the classic British stoicism, writing in his journal (recovered along with his body) that he fully understood that he had taken a risk and that if "things have come out against us . . . we have no cause for complaint." Yet one can't help but think that his comment that Antarctica was an "awful place" could equally apply to North Africa during the Cenomanian.

Those wetlands and reefs were positively teeming with predators. Every which way you look, there seem to be ever more claws and teeth. One of the criticisms often leveled against the "lost world" type of dinosaur movie is that there are usually too many predators with no visible food chain to support them, yet that is exactly the situation we have in North Africa.

That point must be understood. It's not the number of carnivore species that makes this region unique, because the Morrison Formation also has a diverse theropod assemblage. It's the absence of herbivores that stands out.

This strange ecological pattern has been known for a long time. Stromer (in Nothdurft et al. 2002) referred to it as his "three theropod riddle": How were three species of carnivore supporting themselves in the near absence of herbivores? With the number of known theropods, sharks, and crocodylomorphs only increasing, predators have become so overabundant that some workers have refused to accept it, assuming that there must be an error in our analysis or with the lithostratigraphy (McGowan and Dyke 2009; Ibrahim et al. 2020a).

Yet despite the denial of some workers, the pattern is undoubtedly genuine. North Africa was hell on prehistoric earth. This leads us back to the question that bedeviled Stromer: How were all these carnivores supporting themselves?

Predator/Prey Ratios

Let's take a look at the precise numbers we're dealing with. There are at least three species of herbivore in Egypt (two titanosaurs and a third poorly known sauropod), compared to an estimated six theropod species (see fig. 6.10). Poor, but not as unbalanced as initially supposed; the ratio was as bad as seven to two in the earliest studies by Stromer (1931, 1934a, 1936). In Morocco, we have at least four herbivore species (an ornithopod and three sauropods) and an estimated six theropod species. However, there were clearly more sauropod taxa present in Morocco than *Rebbachisaurus* and unnamed titanosaurs; we just don't know how many.

The biomass of each species is also a cause for concern. One prey species, such as wildebeest, can support a large number of carnivore species because it exists in large enough numbers, yet the number of herbivore fossils from North Africa are limited. Carnivore remains are far more common across the board, although Morocco fares better than Egypt, as fragmentary sauropod specimens are found more frequently. However, this might be due to the currently poor sample size for Bahariya. Regardless, North Africa does not appear to have hosted the large herbivore herds seen elsewhere.

It's been suggested that their rarity was due to the large number of theropod species acting as a deterrent (Nash 2012), similar to Shark Bay, Australia, where shark populations keep aquatic herbivores away (Wirsing

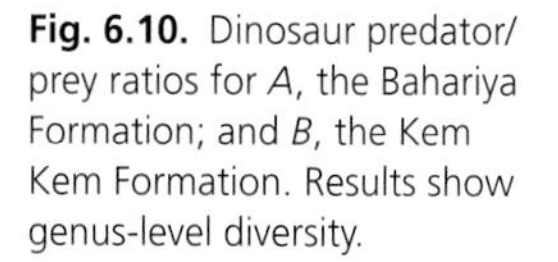

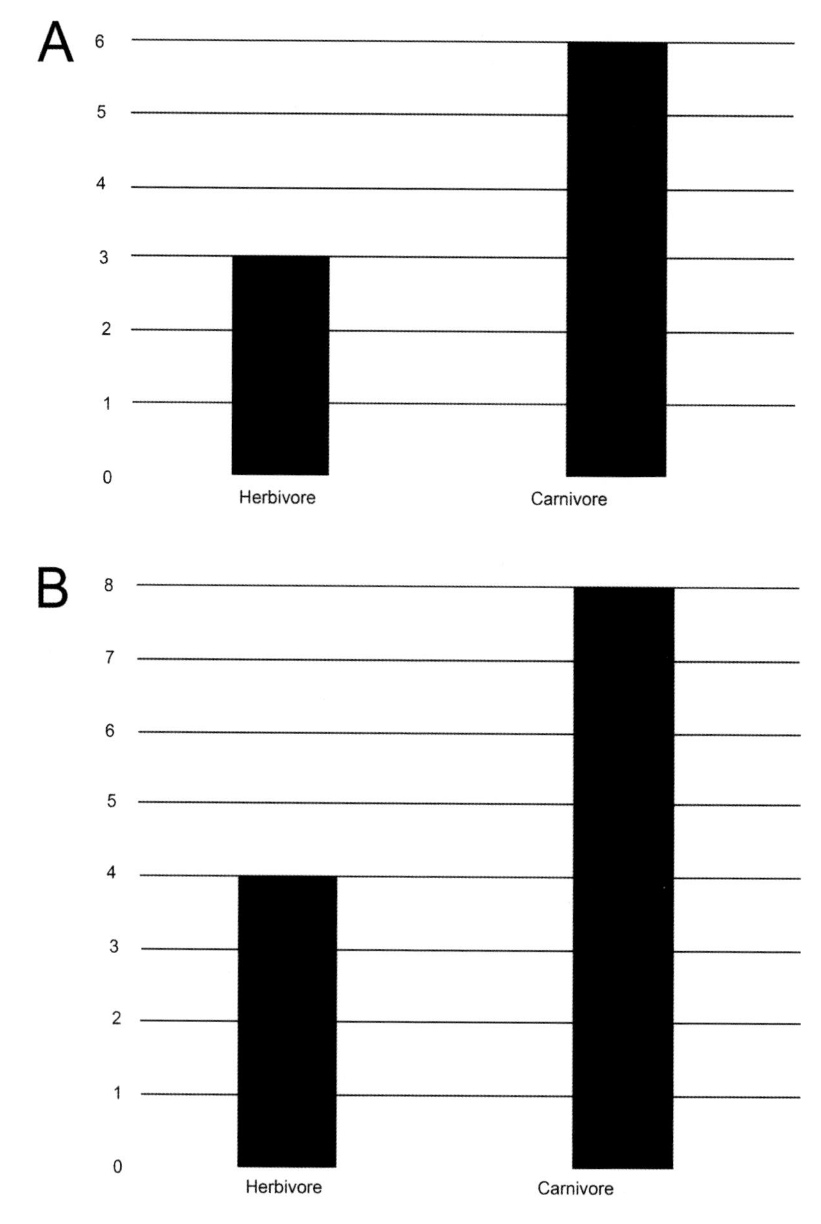

Fig. 6.10. Dinosaur predator/prey ratios for *A*, the Bahariya Formation; and *B*, the Kem Kem Formation. Results show genus-level diversity.

et al. 2007; Wirsing and Ripple 2011). While this may be true for the Kem Kem Formation (although it must be noted that the lack of abundant vegetation on the open floodplain (Läng 2014) would be a more plausible explanation for their rarity), it's doubtful that this applies to Bahariya, with its abundant and diverse flora (Baioumi et al. 2012; Moustafa and Lashin 2012).

A more likely explanation would be North Africa's distinct climate, as ground-feeding herbivores become rarer as the environment becomes more arid (Noto and Grossman 2010). This would also explain the dominance of high browsing titanosaurids in this ecosystem. Further evidence of this comes from a recent study that shows that titanosaurid teeth are more common than those of the rebbachisaurids in the Continental Intercalaire, but the sample size (eighteen) is too small to say whether this is a genuine ecological signal or a coincidence (Holwerda et al. 2018).

As for how many individuals there were of each theropod species, a specimen count—based on the number of identifiable specimens—for Bahariya shows *Carcharodontosaurus* and *Bahariasaurus*, known only from their holotype and paratype specimens (see fig. 6.11). *Spinosaurus* is known from at least one specimen, two if "*Spinosaurus B*" belongs to that species (Stromer 1915, 1931, 1934a, 1936).

If we include specimens from North Africa in its entirety, the result stays the same for the *Bahariasaurus*, while the number of *Carcharodontosaurus* specimens increases to seven and *Spinosaurus* sp. is now known from an estimated thirty-one specimens (Ibrahim et al. 2014b; Larkin and Longrich 2018). The abelisaurids and dromaeosaurids are not included, because the fragmentary nature of their remains makes it impossible to count the number of species and individuals present accurately.

Both counts show that *Spinosaurus* was the most common taxon (albeit slightly in the case of Bahariya alone), irrespective of whether or not we are dealing with two separate species or two separate genera. This further validates the argument that the spinosaurids were the only year-round occupants of this environment (Fanti et al. 2014), with the others being commuter predators entering the mangroves and river networks to hunt.

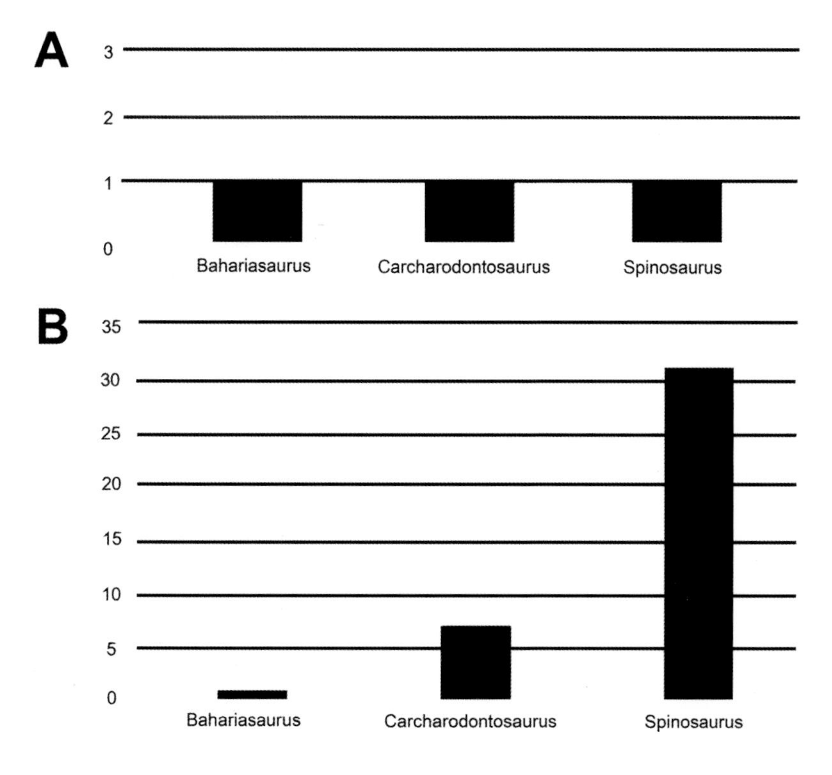

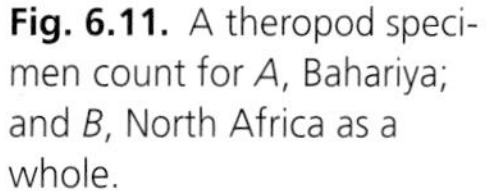
Fig. 6.11. A theropod specimen count for *A*, Bahariya; and *B*, North Africa as a whole.

Another study, from Algeria, underlines the abundance of spinosaurids in this ecosystem, with them making up 94% of all dinosaur remains found. These Algerian specimens are not included in the above specimen count, as they are known solely from teeth (Benyoucef et al. 2016), and spinosaurids appear to have replaced their teeth at a much faster rate than other large theropods, which may result in an overabundance of teeth (Heckeberg and Rauhut 2020).

This offsets the overabundance of theropods, as three of the five named taxa are remarkably rare, which seems to be true of the unnamed abelisaurids and dromaeosaurids as well. This pattern is also seen in the Morrison Formation, which also has a large number of theropod species, but 75 percent of all theropod remains belong to just one species, with the rest in the minority (Foster 2007). This dominance of piscivorous theropods in North Africa during the Cenomanian validates Russell's original hypothesis that North Africa had a food chain with top carnivores feeding directly on fish.

The current predator/prey ratio is only provisional and could change once we have a greater resolution on abelisaurid and dromaeosaurid taxonomy.

While many have noted the abundance of large carnivorous dinosaurs, few have noted the equally large number of large predatory fish (see fig. 6.12). The elasmobranchs were incredibly diverse. Weiler (1935) noted that a large number of Bahariyan fish species possessed dentition for grinding shells. This diversity inhabiting a similar niche must have been supported by high crab, bivalve, and gastropod mollusc population densities/species diversity. Sheaves (2005) has also shown that fish

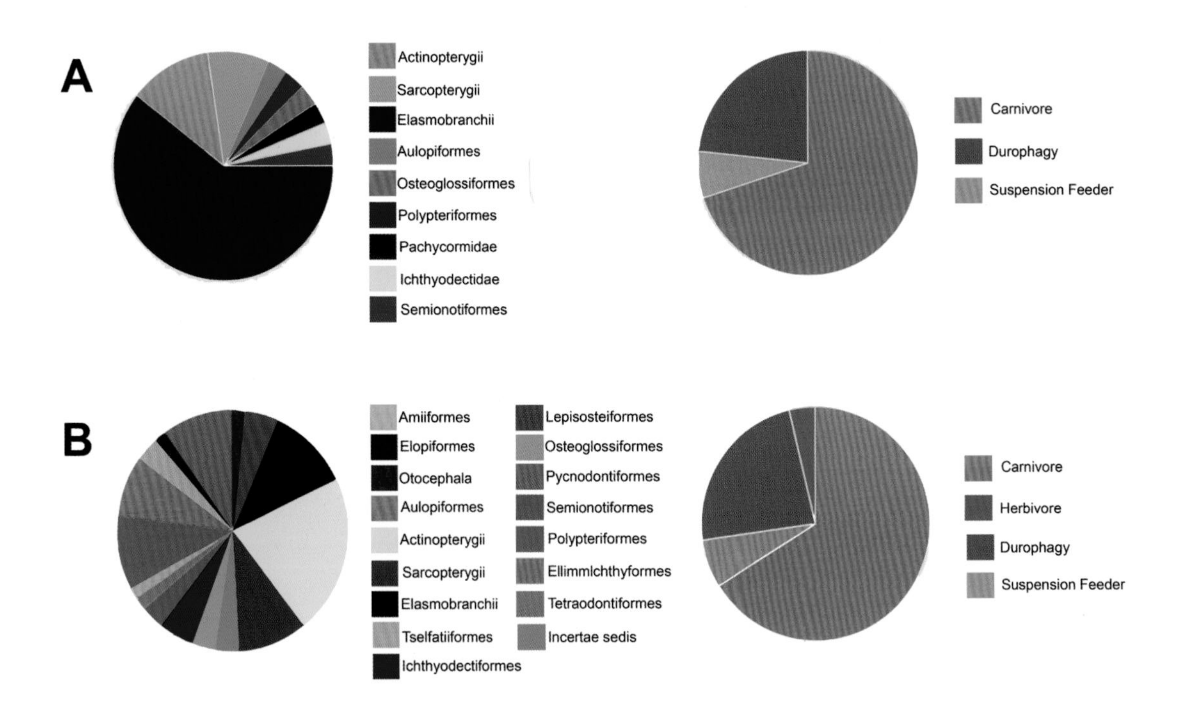

Fig. 6.12. Niche partitioning among North African fish. *A*, the genus-level diversity for the Bahariya Formation and ecological niches; and *B*, the genus-level diversity for the Kem Kem Formation and ecological niches. As many specimens still wait description, only named taxa are included.

populations are larger when the mangroves are always accessible, as was the case in Bahariya with its meso-tidal conditions. Morocco shows the same pattern, with a high diversity of durophagous and carnivorous fishes (Cavin et al. 2015).

This combination of high-density invertebrate populations and an extensive stable environment provided a broad base for various specialized fishes, which in turn sustained a diverse population of aquatic predators (Nash 2012). It must also be noted that such a carnivore-dominated environment (see fig. 6.13) appears to be normal for a healthy tropical oceanic ecosystem (Sandin et al. 2008).

This may explain the prevalence of huge taxa in North Africa. While large dinosaurs are common in most Mesozoic habitats, North Africa seems unique in that everything from the fish to the pterosaurs appears to have evolved into giants. Nothdurft and colleagues (2002) suggest that this was the result of the high productivity of this habitat. I suspect that they were partly right, although I think the nature of this environment, being water based, played a role as well.

Unlike on land, where the biome is based on plants, aquatic biomes are based on phytoplankton, bacterioplankton, and zooplankton—all far more productive in terms of biomass than land plants because of their faster reproductive rates. The abundance of protein-rich creatures and rarity of marine plants (excluding the phytoplankton) thus explains why most marine vertebrates are carnivores (Tucker and Rogers 2014).

Large predators usually feed on prey smaller than themselves; this creates a selective pressure on prey to evolve larger sizes to make capture difficult. This further encourages the development of larger sizes among predators, which concurs with studies suggesting that spinosaurids often preyed on fishes capable of offering a sustained fight (Therrien et al. 2005). So the greater the number of predators, the greater the eventual disparity in body sizes between the trophic levels (Tucker and Rogers 2014).

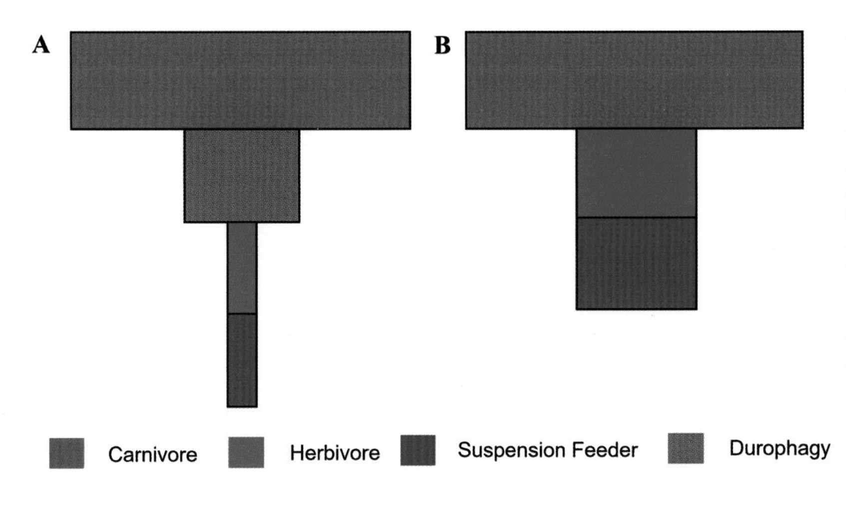

Fig. 6.13. The food pyramid for the marine ecosystem of *A*, Bahariya; and *B*, the modern Northern Line Islands. While a direct comparison is impossible—*A* represents a species count while *B* represents species biomass—both ecosystems show the same general pattern, with a top-heavy pyramid skewered toward carnivores. Adapted from Sandin et al. (2008) with slight modification.

Niche Partitioning

It's no surprise that North Africa's incredible diversity of large fishes coincides with the presence of large piscivores. Russell (1996) suggested that the other theropods may have fed directly on fish. While they undoubtedly would have fed on fish when the opportunity allowed, the presence of large sauropods and ornithopods gave other theropods a greater choice in prey selection (Cavin et al. 2010).

Of the eight theropod species, only four—*Carcharodontosaurus*, *Bahariasaurus*, *Sauroniops*, and any large abelisaurids—would have been predators of large game. Despite their huge size, *Spinosaurus* and *Sigilmassasaurus* were not predators of large terrestrial prey and would only have fed opportunistically on smaller terrestrial animals (Charig and Milner, 1997; Amiot et al. 2010a; Hassler et al. 2018). *Deltadromeus*, *Rugops*, and the dromaeosaurids would also have fed on smaller animals and scavenged large game when it was available (see fig. 6.15).

Ibrahim and colleagues (2020a) and Smyth, Ibrahim, and Martill (2020) snidely ask whether their critics believe that two spinosaurids could share the same niche and ecosystem—to which I ask, why not? They certainly never provided any arguments to prove otherwise. And two spinosaurid taxa definitely coexisted in Brazil (Medeiros 2006; Kellner et al. 2011), and possibly in Niger, if *Cristatusaurus* is valid (Sales and Schultz 2017). And having multiple large theropods with similar niches seems to have been the norm for many Mesozoic ecosystems (Foster 2007).

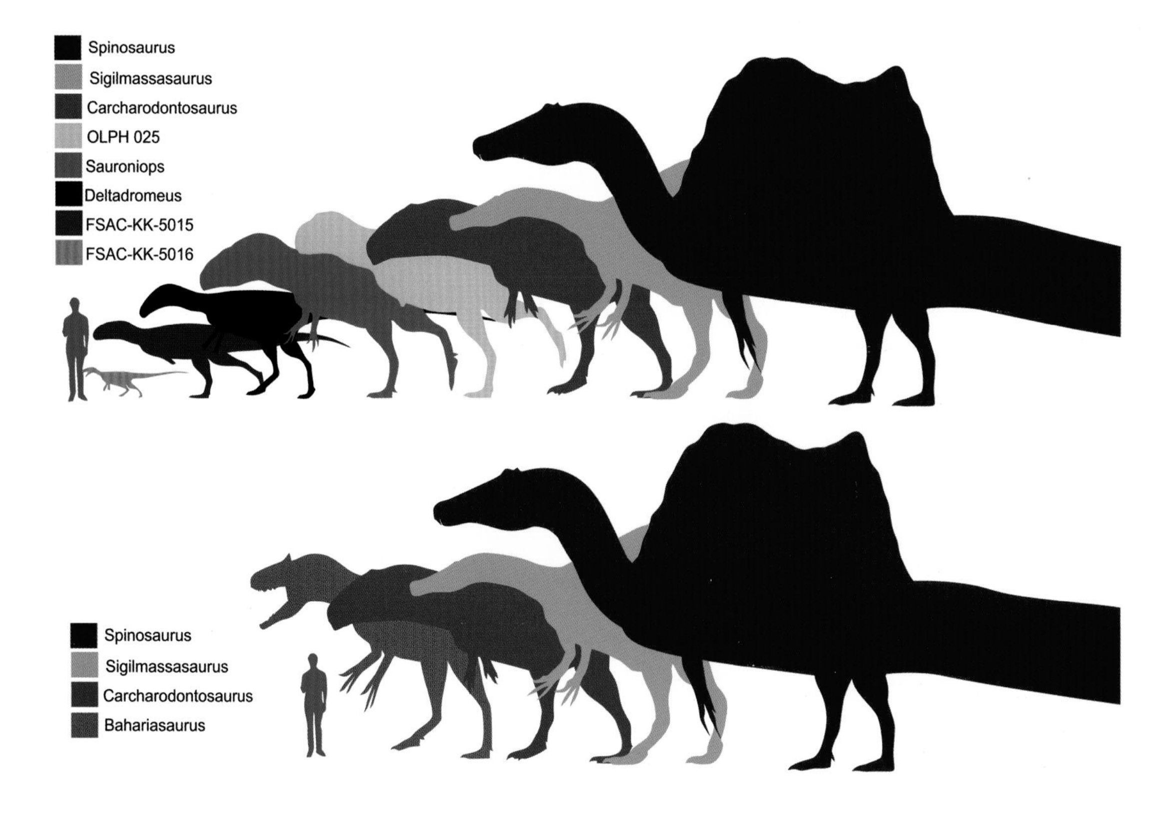

Fig. 6.14. A scale chart for North African theropods from Morocco and Egypt. Egypt shows the pattern seen in the "three theropod riddle," with all four named theropods being giants. This is likely a result of poor sampling, as Morocco shows the typical structure we see in most ecosystems, with a gentle gradient from small to large species. Size estimates are conservative. Artwork modified from Slate Weasel (public domain) and KoprX (CC 4.0). See Wikimedia Commons.

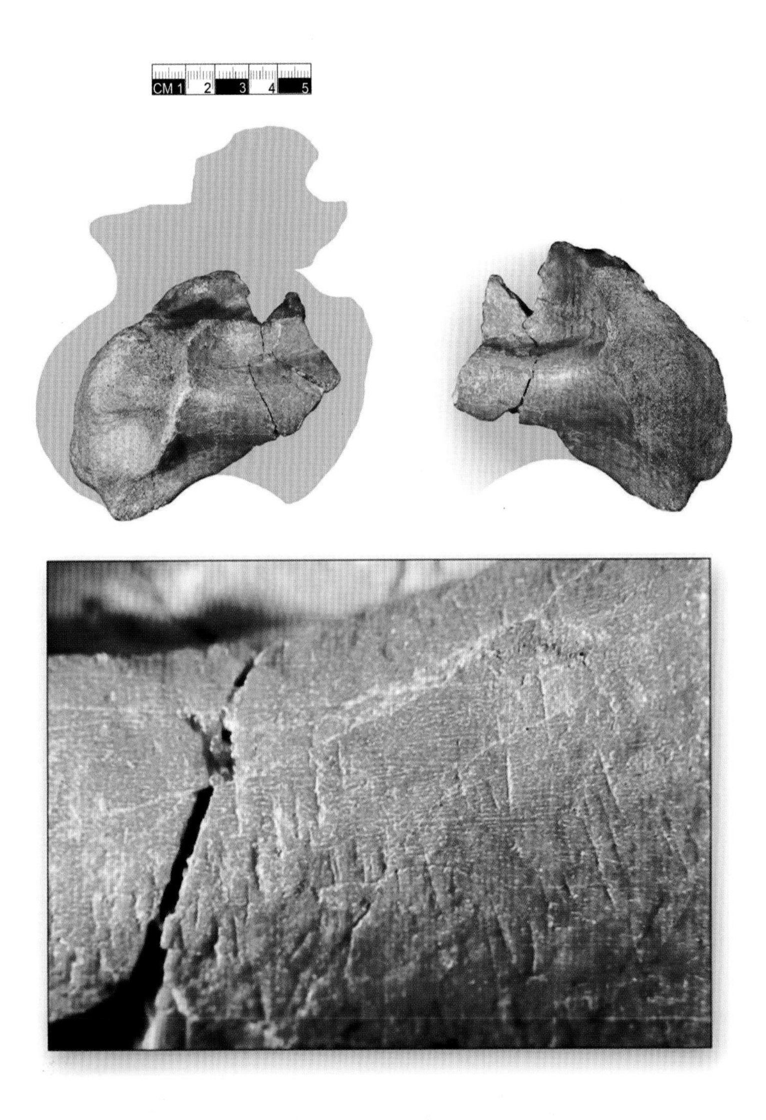

Fig. 6.15. A dinosaur (*Sigilmassasaurus*?) vertebra that has clearly been scavenged after death. The shape of the tooth marks suggests that two different animals, both of which were small in size, as well as an invertebrate fed on this bone. Image provided by Olof Moleman.

Even amalgamating *Spinosaurus* and *Sigilmassasaurus* still fails to get the neatly bifurcated ecological roles many seem to want, as the huge spinosaur-mimicking crocodilian (Larsson et al. 2015) would also occupy the same ecological role as the *Spinosaurs*. This suggests that the issue of niche partitioning is not as clear-cut as some would like it to be.

This brings me to the topic of crocodile abundance in North Africa. Specifically, how was the ecosystem divided between them? Excluding herbivorous forms, a species count reveals that the crocodilians have higher levels of diversity than the theropods, both by country and across North Africa in its totality. The sole exception is Egypt, probably because of limited sampling. We must also remember that large numbers of crocodilian species in the tropics appear to have been normal throughout history (Scheyer et al. 2013).

The crocodylomorphs also show distinct niche partitioning, far more than the theropods. We have piscivores (*Elosuchus*), herbivores (unnamed notosuchian), terrestrial predators (*Hamadasuchus*), durophages (*Lavocatchampsa*), and spinosaurid mimics (unnamed neosuchian). The platyrostral skulled crocodilians (*Stomatosuchus*) were also specialized, either for feeding on a specific type of prey or for hunting in a unique way.

However, only *Elosuchus* can be considered common (see fig. 6.16), as it is known from over fifty specimens, more than all the theropods combined! Was this taxon superabundant, or is this merely a preservational bias? It's interesting to note that we see equally high population densities in extant crocodilians of large sizes (Magnusson and Lima 1991; Fukuda et al. 2011). Such high population densities are a result of being ectothermic. A cold-blooded animal needs less food and eats less frequently than a similarly sized endotherm, enabling large numbers to live in surprisingly small regions (Magnusson and Lima 1991).

Niche partitioning may also occur within a species. Behaviorally, the presence of *Spinosaurus* specimens of various ages, preserved in the same sediments, suggest that if *Spinosaurus* was not gregarious, then it was at least tolerant of others of its kind. However, the disparity in size between the adults and juveniles suggests that their prey selection changed as they aged. The larger individuals would have fed on larger prey (see fig. 6.17), thus avoiding any competition with their own young (Lakin and Longrich 2019).

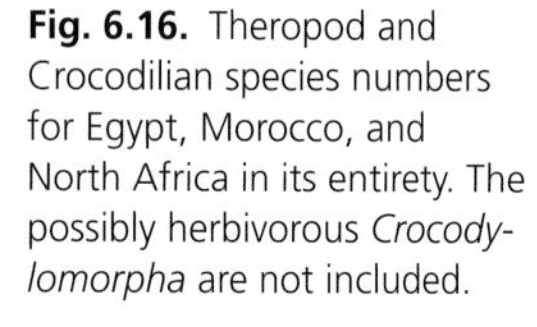

Fig. 6.16. Theropod and Crocodilian species numbers for Egypt, Morocco, and North Africa in its entirety. The possibly herbivorous *Crocodylomorpha* are not included.

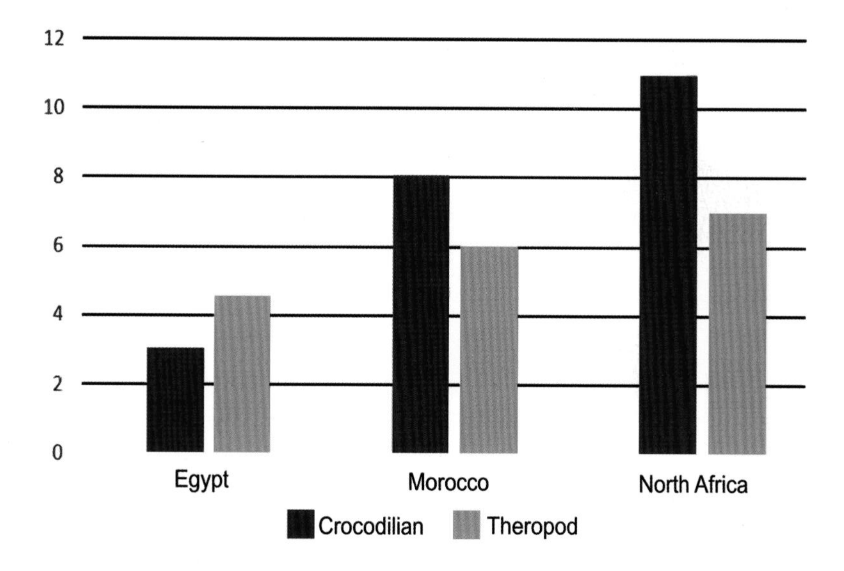

Regardless, the aquatic base of this ecosystem was broader than supposed. This may explain a curious detail noted for the piscivorous pterosaurs. Most assume that ornithocheirids spent their lives out at sea, yet studies by Amiot and colleagues (2010b) suggest that at least one of the Moroccan species may have been different. Carbon isotope studies on isolated ornithocheirid teeth shows that the δ18Op values are within the range of those from terrestrial animals (18.3‰ to 22.9‰), suggesting that they drank from mostly terrestrial sources. Even more telling is that the

δ13Oc values are low (−11.3‰ to −7.5‰). This shows a diet of freshwater fish, as marine fish would have a more positive and variable δ13Oc value.

We even see potential niche partitioning among the fish themselves. At least four pycnodonts taxa—*Coelodus*, *Neoproscinetes*, *Macromesodon*, and *Agassizilia*—each have its own distinct dental arrangements, suggesting that each fed either on different prey or in a unique way (Cooper and Martill 2020).

Of course, niche partitioning did not always work. One example comes from the fish *Tribodus* and *Pseudohypolophus* (Vullo and Neraudeau 2008). When the two ranges did overlap, *Pseudohypolophus* always seems to have replaced *Tribodus*. As a result, the *Tribodus* seems to be a relic taxon, with its range restricted to the tropics, where the cold-water-adapted *Pseudohypolophus* couldn't enter (Vullo and Neraudeau 2008).

Fig. 6.17. A *Spinosaurus* attacking a *Laganosuchus*. While the interaction between North African theropods and crocodilians is poorly understood, age would presumably play a role: while *Spinosaurus* would be prey for *Laganosuchus* when young, the crocodilian would eventually become prey itself as the *Spinosaurus* outgrew it. Artwork by Christopher DiPiazza.

The March of The Oysters

7

North Africa's proximity to both Europe and South America, plus the presence of the Trans-Saharan Seaway and the Tethys Sea, with their major currents, made cosmopolitanism easy. As a result, North Africa was a major evolutionary trading post, especially for marine taxa, as the native fauna from old Gondwana now rubbed shoulders with new arrivals from Europe, the Middle East, and the Americas (Garassino et al. 2008a, 2008b; Vullo and Courville 2014).

It must be remembered that these migrations were not planned events. The species involved did not have any destination in mind. These events were merely a result of animals continually searching for new territories, avoiding competition, and finding new resources (Hanski 1999; Lieberman 2005). These dispersals were also multigenerational events, taking hundreds of years to complete.

Climate and environment also play a role, as a species can only colonize a region that can support it. Likewise, the presence of an ocean will stop any terrestrial animal that is too large to raft over and is incapable of either flight or swimming, while the formation of a land bridge will lead to an exchange of fauna and flora (Karanth 2006).

Bivalves

The best known of these dispersal events occurred among the bivalves. These dispersals seem to have occurred in three distinct waves throughout the Cenomanian (see fig. 7.1). The first was from North America into Europe. During the second wave, many species crossed the Tethys Sea

Fig. 7.1. The migration of Cenomanian bivalves. The solid red line indicates the first migration, the dashed black line indicates the second dispersal event, and the dotted green line indicates the third and final dispersal event (both possible routes shown). Adapted from Abdelhady (2002).

and entered North Africa. The third was from North Africa, either across the Atlantic or through the Trans-Saharan Seaway, into South America (Abdelhady 2002).

While bivalves, in general, were spreading out, the rudists, despite being the dominant reef-building group for much of world, are only known from eastern and central North Africa in Cenomanian times. In the west, they only managed to get as far as southern Morocco (Chikhi-Aouimeur 2008).

Their inability to invade this part of the reef system seems to have been due to a wide variety of factors, such as the rift system that developed in this area and the direction of the prevailing currents, which had a clockwise flow (Ayoub-Hannaa et al. 2014). It's possible that the generally cooler waters and lower salinity in western North Africa also worked against them (Barron 1995).

Ostracods

The oysters weren't the only creatures on the move by the Cenomanian. Ostracod population dynamics in the region also seem to be the result of two dispersal events. The first migration route was to the southwest, from the Middle East along Africa's northern coastline (Elewa and Mohamed 2014). Gebhardt (1999) also suggests that this route turned south, traveling down between Africa and South America and into the South Atlantic. This route appears to have lasted from the beginning of the Cenomanian right to the Turonian age, when decreasing oxygen levels in the ocean water closed it (Elewa and Mohamed 2014).

The second dispersal event occurred at the same time, but this time the route was along Africa's eastern coastline, southward toward Madagascar. This dispersal event began in the Albian age and continued until the late Cenomanian (Elewa and Mohamed 2014).

Fig. 7.2. The migration of Cenomanian ostracods. The solid red line indicates the first migration, the dashed black line indicates the second dispersal event, and the dotted green line indicates the final dispersal event. ASP is the Asian Province, WSTP is the West-South Tethyan Province, ESTP is the East-South Tethyan Province, and WAP is the West African Province. *L* shows the proximal location of Libya. Adapted from Elewa and Mohamed (2014).

These migrations ultimately resulted in three ostracod biozones developing in North Africa (Gebhardt 1999; Elewa and Mohamed 2014): the Asian Province (ASP), the West-South Tethyan Province (WSTP), and the East-South Tethyan Province (ESTP).

Instead of having affinities with either the ESTP or WSTP, Libya shares fauna with the ASP, despite the distance separating them. Elewa and Mohamed (2014) suggest that this could be the result of Libya having a relic fauna while the rest of the Tethyan Province has an ostracod fauna of a more recent origin.

The Trans-Saharan Seaway had no species in common with North Africa until the middle to late Cenomanian (Gebhardt 1999). This resulted in the development of another biozone: the West African Province (WAP). This delayed migration may have been due to the source fauna being composed of deep-water species, yet many shallow-water species also failed to penetrate this seaway even though the northern part of it connected the ESTP and the WSTP (Elewa and Mohamed 2014).

This may have been because the Trans-Saharan Seaway did not reach its maximum extent until the middle to late Cenomanian, when rising sea levels finally flooded the Benue Trough and established a connection between the South Atlantic and the Tethys Sea. This finally led to a faunal interchange between the WAP and the Tethyan Province (Gebhardt 1999).

These dispersal events were caused primarily by changes in sea level. Rising sea levels at the start of the Cenomanian forced deep marine ostracods to the southwest, following the edge of North Africa's outer shelf. During the middle and late Cenomanian, fluctuating sea levels resulted in shallow-water species becoming dominant. This led to a final wave of migration (see fig. 7.2), primarily southward along Africa's eastern coastline (Elewa and Mohamed 2014).

Echinoids

Echinoids underwent three dispersal events (Roney 2013). In the late Aptian/early Cenomanian, North Africa was still joined to northern South America, cutting the Atlantic in two. The Trans-Saharan Seaway also didn't exist at that point. For these reasons, the echinoids dispersed down the east coast of Africa, through the Mozambique Channel, and around the Cape to enter the South Atlantic (Néraudeau and Mathey 2000; Roney 2013).

The second dispersal event occurred when Africa finally separated from South America in the latter half of the Cenomanian, uniting the North and South Atlantic. This allowed the second wave of migration to flow westward into the Atlantic through the Caribbean Sea, which connected the Tethys, Atlantic, and Western Interior Seaway (Néraudeau and Mathey 2000; Roney 2013).

Fig. 7.3. The migration patterns for Cenomanian echinoids. The solid red line indicates the first migration, slashed black line indicates the second dispersal event and the dotted green line indicates the third and final dispersal event. Adapted from Néraudeau and Mathey 2000 and Roney 2013.

The final dispersal (see fig. 7.3) was into the Atlantic via the Trans-Saharan Seaway (Néraudeau and Mathey 2000; Roney 2013). This dispersal event was the most limited because the Benue Trough was above sea level at various points in the Cenomanian, periodically blocking the route (Reyment 1980).

Lizards

An unusual migration took place among the polyglyphanodontian lizards (see fig. 7.4), with the genus *Bicuspidon* being found in the US, Morocco, and Hungary. How one small lizard managed to cover so much ground needs some explaining. Given their prevalence in the US, it's supposed that the Polyglyphanodontia originated in North America (Vullo and Rage 2018).

Thus, the African species of *Bicuspidon* must have been the result of a rapid dispersal out of North America in the early Cenomanian. One proposal is that it traveled from North America into Europe and then on into Africa. This is extremely unlikely, as the oldest known *Bicuspidon* specimens from Hungary postdate the Moroccan and American specimens by millions of years (Vullo and Rage 2018).

A second route would have brought the genus from North America into South America, and then on into Africa and through to Europe (Vullo and Rage 2018). While this route is more likely than the first proposal, no confirmed polyglyphanodontians, let alone *Bicuspidon* itself, are known from South America.

The third theory is that the genus traveled to Africa, across the Caribbean Sea, on rafts of vegetation. After making landfall, the taxon then migrated on into Europe. The problem with this suggestion is that the prevailing currents seemed to flow in the opposite direction for the most part, away from Africa and toward the Caribbean (Vullo and Rage 2018).

Fig. 7.4. The possible migration patterns for Cenomanian polyglyphanodontian lizards. The solid red line indicates the route from the Americas to Africa and finally into Europe. The dashed black line indicates the second possible route from America to Europe and then Africa. The dotted green indicates the third possible route, rafting from America to Africa across the Caribbean and then island hopping to Europe. Adapted from Vullo and Rage 2018.

Given this, it's even been suggested that we may have the situation backward and that the Polyglyphanodontia may have had an African origin, as the American and African species of *Bicuspidon* are both Cenomanian in age, so it's difficult to tell which came first (Vullo and Rage 2018). Ultimately, the question of polyglyphanodontian dispersal will remain unanswered until we have a better understanding of South American lizards from this interval and better evidence of early polyglyphanodontians from any continent.

Dinosaurs

Faunal interchange is more difficult to demonstrate on land. Partly this is because most dinosaur groups had already achieved a near pan-global distribution. Indeed, the only Cenomanian migration I can think of is among the rebbachisaurids (see fig. 7.5). This group is assumed to have evolved in the Jurassic (133–127 mya), given that it's more primitive than the other Flagellicaudatans (Wilson and Allain 2015). The discovery of the basal rebbachisaurid *Maraapunisaurus fragillimus* from the Jurassic of the US supports that theory (Carpenter 2018) and suggests a North American origin for this group.

Traditionally, the view has been that they dispersed from South America into Africa and then Europe (Fanti et al. 2015). However, the evidence is beginning to suggest that the situation was actually reversed. As previously stated, the oldest rebbachisaurid known is from North America, and the next oldest are from Europe: *Xenoposeidon* from Great Britain (the second oldest rebbachisaurid known as only *M. fragillimus* predates it), followed by *Histriasaurus* from Croatia (Sereno et al. 2007; Fernández-Baldor et al. 2011; Taylor 2018). The first records from Africa and South America only occur well after them. This suggests a migration from North America into Europe, and then on through to Africa and South America (Carpenter 2018).

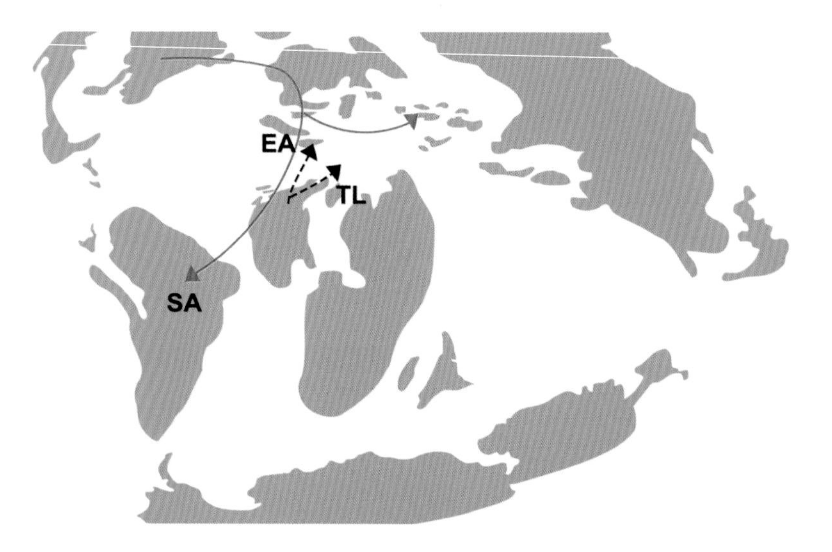

Fig. 7.5. The migration patterns for the rebbachisaurids. The solid red line indicates the first wave of migration, while the dashed black indicates the second. *SA*, South American clade; *EA*, Euro-Africa clade; and *TL*, African lineage. Adapted from Fanti et al. (2015).

However, some still support a South American or even an African origin for this group (Canudo et al. 2018). The presence of basal rebbachisaurids (within the family Rebbachisauridae but outside the clade Khebbashia) in South America does complicate things, especially as they appear so late in the fossil record, occurring at the same time as the derived African/European forms. More work needs to be done with the early rebbachisaurids, preferably with new specimens filling in the ghost lineage, before we can deduce their actual paleobiogeographical pattern.

Regardless, there appears to have been at least two dispersal events, although given the confusing paleobiogeographical signals I personally suspect that there could have been more. The first took place roughly 130 mya when the rebbachisaurids first left the Americas, irrespective of whether they originated in the north or south (Fanti et al. 2015). This initial migration produced a host of basal species through Europe (Sereno et al. 2007; Fernández-Baldor et al. 2011; Taylor 2018; Canudo et al. 2018).

The second took place during the Aptian and continued through the Cenomanian age. This dispersal led to the evolution of a "Euro-African" subclade that includes our very own *Rebbachisaurus*, with one branch (the *Tataouinea* lineage) found only in North Africa.

The titanosaurids might also have migrated using the same route. Ibrahim and colleagues (2016) note that FSAC KK-7000 is more closely related to South American and European titanosaurids than to other African sauropods, suggesting it might be the product of a recent faunal interchange. The tooth morphology of North African sauropods also suggests a close taxonomic relationship between African and European sauropods (Holwerda et al. 2018). This also suggests that sub-Saharan Africa was increasingly isolated from the mainstream sauropod evolution by the Cenomanian (Ibrahim et al. 2016; Gorscak 2017).

You'd think that a biome this complex would function like a well-oiled machine. Unfortunately, after centuries of stability, it ultimately all collapsed into ruin.

The Cenomanian Mass Extinction

8

Foster (2007) once bewailed the fact that we will never know the end of the story for the Morrison Formation because the rocks recording the end of the Morrison have not been preserved. Such is not the case with the Cenomanian of North Africa, where we have the complete series of sediments. Indeed, the end of this era is hard to miss as it coincided with a mass extinction.

There have been five mass extinctions. The end of the Cenomanian saw an extinction event that, while not severe enough to be considered a *true* mass extinction, still cleared the decks on a global scale (Benyoucef et al. 2012).

Unlike the quick and catastrophic mass extinction at the end of the Mesozoic that wiped out the terrestrial dinosaurs (Jablonski and Chaloner 1994), the Cenomanian extinction appears to have been a protracted affair, taking place over a span of 1.46 million years, with the last 520,000 years of that period being the worst (Kauffman 1995). This extinction is also unique in that it affected mostly marine life. Before this extinction, almost all the marine invertebrates in this region were generalists. There was also great diversity, with many taxa having multiple species (Gebhardt 1999). There had been minor die-offs throughout the Cenomanian, but life had always recovered rapidly (Pandey et al. 2011). Not this time.

In many ways, North Africa was a victim of its own success. Toward the end of the Cenomanian, the swamps and wetlands became overproductive. The overabundance of nutrients and iron being delivered into this ecosystem resulted in the eutrophication of the waters, triggering red tides (Martin and Fitzwater 1988; Frost 1996; Boyd et al. 2000; Tanner and Khalifa 2009; Davesne et al. 2018).

While the exact effects vary depending on the taxa in question (Lyons et al. 2014), generally such algal blooms initially boost productivity but lead to a decline in species richness in the long term (see fig. 8.1). They choke the waterways, blocking out the light and reducing water quality through sheer weight in numbers, thus preventing both coral and plant growth. The decay process also absorbs all the dissolved oxygen in the water as all those dead bodies pile up (Dolbeth et al. 2003; Worm and Lotze 2006; Lyons et al. 2014).

The Cenomanian overproductivity of plankton and aerobic bacteria was all part of a larger global extinction (Sinton and Duncan 1997; Kerr 1998; Huber et al. 1999, 2002; Leckie et al. 2002; Meyer and Kump 2008; Yilmaza et al. 2010). The waters of the Tethys Sea had always been

B

Fig. 8.1. *A*, an algal bloom off the coast of Britain, as seen from space, showing how extensive these events can be. *B*, a pond where an algal bloom has killed off much of the aquatic life by removing oxygen from the water, blocking sunlight, and producing natural toxins. Credit NEODAAS and Lamiot / CC-BY-SA-3.0.

oxygen-poor throughout the Cenomanian, but things appear to have reached a tipping point by the end of this age. Why this happened is still debated. Some authors speculate that the ball started rolling 500,000 years prior, triggered by a period of underwater volcanism in the Americas and the Caribbean (Kauffman 1995; New Scientist 2008; van Helmond et al. 2014).

The result, when coupled with failing ocean currents (Kauffman 1995), was an expansion of the oxygen minimum zone, which is usually found at depths of 200 to 1,000 meters (656 to 3280 feet). During the Cenomanian, it rose to within just 100 to 200 meters (328 to 656 feet) of the surface (Schlanger et al. 1987). This created a vast dead zone where little could survive, except for some kinds of plankton and aerobic bacteria. This would have made matters even worse.

However, there are other proposed triggers for this extinction event. A second theory is that the oceans were poisoned by an increase in the levels of trace elements, caused by the release of toxic chemicals and gasses in the deep ocean as a result of volcanic activity. These would have then risen through the water column, becoming more and more concentrated (Kauffman 1995).

Sediments with raised levels of trace elements are known from marine deposits on four different continents. All coincide with both the Cenomanian extinction and the development of a superswell or superplume (an upwelling of molten rock rising to the surface from the planet's mantle) in the Pacific (Kauffman 1995).

It's been noted that the mass die-offs during the Cenomanian-Turonian interval coincide with a rise in iridium levels, which may suggest a series of asteroid impacts. Indeed, four impact craters—the Steen River Crater, Logoisk Crater, Deep Bay Crater, and the West Hawk Crater—have been found with dates that coincide with the Cenomanian/Turonian boundary (Kauffman 1995).

Deducing which of the proposed explanations was the causal factor in this extinction event is difficult, as each mechanism has evidence in its favor. The first two, though, are not mutually exclusive, as both are a result of a similar process: an upsurge in volcanic activity. Likewise, the existence of impact craters is also indisputable. So it's possible, even likely, in my opinion, that all three played a role, as they seem to have occurred over roughly the same period.

Regardless of the cause, the creation of the marine dead zone coincided with the release of SO_2, H_2S, and CO_2. For organisms that produce calcium shells (molluscs, crustaceans, echinoderms, coral, etc.) this was a disaster, preventing them from forming their exoskeletons. And as animals with carbonate shells die off, the decaying bodies can release carbon dioxide at specific water depths. This would have led to a runaway greenhouse effect (Nienhuis et al. 2010; National Research Council 2011).

As mentioned previously, echinoderms fix an estimated 0.1 gigatonnes of calcium carbonate a year worldwide (Uthicke et al. 2009; Kaplan 2010).

In the Sinai, echinoids were so prevalent during the mid-Cenomanian that they, along with the gastropods, replaced the oysters as the main reefal components in these sediments (Gertsch et al. 2010a).

Yet at the Cenomanian/Turonian boundary, their population crashed. In the mid-Cenomanian sediments of the Sinai Peninsula, echinoids composed 14.77% of all the fossils found in the East Themed area and 23.25% of those at Wadi Quseib (Ayoub-Hannaa and Fürsich 2011). Yet by the late Cenomanian, they composed a mere 1.36% (Ayoub-Hannaa 2011) and were absent in some regions (Gertsch et al. 2010a).

So the decline of this group, which is also recorded in parts of the Americas (Aguilera-Franco and Allison 2005), would have left more carbonate material in the sea to break down. This became a vicious cycle, with populations declining because of rising CO_2 levels, and their absence leading in turn to yet higher CO_2 levels.

This led to the creation of vast beds of black shale, formed from unoxidized carbon and produced by the stagnant, oxygen-poor ocean waters. These beds can be found the world over, but the Tethys and Atlantic Oceans were hit exceptionally hard (Schlanger et al. 1987).

That conditions in the Galala Sea were becoming harsher is also proved by the presence of monospecific bonebeds, composed almost entirely of gastropod molluscs and certain types of benthic foraminifera (Gertsch et al. 2010a; Ayoub-Hannaa and Fürsich 2011; Contessi and Fanti 2012a). This is a result of the environment being so hostile that most species can't survive in it, but those that can survive proliferate extensively because they face little in the way of competition or predation.

Mekawy (2010) suggests that the late Cenomanian also saw the rise of predators evolved specifically to feed on shelled creatures. He cited a general increase in size and shell thickness, as well as a tendency to form ever tighter colonies to make extraction difficult. While I agree that an increase in diversity of durophagous predators was a factor, the degrading environment must have also played a role.

This was compounded by the continued rise in sea levels at the end of the Cenomanian to the highest levels they have ever reached during the whole of the Phanerozoic (Abdelhady 2003). Morocco, Tunisia, Egypt, and Algeria were almost completely flooded (Beadnell 1902; Cavin et al. 2010; Martill et al. 2011; Benyoucef et al. 2012). This led to extinctions on land, as living space became more restricted, and to the loss of extensive mangrove and reefal systems, which would be drowned as the waters deepened.

As the water got deeper and more acidic and ocean circulation failed (Kauffman 1995; Dhondt et al. 1999; Ayoub-Hannaa and Fürsich 2012), only those species capable of living in such a harsh environment survived (Elewa and Mohamed 2014). The Ichthyosauria never recovered (Fischer et al. 2016), although the pliosaurids and plesiosaurids pulled through. Ammonite numbers and diversity also fell, and by the Turonian, they were mostly restricted to deep-water environments (Gertsch et al. 2010a).

On land, the extinction spelled the end for both the spinosaurids and the rebbachisaurids. The decline of the spinosaurids is hardly surprising. They were, as far as we know, all multiton animals that could only evolve because of the presence of equally giant fish and wetlands on a continental scale. Once both of those went, the spinosaurids soon followed.

The fall of the rebbachisaurids is more difficult to explain. While competition with titanosaurids does not seem to have been a factor (Barrett and Upchurch 2005), we must also remember that the Diplodocoidea were already in decline. The rebbachisaurids were among the last surviving members of this group. Personally, I suspect that overspecialization, followed by habitat loss, played a role.

The fall in sauropod diversity might explain why carcharodontosaurids also took a hit (Novas, de Valais, et al. 2005). The absence of prey would knock the props out from under the top predators, although the carcharodontosaurids did survive and limped on for a while (Candeiro et al. 2012).

Of course, there were also creatures that benefited from this disaster. The Ammonite *Pseudaspidoceras* flourished during and after this crisis (Ifrim 2013). The abelisaurids also came out on top, replacing the carcharodontosaurids and subsequently dominating Gondwana right up until the end of the Mesozoic (Novas, de Valais, et al. 2005).

This all brings us back to the concept of widespread trophic cascades. Such feedback loops can be harmful as well as positive. In this scenario, once one thing went wrong, it had a knock-on effect, and everything else that could go wrong went wrong.

For example, in Yellowstone, the extirpation of wolves led to a decline in the numbers of bison, six species of songbird, and many species of both woody and herbaceous plants. This was due to elk numbers exploding once their main predator was removed. The increase in elk resulted in the overgrazing of many plant species, which subsequently put pressure on the other large herbivores, as they had to compete for diminishing resources. Smaller vertebrates, many of which only utilize one specific species of tree, also declined (Baril 2009; Ripple and Beschta 2012).

This also had a knock-on effect in the riparian ecosystems. The overgrazing of willow and aspen trees deprived the beavers of both their preferred food and the material for the construction of their dams. The crash in the beaver population resulted in a fall in both nutrient recycling rates and species diversity, as the marshlands drained away without beaver dams to maintain them (Ripple and Beschta 2012). It even affected scavengers like ravens, bald eagles, and fifty-seven species of beetle, as these are partially subsidized by scavenging wolf kills and the amount of available carrion dropped from an estimated 1524 kg to 458 kg in the winter of 1990–91 (Wilmers et al. 2003).

The trophic cascade at work during the Cenomanian extinction event is still poorly understood. It appears to have occurred in at least five pulses, with each specific group succumbing in sequence instead of all declining simultaneously (Kauffman 1995; Abdelhady 2003; Gertsch et al. 2010b; van Helmond et al. 2014; Fischer et al. 2016). Such negative cascades throughout the system would have continued until events finally peaked and the loop eventually began to reverse itself.

This would have been when oxygen levels in the ocean began to increase again. This is reflected in the fossil record by the decline of monospecific assemblages and increasing diversity on both a family and species level, as various groups recolonize regions where they had been extirpated, and renewed speciation occurs among survivors (Gertsch et al. 2010a; van Helmond et al. 2014). The pattern of recovery on land is more difficult to deduce, but the general outline was probably the same, although some groups, like the carcharodontosaurids, never regained their former diversity.

Of course, there are still some questions about what happened. Gale and colleagues (2000) maintain that there was no extinction at all, basing this opinion on studies of the Anglo-Paris Basin. While acknowledging the fall in productivity in the surface waters, they maintain that there was no decline in biodiversity, suggesting a series of migrations/immigrations that give the illusion of extinctions. Monnet (2009) makes a similar point concerning the ammonite fauna of Tunisia, which appears to have increased in diversity during this extinction event.

While I accept that the damage inflicted in the Anglo-Paris Basin was milder than elsewhere, the evidence for the extinction is too overwhelming to ignore (Abdelhady 2003). I suspect that the conditions here were a local phenomenon which created a refuge for marine species. Gertsch and colleagues (2010b) and Keller and Pardo (2004) also documented oxygen-rich areas in North Africa created by localized climatic and geographical conditions. At Azazoul, in southwestern Morocco, the shallow waters retained their diverse reefs throughout this extinction. This reef system didn't succumb until conditions finally became anoxic in the early Turonian (Gertsch et al. 2010b).

Such localized conditions may explain why ammonite diversity appears to have increased in Tunisia during this extinction event when it fell elsewhere in North Africa (Gertsch et al. 2010a). Van Helmond and colleagues (2014) notes that the Atlantic and Tethys were hit hardest out of all the world's oceans because of their restricted basins. Tunisia at the time would have been at the mouth of the Trans-Saharan Seaway, which might have offered some form of protection. The ammonites of the Western Interior Seaway also appear to have come through this event with minor loses (Monnet 2009). What protection these epicontinental seas offered, if any, is unclear, especially as these waters were just as anoxic as those recorded elsewhere. Regardless, these localized conditions in parts of Africa do not change the trend seen on the global level.

For the rest of the ecosystem, the OAE 2 event, as it's known, spelled the end (see plate 8). It takes somewhere between 5 and 30 million years for the planet to recover from a mass extinction (Quammen 1998). The OAE 2 event was relatively minor, so recovery would have been quick, but the subsequent ecosystems were never as diverse as they had been (Abdelhady 2003).

Perhaps we can take some solace from the excellent example the Cenomanian extinction provides us within an age of renewed global warming. At the end of the Cenomanian, extinctions linked to global warming and rising sea levels were caused by natural events no one could have prevented. Now it's caused by us, and I'm constantly shocked by how few people seem to care.

But even if there is no hope for these wonderful, long-ago wetlands, there is still hope for many modern creatures. Just because *Elosuchus* was doomed, there is no need to see the gharial go the same way. An estimated 99% of all life on earth is already extinct (Newman 1997), so, after already losing so many wonders like the Cenomanian wetlands, do we really want to lose any more?

Plate 8, Overleaf A malnourished Spinosaurus aegyptiacus sits on a sandbank while watching the sun set, after spending two months on the traditional lekking grounds futilely calling for a mate. He's completely unaware that he's the last of his kind. With his passing, the world loses not only his species but the entire spinosaurid race.

"These are the closing days of my era."

—**ELDER SCROLLS**

Appendix

Table A1. Data taken from Moustafa and Lashin (2012), Baioumi et al. (2012), and Tahoun et al. (2013)

POLLEN AND MIOSPORES

Marchantiophyta
Aequitriradites spinulosus

Polypodiopsids
Cicatricosisporites orbiculatus
Concavissimisporites variverrucatus and *Concavissimisporites punctatus*
Crybelosporites pannuceus
Cyathidites sp.
Deltoidospora sp.
Dictyophyllidites harrisii
Matonisporites simplex
Impradecispora apiverrucata
Scortea hamoza and *Scortea tecta*
Pilosisporites trichopapillosus

Gymnosperms
Alaticolpites limai
Appendicisporites sp.
Araucariacites australis & *Araucaiacites* sp.
Classopollis torosus, *Classopollis brasiliensis*, and *Classopollis* sp.
Dicheiropollis etruscus

Angiosperms
Afropollis jardinus, *Afropollis kahramanensis*, and *Afropollis operculatus*
Cretacaeiporites densimurus and *Cretacaeiporites scabratus*
Retimonocolpites sp.
Retitricolpites sp.
Tricolpites sp.
Elateroplicites sp.
Elaterocolpites castelainii
Elaterosporites verrucatus and *Elaterosporites klaszii*
Ephedripites sp.
Galeacornea causea
Spheripollenites psilatus

Table A2. Data taken from Gertsch et al. (2010a), Cavin et al. (2010), Lézin et al. (2012), Garassino et al. (2008a), and Shahin and El Baz (2014)

MICROFAUNA AND MEIOFAUNA

Dinoflagellates

Botryococcus sp.
Coronifera oceanica
Cyclonephelium vannophorum and *Cyclonephelium edwardsii*
Dinopterygium cladoides
Exochosphaeridium sp.
Florentinia mantlii, *Florentinia cooksoniae*, and *Florentinia* sp.
Mudrongia simplex
Palaeoperidinium cretaceum
Pediastrum sp.
Peseudoceratium securigerum and *Peseudoceratium anaphrissum*
Scenedesmus sp.
Subtilisphaera perlucida and *Subtilisphaera senegalensis*
Xenascus cerartoides
Xiphophoridium alatum

Foraminifera

Ammobaculites plummerae
Anaticinella planoconvexa and *Anticinella* sp.
Asterohedbergella asterospinosa
Charentia cuvillieri
Coryphostoma plaitum
Cuneolina pavonia
Daxia cenomana
Dentolina sp.
Dicarinella hagni and *Dicarinella algeriana*
Favusella washitensis
Fursenkoina nederi
Gaudryina sp.
Gavelinella sandidgei
Globigerinelloides bentonensis, *Globigerinelloides ultramicra*, and *Globigerinelloides* sp.
Guembelitria cenomana
Hedbergella delrioensis, *Hedbergella simplex*, *Hedbergella planispira*, and *Hedbergella* sp.
Heterohelix moremani, *Heterohelix globulosa*, *Heterohelix Americana*, and *Heterohelix* sp.
Merlingina cretacea
Orbitolina concava
Pleurostomella sp.
Praebulimina aspra and *Praebulimina nannina*
Praeglobotruncana stephani, *Praeglobotruncana gibba*, and *Praeglobotruncana kalaati*
Praealveolina cretacea and *Praealveolina tenuis*
Pseudotextularia sp.
Pyramidina prolixa
Rotalipora greenhornensis and *Rotalipora cushmani*
Thomasinella aegyptia
Whiteinella archeocretacea, *Whiteinella brittonensis*, *Whiteinella inornata*, *Whiteinella praehelvetica*, *Whiteinella baltica*, and *Whiteinella* sp.

Table A3. Data taken from Ayoub-Hannaa (2011) and Ayoub-Hannaa and Fürsich (2012b)

ECHINOIDS

Echinothurioida
Archiacia pescameli, *Archiacia aegyptiaca*, *Archiacia araidahensis*, *Archiacia Saadensis*, and *Archiacia palmata*
Gentilia syriensis
Hemiaster syriacus, *Hemiaster pseudofourneli*, *Hemiaster pseudofourneli*, *Hemiaster fourneli* (var. *ambiguous* and *refanensi*), and "*Hemiaster*" *heberti turonensis*
Pygurus subproductus

Salenioida
Holosalenia batnensis

Acroechinoidea
Orthopsis ruppelii, *Orthopsis ovata*, and *Orthopsis miliaris*

Phymosomatoida
Aegyptiaris halalensis
Phymosoma sinaeum

Echinacea
Micropedina olisiponensis, *Micropedina cotteaui*, and *Micropedina simplex*
Pedinopsis sinaica, *Pedinopsis meridanensis*, and *Pedinopsis desori*

Phymosomatoida
Tetragramma marticense and *Tetragramma variolare*

Arbacioida
Codiopsis sp.
Goniopygus menardi and *Goniopygus coquandi*

Cidaroida
Sinaecidaris gauthieri and *Sinaecidaris* sp.

Diadematoida
Heterodiadema libycum and *Heterodiadema ouremense*
Notatudiadema rekeibensis

Holectypoida
Coenholectypus excisus, *Coenholectypus larteti*, *Coenholectypus pulvinatus*, *Coenholectypus turonensis*, and *Coenholectypus neocomiensis*
Anorthopygus orbicularis and *Anorthopygus atavus*

Spatangoida
Mecaster batnensis, *Mecaster cubicus*, *Mecaster semicavatus*, *Mecaster batnensis*, *Mecaster cubicus*, and *Mecaster pseudofourneli*

Hemicidaroida
Heterodiadema libycum and *Heterodiadema ouremense*
Micropedina olisiponensis, *Micropedina cotteaui*, and *Micropedina simplex*
Pedinopsis sinaica, *Pedinopsis meridanensis*, and *Pedinopsis desori*
Tetragramma marticense and *Tetragramma variolare*

Cassiduloida
Petalobrissus cf. *Pygmaeus*

Table A4. Data taken from Gertsch et al. (2010a), Ayoub-Hannaa (2011), and Ayoub-Hannaa and Fürsich (2012b)

GASTROPODS

Neogastropoda
Fasciolaria gen. et sp. indet.
Volutoderma elleryi

Stromboidea
Aporrhais blanckenhorni, *Aporrhais dutrugei*, *Aporrhais* sp. 1, and *Aporrhais* sp. 2
Checchiaia sanfilippoi
Columbellina cf. *fusiformis*
Pterocera incerta
Pterodonta cf. *deffisi*, *Pterodonta gigantica*, & *Pterodonta* sp.
Strombus mermeli
Tylostoma globosum, *Tylostoma pallaryi*, *Tylostoma cossoni*, and *Tylostoma* sp.

Eucycloidea
Calliomphalus orientalis

Sorbeoconcha
Cimolithium tenouklense and *Cimolithium* sp.
Diozoptyxis blancheti
Pyrazus valeriae

Caenogastropoda
Ampullina dupinii and *Ampullina* sp. (*Ampullina uchauxiensis*?).
Campanile ganesha and *Campanile* sp.
Harpagodes nodosus, *Harpagodes heberti*, *Harpagodes* cf. *heberti*, and *Harpagodes* sp.
Turritella calava

Trochoidea
Pchelinsevia coquandian

Cephalaspidea
Cylichna sp.

"**Opisthobranchia**"
Mrhilaia nerineœformis, *Mrhilaia haugi*, and *Mrhilaia* sp.
Neoptyxis olisiponensis
Sogdianella laevis laevis

Heterobranchia
Nerinea sp.

Table A5. Data taken from Ayoub-Hannaa (2011) and Ayoub-Hannaa and Fürsich (2012b)

BIVALVES

Rudist

Apricardia matheroni and *Apricardia noncarinata*
Bournonia africana
Durania arnaudi, *Durania blayaci*, and *Durania* sp.
Eoradiolites lenis externus and *Eoradiolites lenis liratus*
Ichthyosarcolites triangularis and *Ichthyosarcolites* sp.
Praeradiolites biskraensis and *Praeradiolites* sp.
Radiolites sauvagesi
Sauvagesia sharpie and *Sauvagesia* sp.
Vaccinites rousseli

Oyster

Ambigostrea pseudovillei
Ceratostreon flabellatum
Costagyra olisiponensis
Courvostrea rouvillei
Exogyra delettrei, *Exogyra conica*, *Exogyra olisiponensis*, *Exogyra columba*, and *Exogyra* sp.
Gyrostrea delettrei and *Gyrostrea* cf. *anubis*
Ilymatogyra africana (*forma typica* and *forma crassa*)
Lopha syphax
Nayadina gaudryi
Ostrea flabellata, *Ostrea suborbiculata*, *Ostrea Bouvilleii*, and *Ostrea* sp.
Phelopteria atra and *Phelopteria gravida*
Pycnodonte vesicularis and Pycnodonte (Phygraea) vesicularis vesiculosa
Rastellum carinatum and *Rastellum* sp.
Rhynchostreon suborbiculatum

Cockle

Cardium sp.
Granocardium productum, *Granocardium desvauxi*, and *Granocardium carolinum*
Protocardia hillana and *Protocardia* sp.
Pseudolima itieriana

Mussel

Modiolus aequalis
Mytilus cf. *Galliennei*

Scallop

Chondrodonta joannae
Neithea coquandi, *Neithea quinquecostata*, *Neithea dutrugei*, *N. hispanica*, and *Neithea* sp.
Neithella notabilis
Plicatula auressensis, *Plicatula fourneli*, *Plicatula ferryi*, and *Plicatula ferryi*

Clam

Arctica picteti, *Arctica cordata*, *Arctica humei*, *Arctica roudaireia*, and *Arctica inornata*
Cardita? nicaisei
Clisocolus corrugatus
Corbula sp.
Crassatella (*Rochella*) *tenuicostata*
Cyprina roudaireia
Glossus aquilinus
Inoceramus atlanticus
Lucina fallax
Maghrebella cf. *forgemoli* and *Maghrebella* sp.
Meretrix desvauxi and *Meretrix* sp.
Nucula margaritifera and *Nucula neckeriana*
Parasea faba faba
Pholadomya pedernalis and *Pholadomya vignesi*
Poromya? Ligeriensis
Pterotrigonia scabra
Sphaera cf. *corrugata*
Tenea delettrei
Trigonia ethra and *Trigonia auressens*
Venericardia? Forgemoli
Veniella cf. *trapezoidalis*

Table A6. Data taken from Ismail and Soliman (1997), Gebhardt (1999), Brusca and Brusca (2003), Lézin et al. (2012), and Elewa and Mohamed (2014)

OSTRACODS

Podocopida

Amphicytherura sexta
Anticythereis gaensis
Bairdia bassiounii, *B. elongata*, and *Bairdia* sp. 1 and sp. 2
Brachycythere ledaforma porosa
Bythoceratina avnonensis and *Bythoceratina tamarae*
Bythocypris eskeri
Cythereis bicornis levis, *Cythereis canteriolata*, *Cythereis namousensis*, and *Cythereis algeriana*
Dolocytheridea atlastica
Eocytheropteron? cf. *punctata*
Fabanella sp. 1 and sp. 2
Looneyella sohni
Loxoconcha clinocosta and *Loxoconcha fletcheri*
Metacytheropteron berbericum
Neocythere? *mackenziei*
Ovocytheridea caudata, *Ovocytheridea producta*, and *Ovocytheridea reniformis*
Paracypris mdaouerensis, *Paracypris triangularis*, *Paracypris dubertreti*, *Paracypris acutocaudata*, and *Paracypris angusta*
Peloriops pustulata and *Peloriops aegyptiaca*
Pterigocythere raabi
Veeniacythereis jezzineensis and *Veeniacythereis streblolophata schista*
Xestoleberis obesa

Platycopida

Cytherella ovata, *Cytherella paenovata*, *Cytherella parallela*, *Cytherella aegyptiensis*, *Cytherella sulcata*, *Cytherella gigantosulcata*, *Cytherella postangulata*, and *Cytherella* sp.

Literature Cited

Abdallah, A. M., and A. Adindani. 1963. "Stratigraphy of Upper Paleozoic Rocks, Western Side of the Gulf of Suez." Geological Survey and Mineral Research of Egypt, Cairo, no. 25.

Abdel-Gawad, G. I., H. A. El-Sheikh, M. A. Abdelhamid, M. K. El-Beshtawy, M. M. Abed, F. T. Fürsich, and G. M. El Qot. 2004. "Stratigraphic Studies on Some Upper Cretaceous Successions in Sinai, Egypt." *Egyptian Journal of Paleontology* 4:263–303.

Abdelhady, A. A. 2002. "Stratigraphical and Paleontological Studies on the Upper Cretaceous Strata, North Wadi Qena, Eastern Desert, Egypt." Master's thesis, Minia University.

———. 2003. "Cenomanian/Turonian Mass Extinction of Macroinvertebrates in the Context of Paleoecology: A Case Study from North Wadi Qena, Eastern Desert, Egypt." In *Mass Extinction*, edited by A. M. T. Elewa, 103–127. Heidelberg: Springer.

———. 2007. "Stratigraphical and Paleontological Studies on the Upper Cretaceous Strata, North Wadi Qena, Eastern Desert Egypt." Master's thesis, Minia University.

Abdelsalam, M. G., J-P. Liegeois, and R. J. Stern. 2002. "The Saharan Metacraton." *Journal of African Earth Sciences* 34(3):119–36.

Abuelkheir, G., and M. K. Abdelgawad. 2017. "Discovery of Testudines Material from the Cenomanian Maghrabi Formation, South Western Desert, Egypt." Abstract presented at the 71st Annual Meeting of the Society of Vertebrate Paleontology, Las Vegas.

Agassiz, L. 1833–44. *Recherches sur les Poissons Fossiles*. Neuchatel.

Agnoli, G. L., and P. Rosa. 2022. "*Biology of Chrysididae*." Chrysis.net. Accessed January 10, 2022. https://www.chrysis.net/chrysididae/biology-of-chrysididae/.

Agnolín, F. L., and F. E. Novas. 2013. *Avian Ancestors: A Review of the Phylogenetic Relationships of the Theropods Unenlagiidae, Microraptoria, Anchiornis and Scansoriopterygidae*. SpringerBriefs in Earth System Sciences. Dordrecht: Springer.

Aguilera-Franco, N., and P. Allison. 2005. "Events of the Cenomanian-Turonian Succession, Southern Mexico Eventos de una Sucesión del Cenomaniano-Turoniano del Sur de México." *Journal of Iberian Geology* 31:25–50.

Ahyong, S. T., K. Baba., E. Macpherson, and G. C. B. Poore. 2010. "A New Classification of the Galatheoidea (Crustacea: Decapoda: Anomura)." *Zootaxa* 2676:57–68.

Alekseev, P. I. 2009. "Genus *Liriodendrites* in Cretaceous and Early Paleogene Floras of Northern Asia." *Paleontological Journal* 43(10):1181–89.

Ali, I. G., G. Sheridan., J. R. J. French, and B. M. S. Ahmed. 2013. "Ecological Benefits of Termite Soil Interaction and Microbial Symbiosis in the Soil Ecosystem." *Journal of Earth Sciences and Geotechnical Engineering* 3(4):63–85.

Allain, R., T. Xaisanavong., P, Richir, and B. Khentavong. 2012. "The First Definitive Asian Spinosaurid (Dinosauria: Theropoda) from the Early Cretaceous of Laos." *Naturwissenschaften* 99(5):369–77.

Alongi, D. M. 1987. "Intertidal Zonation and Seasonality of Meiobenthos in Tropical Mangrove Estuaries." *Marine Biology* 95:447–58.

Alvin, K. L. 1971. "*Weichselia reticulata* (Stokes et Webb) Fontaine from the Wealden of Belgium. Institut royal des sciences naturelles de Belgique." *Mémoires*, no. 166, 3–39.

Aly, M. F., and G. I. Abdel-Gawad. 2001. "Upper Cenomaniane/Lower Turonian Ammonites from North and Central Sinai, Egypt." *El-Minia Science Bulletin*, no. 13, 17–60.

Aly, M. F., A. Smadi, and H. A. Azzam. 2008. "Late Cenomanian-Early Turonian ammonites of Jordan." 7th International Symposium on the Cretaceous, Switzerland.

Amiot, R., E. Buffetaut, C. Lécuyer, X. Wang, L. Boudad, Z. Ding, F. Fourel, S. Hutt, F. Martineau, M. A. Medeiros, J. Mo, L. Simon, V. Suteethorn, S. Sweetman, H. Tong, F. Zhang, and Z. Zhou. 2010a. "Oxygen Isotope Evidence for Semi-aquatic Habits among Spinosaurid Theropods." *Geology* 38(2):139–42.

Amiot, R., E. Buffetaut, H. Tong, L. Boudad, and L. Kabiri. 2004. "Isolated Theropod Teeth from the Cenomanian of Morocco and Their Palaebiogeographical Significance." *Revue de Paléobiologie* 9:143–49.

Amiot, R., X, Wang, C. Lécuyer, E. Buffetaut, L. Boudad, L. Cavin, Z. Ding, F. Fluteau, A. W. A. Kellner, H. Tong, and F. Zhang. 2010b. "Oxygen and Carbon Isotope Compositions of Middle Cretaceous Vertebrates from North Africa and Brazil: Ecological and Environmental Significance." *Palaeogeography, Palaeoclimatology, Palaeoecology* 297:439–51.

Andrews, C. W. 1911. "Description of a New Plesiosaur (*Plesiosaurus capensis*, Sp. Nov.) from the Uitenhage Beds of Cape Colony." *Annals of the South African Museum* 7(4):309–22.

Anfinson, O. A., M. G. Lockley, S. H. Kim, K. S. Kimd, and J. Y. Kim. 2009. "First Report of the Small Bird Track *Koreanaornis* from the Cretaceous of North America: Implications for Avian Ichnotaxonomy and Paleoecology." *Cretaceous Research* 30(4):885–94.

Apesteguía, S. 2005. "Evolution of the Titanosaur Metacarpus." In *Thunder-Lizards: The Sauropodomorph Dinosaurs*, edited by V. Tidwell and K. Carpenter, 321–45. Bloomington: Indiana University Press.

Apesteguía, S., J. D. Daza., T. R. Simões, and J-C. Rage. 2016. "The First Iguanian Lizard from the Mesozoic of Africa." *Royal Society, Open Science* 3:160–462.

Apesteguía, S., P. A. Gallina, and A. Haluza. 2010. "Not Just a Pretty Face: Anatomical Peculiarities in the Postcranium of Rebbachisaurids (Sauropoda: Diplodocoidea)." *Historical Biology* 22(1-3):165–74.

Aragão, L. E. O. C., Y. Malhi, D. B. Metcalfe, J. E. Silva-Espejo, E. Jimenez, D. Navarrete, S. Almeida, A. C. L. Costa, N. Salinas, O. L. Phillips, L. O. Anderson, E. Alvarez, T. R. Baker, P. H. Goncalvez, J. Huaman-Ovalle, M. Mamani-Solorzano, P. Meir, A. Monteagudo, S. Patino, M. C. Penuela, A. Prieto, C. A. Quesada, A. Rozas-Davila, A. Rudas, J. A. Silva, and R. Vasquez. 2009. "Above- and Below-Ground Net Primary Productivity across Ten Amazonian Forests on Contrasting Soils." *Biogeosciences* 6(12):2759–2778.

Arambourg, C. 1940. "Le groupe des Ganopristinés." *Bulletin de la Société Géologique de France*, série 5:127–47.

——. 1943. "Note préliminaire sur quelques poissons fossiles nouveaux. I. Les poissons du Djebel Tselfat (Maroc)." *Bulletin de la Société Géologique de France*, série 5:281–88.

——. 1954. "Les Poissons crétacés du Jebel Tselfat (Maroc)." *Editions du Service Géologique du Maroc* 118:1–188.

——. 1968. "A propos du genre *Clupavus Aramb* (Rectification de nomenclature)." *Bulletin du Muséum national d'Histoire naturelle* 39:1236–1236.

Arambourg, C., and L. Joleaud. 1943. "Vertébrés fossiles du bassin du Niger." *Bulletin du Service des Mines de l'Afrique Occidentale Française* 7:27–84.

Arden, T. M. S., C. G. Klein, S. Zouhri, and N. R. Longrich. 2018. "Aquatic Adaptation in the Skull of Carnivorous Dinosaurs (Theropoda: Spinosauridae) and the Evolution of Aquatic Habits in Spinosaurids." *Cretaceous Research* 93:275–84.

Arnold, C. A., and J. S. Lowther. 1955. "A New Cretaceous Conifer from Northern Alaska." *American Journal of Botany* 42(6):522–28.

Averianov, A. 2014. "Review of Taxonomy, Geographic Distribution, and Paleoenvironments of Azhdarchidae (Pterosauria)." *ZooKeys* 432:1–107.

Averianov, A., G. Dyke, I. Danilov, and P. Skutschas. 2015. "The Paleoenvironments of Azhdarchid Pterosaurs Localities in the Late Cretaceous of Kazakhstan." *ZooKeys* 483:59–80.

Axsmith, B. J., and B. F. Jacobs. 2005. "The Conifer *Frenelopsis ramosissma* (Cheirolepidiaceae) in the Lower Cretaceous of Texas." *International Journal of Plant Sciences* 166(2):327–37.

Ayoub-Hannaa, W. S. 2011. "Taxonomy and Palaeoecology of the Cenomanian-Turonian Macroinvertebrates from Eastern Sinai, Egypt." Doctoral diss., Bayerischen Julius-Maximilians-Universität Würzburg.

Ayoub-Hannaa, W. S., and F. T. Fürsich. 2011. "Revision of Cenomanian-Turonian (Upper Cretaceous) Gastropods from Egypt." *Zitteliana* 51:115–52.

——. 2012a. "*Apricardia noncarinata* N. Sp. (Bivalvia, Requieniidae) from the Cenomanian (Upper Cretaceous) of Central-East Sinai, Egypt." *Neues Jahrbuch für Geologie und Paläontologie-Abhandlungen Band* 263 (1): 75–84.

——. 2012b. "Palaeoecology and Environmental Significance of Benthic Associations from the Cenomanian-Turonian of Eastern Sinai, Egypt." *Beringeria* 42:93–138.

Ayoub-Hannaa, W. S., F. T. Fürsich, and G. M. El Qot. 2014. "Cenomanian-Turonian Bivalves from Eastern Sinai, Egypt." Palaeontographica Abteilung A Band 301 Lieferung 3-6:63–168.

Báez, A. M. 1981. "Redescription and Relationships of *Saltenia ibanezi*, a Late Cretaceous Pipid Frog from Northwestern Argentina." *Ameghiniana* 18(3-4):127–54.

Bailey, J. B. 1997. "Neural Spine Elongation in Dinosaurs: Sailbacks or Buffalo-Backs?" *Journal of Paleontology* 71(6):1124–46.

Baioumi, A., A. El Hakam, M. M. M. Mandur, and T. F. Moustfa. 2012. "Aptian-Cenomanian Palynozonation and Paleoecology from Horous -1 Well, Northern Western Desert, Egypt." *Journal of Applied Sciences Research* 8:1490–501.

Baioumy, H. M., and S. N. Boulis. 2012. "Glauconites from the Bahariya Oasis: An Evidence for Cenomanian Marine Transgression in Egypt." *Journal of African Earth Sciences* 70:1–7.

Ballesteros, J. A., and P. P. Sharma. 2019. "A Critical Appraisal of the Placement of Xiphosura (Chelicerata) with Account of Known Sources of Phylogenetic Error." *Systematic Biology* 68(6):896–917.

Banerjee, S. 1980. "Lexique stratigraphique international, v. IV, Afrique, part IV, Libye, Cong." Geol. International commission de stratig., Rech. Sci., Paris.

Bannikov, A. F., and F. Bacchia. 2000. "A Remarkable Clupeomorph Fish (Pisces, Teleostei) from a New Upper Cretaceous Marine Locality in Lebanon." *Senckenbergiana lethaea* 80:3–11.

Barhoumi, Z., W. Djebali, C. Abdelly, W. Chaïbi, and A. Smaoui. 2008. "Ultrastructure of *Aeluropus littoralis* Leaf Salt Glands under NaCl Stress." *Protoplasma* 233(3):195–202.

Baril, L. M. 2009. "Change in Deciduous Woody Vegetation, Implications of Increased Willow (*Salix* spp.) Growth for Bird Species Diversity and Willow Species Composition in and around Yellowstone National Park's Northern Range." Master's thesis, Montana State University.

Barrett, P. M., and P. Upchurch. 2005. "Sauropodomorph diversity through time." In *The Sauropods Evolution and Paleobiology*, edited by K. Curry Rogers and J. Wilson. Berkeley: University of California Press: 125–156.

Barron, E. J. 1995. "Tropical Climate Stability and Implications for the Distribution of Life." In *Effects of Past Global Change on Life*, National Research Council. Washington, DC: National Academies Press.

Bates, R. L., and J. A. Jackson. 1980. *Glossary of Geology*. Falls Church, VA: American Geological Institute.

Barbacka, M., and E. Bodor. 2008. "Systematic and Palaeoenvironmental Investigations of Fossil Ferns *Cladophlebis* and *Todites* from the Liassic of Hungary." *Acta Palaeobotanica* 48(2):133–49.

Baron, M. G., D. B. Norman, and P. M. Barrett. 2017. "A New Hypothesis of Dinosaur Relationships and Early Dinosaur Evolution." *Nature* 543:501–506.

Baron-Szabo, R. C. 2006. "Corals of the K/T Boundary: Scleractinian Corals of the Suborders Astrocoeniina, Faviina, Rhipidogyrina, and Amphiastraeina." *Journal of Systematic Palaeontology* 4(1):1–108.

Barr, F. T., and A. A. Weegar. 1972. "*Stratigraphic Nomenclature of the Sirte Basin, Libya*." Tripoli: Petroleum Exploration Society of Libya.

Barroso-Barcenilla, F. 2004. "Acanthoceratidae and Ammonite Zonation of the Upper Cenomanian and the Lower Turonian in the Puentedey Area, Basque-Cantabrian Basin, Spain." *Coloquios de Paleontología* 54:83–114.

Barthoux, J., and P. H. Fritel. 1925. "Flore crétacée du Grès de Nubie." *Mémoires présentés à l'Institut d'Egypte* 7 (2): 65–119.

Beadnell, H. J. L. 1902. *The Cretaceous Region of Abu Roash, near the Pyramids of Giza*. Giza: Egyptian Survey Department.

Becker, M. A., and J. A. Chamberlain Jr. 2012. "*Squalicorax* Chips a Tooth: A Consequence of Feeding-Related Behavior from the Lowermost Navesink Formation (Late Cretaceous: Campanian-Maastrichtian) of Monmouth County, New Jersey, USA." *Geosciences* 2(2):109–29.

Bègue, J-F., M. Sanchez, C. Claire Micheneau, and J. Fournel. 2014. "New Record of Day Geckos Feeding on Orchid Nectar in Reunion Island: Can Lizards Pollinate Orchid Species?" *Herpetology Notes* 7:689–92.

Belvedere, M., J. Nour-Eddine, Breda, G. Gttolin, H. Bourget, F. Khaldoune, and G. Dyke. 2013. "Vertebrate Footprints from the Kem Kem Beds (Morocco): A Novel Ichnological Approach to Faunal Reconstruction." *Palaeogeography, Palaeoclimatology, Palaeoecology* 383–84:52–58.

Benedict, J. E. 1902. "Descriptions of a New Genus and Forty-Six New Species of Crustaceans of the Family Galatheidae, with a List of the Known Marine Species." *Proceedings of the United States National Museum* 26(1311): 243–334.

Benson, R. B. J., T. H. Rich, P. Vickers-Rich, and M. Hall. 2012. "Theropod Fauna from Southern Australia Indicates High Polar Diversity and Climate-Driven Dinosaur Provinciality." *PLoS ONE* 7 (5):e37122.

Benton, M. J. 2015. "Palaeodiversity and Formation Counts: Redundancy or Bias?" *Palaeontology* 58(6): 1003–10291.

Benton, M. J., S. Bouaziz, E. Buffetaut, D. Martill, M. Ouaja, M. Soussi, and C. Trueman. 2000. "Dinosaurs and Other Fossil Vertebrates from Fluvial Deposits in the Lower Cretaceous of Southern Tunisia." *Palaeogeography, Palaeoclimatology, Palaeoecology* 157(3-4):227–46.

Benyoucef, M. 2017. "Litho and Biostratigraphy, Facies Patterns and Depositional Sequences of the Cenomanian-Turonian Deposits in the Ksour Mountains (Saharan Atlas, Algeria)." *Cretaceous Research* 78:34–55.

Benyoucef, M., A. Biely, Y. Kamoun, and M. Zouari. 1985. "L'Albien moyen supérieur à *Knemiceras* forme la base de la grande ransgression crétacée au Tebaga de Medenine (Tunisie méridionale)." *Comptes Rendus de l'Académie des Sciences, Paris*, série 2 (300):965–68.

Benyoucef, M., C. Meister, M. Bensalah, and F. Z. Malti. 2012. "La plateforme préafricaine (Cénomanien supérieur—Turonien inférieur) dans la région de Béchar (Algérie): stratigraphie, paléoenvironnements et signification paléobiogéographique." *Revue de Paléobiologie, Genève* 33(1):205–18.

Benyoucef, M., M Adaci, C. Meister, E. Läng, F-Z. Malti, K. Mebarki, A. Cherif, D. Zaoui, A. Benyoucef, and M. Bensalah. 2014a. "Le Continental Intercalaire dans la région du Guir (Algérie): nouvelles données paléontologiques, ichnologiques et sédimentologiques." *Revue de Paléobiologie, Genève* 33(1):281–97.

Benyoucef, M., M. Adaci, C. Meister, M. Benyoucef, B. Ferré, E. Läng, L. Calvin, D. Zaoui, D. Desmares, L. Villier, F-Z. Malti, and M. Bensalah. 2014b. "Lithostratigraphic, Palaeoenvironmental and Sequential Evolution of the Cenomanian-Lower Turonian in the Guir Region (Western Algeria)." *Notebooks on Geology* 16:271–96.

Benyoucef, M., E. Läng, L. Cavin, K. Mebarki, M. Adaci, and M. Bensalah. 2015. "Overabundance of Piscivorous Dinosaurs (Theropoda: Spinosauridae) in the Mid-Cretaceous of North Africa: The Algerian Dilemma." *Cretaceous Research* 55:44–55.

Benyoucef, M., C. Meister., K. Mebarki, and M. Bensalah. 2016. "Evolution lithostratigraphique, paleoenvironmentale et sequentielle du Cenomanien-Turonien inferieur dans la region du Guir (Oeust algerien)." *Carnets Geol* 16:271–96.

Benyoucef, M. D. Zaoui, M. Adaci, B. Ferré, C. Meister, A. Piuz, G. M. El Qot, A. Mennad, S. Tchenar, and M. Bensalah. 2019. "Stratigraphic and Sedimentological Framework of the Tinrhert Plateau (Cenomanian-Turonian, SE Algeria)." *Cretaceous Research* 98:95–121.

Berry, E. W. 1914. *The Upper Cretaceous and Eocene Floras of South Carolina and Georgia*. Washington, DC: US Government Printing Office.

Bertrand, J-M., and R. Caby. 1978. "Geodynamic Evolution of the PanAfrican Orogenic Belt: A New Interpretation of the Hoggar Shield (Algerian Sahara)." *Geologische Rundschau* 67(2):357–88.

Bice, K. L., and R. D. Norris. 2002. "Possible Atmospheric CO_2 Extremes of the Middle Cretaceous (Late Albian-Turonian)." *Paleoceanography* 17(4):22-1–22-17.

Blanco, A., and J. Alvarado-Ortega. 2006. "*Rhynchodercetis regio*, Sp. Nov., a Dercetid Fish (Teleostei: Aulopiformes) from Vallecillo, Nuevo León State, Northeastern Mexico." *Journal of Vertebrate Paleontology* 26(3):552–58.

Blanco-Moreno, C., A.-L. Decombeix, and C. Prestianni. 2020. "New Insights into the Affinities, Autoecology, and Habit of the Mesozoic Fern *Weichselia reticulata* Based on the Revision of Stems from Bernissart (Mons Basin, Belgium)." *Papers in Palaeontology* 7(3):1351–1372.

Bodin, S., L. Petitpierre, J. Wood, I. Elkanouni, and J. Redfern. 2010. "Timing of Early to Mid-Cretaceous Tectonic Phases along North Africa: New Insights from the Jeffara Escarpment (Libya-Tunisia)." *Journal of African Earth Sciences* 58(3):489–506.

Bolkhovitina, N. A. 1953. "Spores and Pollen Characteristic of Cretaceous Deposits in the Central Regions of the USSR." *Proceedings of the USSR Academy of Sciences* 61:1–16.

Bommer, C. 1911. "Contribution a` l'étude du genre Weichselian (note pre´liminaire)." *Bulletin de la Société royale de botanique de Belgique* 47:296–304.

Bonachela, J. A., R. M. Pringle, E. Sheffer, T. C. Coverdale, J. A. Guyton, K. K. Caylor, S. A. Levin, and C. E. Tarnita. 2015. "Termite Mounds Can Increase the Robustness of Dryland Ecosystems to Climatic Change." *Science* 347(6222):651–55.

Bonaparte, J. F., F. E. Novas, and R. A. Coria. 1990. "*Carnotaurus sastrei* Bonaparte, the Horned, Lightly Built Carnosaur from the Middle Cretaceous of Patagonia." *Contributions to Science of the Natural History Museum of Los Angeles County* 416:1–42.

Bornhorst, H. L. 1991. "Propagating native *Hibiscus*." *Horticulture Digest* 93:3–5.

Bouchet, P., J-P Rocroi, R. Bieler, J. G. Carter, E. V. Coan, R. Bieler, J. G. Carter, and E. V. Coan. 2010. "Nomenclator of Bivalve Families with a Classification of Bivalve Families." *Malacologia* 52(2):1–184.

Boudad, L., E. Läng, and L. Cavin. 2014. "What Kind of Heritage for the Kem Kem?" La Cinquième Rencontre Internationale Sur la Valorisation et la Préservation du Patrimoine Paléontologique (RIV3P5), Oujda, Maroc.

Bouillon, S., A. V. Borges, E. Castañeda-Moya, K. Diele, T. Dittmar, N. C. Duke, E. Kristensen, S. Y. Lee, C. Marchand, J. J. Middelburg, V. H. Rivera-Monroy, T. J. Smith III, and R. R. Twilley. 2008. "Mangrove Production and Carbon Sinks: A Revision of Global Budget Estimates." *Global Biogeochemical Cyles* 22(2):1–12.

Bourquin, S., R. Eschard, and B. Hamouche. 2010. "High-Resolution Sequence Stratigraphy of Upper Triassic Succession (Carnian-Rhaetian) of the Zarzaitine Outcrops (Algeria): A Model of Fluviolacustrine Deposits." *Journal of African Earth Sciences* 58(2):365–86.

Boyd, P. W., A. J. Watson, C. S. Law, E. R. Abraham, T. Trull, R. Murdoch, D. C. E. Bakker, A. R.

Bowie, K. O. Buesseler, H. Chang, M. Charette, P. Croot, K. Downing, R. Frew, M. Gall, M. Hadfield, J. Hall, M. Harvey, G. Jameson, J. LaRoche, M. Liddicoat, R. Ling, M. T. Maldonado, R. M. McKay, S. Nodder, S. Pickmere, R. Pridmore, S. Rintoul, K. Safi, P. Sutton, R. Strzepek, K.Tanneberger, S. Turner, A. Waite, and J. Zeldis. 2000. "A Mesoscale Phytoplankton Bloom in the Polar Southern Ocean Stimulated by Iron Fertilization." *Nature* 407:695–702.

Brea, M. 2020. "Giant Bones from Egypt: Is *Deltadromeus* Present in Bahariya?." *Mysteries of the Bahariya Oasis* (blog), Accessed December 21, 2020. https://bahariyamysteries.blogspot.com/2021/12/giant-bones-from-egyptis-deltadromeus.html.

Brenner, G. J. 1996. Evidence for the Earliest Stage of Angiosperm Pollen Evolution: A Paleoequatorial Section from Israel." In *Flowering Plant Origin, Evolution & Phylogeny*, edited by D. W. Taylor and L. J. Hickey, 91–115. Boston: Chapman and Hall.

Brichant, A. L. 1952. *A Broad Outline of the Geology and Mineral Possibilities of Libya*. UN Tech. Assistance Program, Rept. AfAC 32 Ta 27.

Brink, K. S., D. Larson, and D. Evans. 2015. "Enamel Microstructure in Ornithocheirid Pterosaurs." Abstract presented at the 75th Annual Meeting of the Society of Vertebrate Paleontology, Dallas.

Brusatte, S. L. 2012. *Dinosaur Paleobiology*. TOPA Topics in Paleobiology. Hoboken, NJ: Wiley-Blackwell.

Brusatte, S. L., and P. C. Sereno. 2007. "A New Species of *Carcharodontosaurus* (Dinosauria: Theropoda) from the Cenomanian of Niger and a Revision of the Genus." *Journal of Vertebrate Paleontology* 27(4):902–16.

Brusca, R. C., and G. J. Brusca. 2003. *Invertebrates*. 2nd ed. Sunderland: Sinauer Associates.

Buckmeier, D. L. 2008. *Life History and Status of Alligator Gar (*Atractosteus spatula*), with Recommendations for Management*. Mountain Home, TX: Heart of Hills Fisheries Science Center, Texas Parks and Wildlife Department.

Buffetaut, E. 1976. "Ostéologie et affinités de *Trematochampsa taqueti* (Crocodylia, Mesosuchia) du Sénonien inférieur d'In Beceten (République du Niger)." *Géobios* 9:143–98.

——. 1982. "Radiation évolutive, paléoécologie et biogéographie des crocodiliens Mésosuchiens." *Memoirs of the Geological Society of France*, no. 142: 88.

——. 1991. "*Itasuchus*." In *Santana Fossils: An Illustrated Atlas*, edited by J. G. Maisey, 348–350. Neptune City: TFH.

——. 1994. "A New Crocodilian from the Cretaceous of Southern Morocco." *Comptes Rendus de l'Académie Des Sciences* 319(2):1563–68.

Buffrénil, V. de., and J-C. Rage. 1993. "La pachyostose vertébrale de *Simoliophis* (Reptilia, Squamata): données comparatives et considerations fonctionnelles." *Annales de Paléontologie, Paris* 79:315–55.

Burch, S. 2013. "The Myological Consequences of Extreme Limb Reduction: New Insights from the Forelimb Musculature of Abelisaurid Theropods." Abstract presented at the 73rd Annual Meeting of the Society of Vertebrate Palaeontology, Los Angeles.

Burness, G. P., and T. Flannery. 2001. "Dinosaurs, Dragons, and Dwarfs: The Evolution of Maximal Body Size." *PNAS* 98(25):14518–14523.

Burollet, P. F. 1960. *Stratigraphic Lexicon of Libya*. Bulletin no. 13. Tripoli: Industrial Research Centre.

Busson, G. 1967. "Le Mésozoïque saharien. 1re partie: l'extrême Sud tunisien." *Centre Recherche Zones Arides, Sérvice géologique, CNRS* 8:1–194.

Callapez, P. M., F. Barroso-Barcenilla, A. F. Soares, M. Segura, and V. F. Santos. 2018. "On the Co-occurrence of *Rubroceras* and *Vascoceras* (Ammonoidea, Vascoceratidae) in the Upper Cenomanian of the West Portuguese Carbonate Platform." *Cretaceous Research* 88:325–336.

Candeiro, C. R. A., S. L. Brusatte, and A. L. Souza. 2017. "Spinosaurid Dinosaurs from the Early Cretaceous of North Africa and Europe: Fossil Record, Biogeography and Extinction." *Anuário do Instituto de Geociências* 40(3):294–302.

Candeiro, C. R. A., P. Currie, and L. Bergqvist. 2012. "Theropod Teeth from the Marília Formation (Late Maastrichtian) at the Paleontological Site of Peirópolis in Minas Gerais State, Brazil." *Revista Brasileira de Geociências* 42(2):323–30.

Candeiro, C. R. A., F. Fanti., F. Therrien, and M. C. Lamanna. 2011. "Continental Fossil Vertebrates from the Mid-Cretaceous (Albian-Cenomanian) Alcântara Formation, Brazil, and Their Relationship with Contemporaneous Faunas from North Africa." *Journal of African Earth Sciences* 60(3):79–92.

Candeiro, C. R. A., L. M. Gil, and P. E. P. de Castro. 2018. "Large-Sized Theropod *Spinosaurus*: An Important Component of the Carnivorous Dinosaur Fauna in Southern Continents during the Cretaceous." *Earth Sciences Bulletin* 189(4–6):1–9.

Canerot, J., B. Andreu, D. Chafiki, Kh. El Hariri, and A. Souhel. 2003. "Mesozoic carbonate platforms and associated siliciclastic spreading." In *North African Cretaceous Carbonate Platform Systems*, edited by E. Gili, M. N. El Negra, and P. W. Skelton, 19-29. NATO Science Series 4. Heidelberg: Springer.

Cannatella, D. C., and L. Trueb. 1988. "Evolution of Pipoid Frogs: Morphology and Phylogenetic Relationships of *Pseudhymenochirus*." *Journal of Herpetology* 22(4):439–56.

Canudo, J. I., J. L. Carballido, A. Garrido, and S. Leonardo. 2018. "A New Rebbachisaurid Sauropod from the Aptian-Albian, Lower Cretaceous Rayoso Formation, Neuquén, Argentina." *Acta Palaeontologica Polonica* 63(4):1–13.

Cappetta, H. 1987. *Handbook of Paleoichthyology*. Vol 3B, *Chondrichthys II, Mesozoic and Cenozoic Elasmobranchii*. Munich: Gustav Fischer.

———. 2012. *Handbook of Palaeoichthyology*. Vol. 3E, *Chondrichthyes, Mesozoic and Cenozoic Elasmobranchii*. Munich: Gustav Fischer.

Cappetta, H., E. Buffetaut., G. Cuny, and V. Suteethorn. 2006. "A New Elasmobranch Assemblage from the Lower Cretaceous of Thailand." *Palaeontology* 49(3):547–55.

Cappetta, H., and G. R. Case. 1975. "Sélaciens nouveaux du Crétacé du Texas." *Géobios* 8:303–7.

———. 1999. "Additions aux faunes de sélaciens du Crétacé du Texas (Albien supérieur—Campanien)." *Palaeo Ichthyologica* 9:1–112.

Carnevale, G., and A. Rindone. 2011. "The Teleost Fish *Paravinciguerria praecursor* Arambourg, 1954 in the Cenomanian of North-Eastern Sicily." *Bollettino della Società Paleontologica Italiana* 50(1):1–10.

Carpenter, K. 2006. "Biggest of the Big: A Critical Re-evaluation of the Mega-sauropod *Amphicoelias fragillimus*." In *Paleontology and Geology of the Upper Jurassic Morrison Formation*, edited by J. R. Foster and S. G. Lucas, 315–324. *Bulletin of the New Mexico Museum of Natural History and Science*, 36.

———. 2018. "*Maraapunisaurus fragillimus*, N.G. (Formerly *Amphicoelias fragillimus*), a Basal Rebbachisaurid from the Morrison Formation (Upper Jurassic) of Colorado." *Geology of the Intermountain West* 5:227–44.

Carrano, M. T., and S. D. Sampson. 2008. "The Phylogeny of Ceratosauria (Dinosauria: Theropoda)." *Journal of Systematic Palaeontology* 6(2): 183–236.

Carrano, M. T., S. D. Sampson, and C. A. Forster. 2002. "The Osteology of *Masiakasaurus knopfleri*, a Small Abelisauroid (Dinosauria: Theropoda) from the Late Cretaceous of Madagascar." *Journal of Vertebrate Paleontology* 22(3):510–34.

Carrano, M. T., and J. Velez-Juarbe. 2006. "Paleoecology of the Quarry 9 Vertebrate Assemblage from Como Bluff, Wyoming (Morrison Formation, Late Jurassic)." *Palaeogeography, Palaeoclimatology, Palaeoecology* 237(2-4):147–59.

Carroll, R. L. 1988. *Vertebrate Paleontology and Evolution*. New York: W. H. Freeman.

Carter, D. O., D. Yellowlees, and M. Tibbett. 2007. "Cadaver Decomposition in Terrestrial Ecosystems." *Naturwissenschaften* 94(1):12–24.

Carvalho, M. S. S. de., and J. G. Maisey. 2008. "New Occurrence of *Mawsonia* (Sarcopterygii: Actinistia) from the Early Cretaceous of the Sanfranciscana Basin, Minas Gerais, Southeastern Brazil." *Geological Society, London, Special Publications* 295:109–44.

Case, G., and H. Cappetta. 2004. "Additions to the Elasmobranch Fauna from the Late Cretaceous of New Jersey (Lower Navesink Formation, Early Maastrichtian)." *Palaeovertebrata* 33(1-4):1–16.

Castro, D. F., C. E. V. Toledo, E. P. Sousa, and M. A. Medeiros. 2004. "Novas ocorrências de *Asiatoceratodus* (Osteichthyes, Dipnoiformes) na Formação Alcântara, Eocenomaniano da bacia de São Luís, MA, Brasil." *Revista Brasileira de Paleontologia* 7(2):245–48.

Catuneanu, O., M. A. Khalifa, and H. A. Wanas. 2006. "Sequence Stratigraphy of the Lower Cenomanian Bahariya Formation, Bahariya Oasis, Western Desert, Egypt." *Sedimentary Geology* 190(1-4):121–37.

Cau, A. 2009. "*Kemkemia auditorei* (Cau & Maganuco, 2009)—Part One: Small, Beautiful and Bizarre." *Theropoda* (blog), July 13, 2009. http://theropoda.blogspot.co.uk/2009/07/kemkemia-auditorei-cau-maganuco-2009.html.

———. 2012a. "Kem Kem Wars—Episode VI: Return of *Kemkemia*; Prologue." *Theropoda* (blog), April 7, 2012. http://theropoda.blogspot.co.uk/2012/04/kem-kem-wars-episode-vi-return-of.html.

———. 2012b. "Kem Kem Wars, Episode VI: *Kemkemia*, Finale (Lio, Agnolin, Cau and Maganuco 2012)." *Theropoda* (blog), April 16, 2012. http://theropoda.blogspot.co.uk/2012/04/kem-kem-wars-episode-vi-kemkemia-finale.html.

———. 2012c. "The Science behind the Art of *Sauroniops*." *Theropoda* (blog), November 10, 2012. http://theropoda.blogspot.co.uk/2012/11/la-scienza-dietro-larte-di-sauroniops.html.

———. 2014. "*Spinosaurus* Revolution, Episode IV: A Solution to All the Puzzles?" *Theropoda* (blog), September 20, 2014. http://theropoda.blogspot.co.uk/2014/09/spinosaurus-revolution-episodio-iv-una.html.

———. 2015a. "Do We Have (at Least) Two Species of Spinosaurinae Cenomanian in North Africa?" *Theropoda* (blog), November 13, 2015. http://theropoda.blogspot.co.uk/2015/11/abbiamo-almeno-due-specie-di.html.

———. 2015b. "A Frontal Attack by the Allies of the West against Mordor." *Theropoda* (blog), June 5,

2015. http://theropoda.blogspot.co.uk/2015/06/un-attacco-frontale-dagli-alleati.html.
———. 2015c. "If '*Spinosaurus C*' Is a Pipe Dream . . . Is a Chimera of What?" *Theropoda* (blog), October 25, 2015. http://theropoda.blogspot.co.uk/2015/10/se-spinosaurus-c-e-una-chimera-e-una.html.
———. 2015d. "*Maroccanoraptor*." *Theropoda* (blog), December 2, 2015. http://theropoda.blogspot.co.uk/2015/12/maroccanoraptor.html.
———. 2016. "The Unexpected (but Not Too) Possibility of Tyrannosauroidi (Giants) African." *Theropoda* (blog) June 18, 2016. http://theropoda.blogspot.co.uk/2016/06/linattesa-ma-non-troppo-possibilita-di.html.
———. 2017. "The Rise of the Bipedal *Spinosaurus*." *Theropoda* (blog), February 25, 2017. http://theropoda.blogspot.co.uk/2017/02/the-rise-of-bipedal-spinosaurus.html.
———. 2018. "The Assembly of the Avian Body Plan: A 160-Million-Year Long Process." *Bollettino della Società Paleontologica Italiana* 57:1–25.
———. 2019. "Is This Abelisauride Ileum from Kem Kem from Abelisauride?" *Theropoda* (blog), April 3, 2019. http://theropoda.blogspot.com/2019/04/questo-ileo-di-abelisauride-dal-kem-kem.html.
Cau, A., F. M. Dalla Vecchia, and M. Fabbri. 2012a. "Evidence of a New Carcharodontosaurid from the Upper Cretaceous of Morocco." *Acta Palaeontologica Polonica* 57(3):661–65.
———. 2012b. "A Thick-Skulled Theropod (Dinosauria, Saurischia) from the Upper Cretaceous of Morocco with Implications for Carcharodontosaurid Cranial Evolution." *Cretaceous Research* 40:251–60.
Cau, A., and S. Maganuco. 2009. "A New Theropod Dinosaur, Represented by a Single Unusual Caudal Vertebra from the Kem Kem Beds (Cretaceous) of Morocco." *Atti della Società Italiana di Scienze Naturali e del Museo di Storia Naturale di Milano* 150(2): 239–57.
Cavin, L. 1999a. "A New Clupavidae (Teleostei, Ostariophysi) from the Cenomanian of Daoura (Morocco)." *Comptes Rendus de l'Academie des Sciences Series IIA Earth and Planetary Science* 329:689–95
———. 1999b. "Occurrence of a Juvenile Teleost, *Enchodus* Sp., in the Fish Gut Contents from the Upper Cretaceous of Goulmima, Morocco." *Special Papers in Palaeontology*, no. 60:57–72.
———. 2017. *Freshwater Fishes, 250 Million Years of Evolutionary History*. Eugene: ISTE.
Cavin, L., and P. M. Brito. 2001. "A New Lepisosteidae (Actinopterygii: Ginglymodi) from the Cretaceous of the Kem Kem Beds, Southern Morocco." *Bulletin de la Société géologique de France* 172:141–50.
Cavin, L., and D. B. Dutheil. 1999. "A New Cenomanian Ichthyofauna from Southeastern Morocco and Its Relationships with Other Early Late Cretaceous Moroccan Faunas." *Geologie en Mijnbouw* 78:261–66.
Cavin, L., and P. L. Forey. 2001. "Osteology and Systematic Affinities of *Palaeonotopterus greenwoodi* Forey 1997 (Teleostei: Osteoglossomorpha)." *Zoological Journal of the Linnean Society* 133(1):25–52.
———. 2004. "New mawsoniid coelacanth (Sarcopterygii: Actinistia) remains from the Cretaceous of the Kem Kem beds, Southern Morocco." In *Mesozoic Fishes*, vol. 3, *Systematics, Paleoenvironments and Biodiversity*, edited by G. Arratia and A. Tintori, 493–506. Munich: Dr. Friedrich Pfeil.
———. 2008. "A New Tselfatiiform Teleost from the Mid-Cretaceous (Cenomanian) of the Kem Kem Beds, Southern Morocco." In *Mesozoic Fishes*, vol. 4, *Homology and Phylogeny*, edited by G. Arratia, H-P. Schultze, and M. V. H. Wilson. Munich: Dr. Friedrich Pfeil. Munich.
Cavin, L., G. Garcia, and X. Valentin. 2020. "A Minute Freshwater Pycnodont Fish from the Late Cretaceous of Southern France: Palaeocological Implications." *Cretaceous Research* 106:104–242.
Cavin, L., and A. Longbottom. 2008. *Fishes and the Break-Up of Pangea*. London: Geological Society Special Publications.
Cavin, L., M. Martin, and X. Valentin. 1996. "Découverted'Atractosteus africanus (Actinopterygii,Lepisosteidae) dans le Campanien-inférieur de Ventabren (Boûches-du-Rhône, France). Implications paléobiogéographiques." *Revue de Paléobiologie* 15:1–7.
Cavin, L., H. Tong, L. Boudad, C. Meister, A. Piuz, J. Tabouelle, M. Aarab, R. Amiot, E. Buffetaut, G. Dyke, S. Hua, and J. Le Loeuff. 2010. "Vertebrate Assemblages from the Early Late Cretaceous of Southeastern Morocco: An Overview." *Journal of African Earth Sciences* 57(5):391–412.
Cavin, L., L. Boudad, E. Läng, J. Tabouelle, and H. Tong. 2013. "The Continental Bony Fish Assemblage from the Cenomanian (Late Cretaceous) of the Ifezouane Formation, SE Morocco." Abstract presented at the 11th Annual Meeting of the European Association of Vertebrate Palaeontologists, Villers-sur-Mer.
Cavin, L., L. Boudad, H. Tong, E. Läng, J. Tabouelle, and R.Vullo. 2015. "Taxonomic Composition and Trophic Structure of the Continental Bony Fish Assemblage from the Early Late Cretaceous of Southeastern Morocco." *PLoS ONE* 10 (5):e0125786.
Cepek, P. 1979. *Geological Map of Libya*. Sheet Al Qaryat ash Sharqryah NH 33–6, scale I:250 000, 32. Tripoli: Industrial Research Centre.

Cerutti-Pereyra, F., M. Thums, C.M. Austin, C.J.A.J.A. Bradshaw, J.D. Stevens, R.C. Babcock, R.D. Pillans, and M. Meekan. 2014. "Restricted Movements of Juvenile Rays in the Lagoon of Ningaloo Reef, Western Australia: Evidence for the Existence of a Nursery." *Environmental Biology of Fishes* 97(4):371–83.

Cervigón, F., R. Cipriani, W. Fischer, L. Garibaldi, M Hendrickx, A. J. Lemus, R Marquez, J.M. Poutiers, and G.R.Y.B. Rodriquez. 1992. *Fichas FAO de identificación de especies para los fines de la pesca. Guía de campo de las especies comerciales marinas y de aquas salobres de la costa septentrional de Sur América.* Food and Agriculture Organization of the United Nations. Rome: FAO.

Chaffey, C. 2002. "A Fern Which Changed Australian History." *Australian Plants Online*. Accessed January 11, 2022. http://anpsa.org.au/APOL26/jun02-6.html.

Chang, K. 2020. "A Strange Dinosaur May Have Swum the Rivers of Africa." *New York Times*, April 29.

Chapman, A. D. 2006. *Numbers of Living Species in Australia and the World*. Canberra: Australian Biological Resources Study.

Charig, A. J., and A. C. Milner. 1997. "*Baryonyx walkeri*, a Fish-Eating Dinosaur from the Wealden of Surrey." *Bulletin of the Natural History Museum of London (Geology)* 53:11–70.

Checa, A., and D. Martin-Ramos. 1989. "Growth and Function of Spines in the Jurassic Ammonite *Aspidoceras*." *Palaeontology* 32(3):645–55.

Chiarenza, A. A. 2016a. "One, No One and One Hundred Thousand: A Phylogenetic Experimental Approach to Test the Systematic Position of the Kem Kem Abelisauridae Dinosauria: Theropoda." Poster presented at the 2016 meeting of the Progressive Palaeontology Association, Oxford.

Chiarenza, A. A. 2016b. Personal communications, discussion with Jamale Ijouiher.

Chiarenza, A. A., and A. Cau. 2016. "A Large Abelisaurid (Dinosauria, Theropoda) from Morocco and Comments on the Cenomanian Theropods from North Africa." *PeerJ* 4:e1754.

Chikhi-Aouimeur, F. 2008. "Distribution of Rudists on the Saharan Platform during Cretaceous Times." Abstract presented at the Eighth International Congress on Rudists, İzmir.

Chikhi-Aouimeur, F., J. M. Pons., E. Vicens, and H. Abdallah. 2008. "Late Cretaceous Rudists of the Gafsa Region, Tunisia." Abstract presented at the Eighth International Congress on Rudists, İzmir.

Christie, A. M. 1955. *Geology of the Garian Area*. UN Tech. Assistance Program, Rept. TAA/Lib/2, Washington, DC.

Choubert, G., 1948. "Essai sur la paleogeographie du Mesocretace marocain."*Jubilaire de la Societe des Sciences Naturelles du Maroc*, Special Volume:307–29.

Choubert, G., L. Clariond, and J. Hindermeyer. 1952. *Livret-guide de l'excursion C 36. Anti-Atlas central et oriental.* Maroc. Algiers: Congres Geologique International, XIXe Session.

Chudeau, R. 1909. "Ammonites du Damergou (Sahara meridional)." *Bulletin de la Société Géologique de France* 4:67–71.

Churcher, C. S. 1995. "Giant Cretaceous Lungfish *Neoceratodus tuberculatus* from a Deltaic Environment in the Quseir (=Baris) Formation of Kharga Oasis, Western Desert of Egypt." *Journal of Vertebrate Paleontology* 15(4):845–49.

Churcher, C. S., and G. De Iuliis. 2001. "A New Species of *Protopterus* and a Revision of *Ceratodus humei* (Dipnoi: Ceratodontiformes) from the Mut Formation of Eastern Dakhleh Oasis, Western Desert of Egypt." *Journal of Palaeontology* 44(2):305–23.

Churcher, C. S., G. De Iuliis, and M. R. Kleindienst. 2006. "A New Genus for the Dipnoan Species *Ceratodus tuberculatus* Tabaste, 1963." *Geodiversitas* 28(4):635–47.

Chure, D. J. 2000. "A New Species of *Allosaurus* from the Morrison Formation of Dinosaur National Monument (Utah-Colorado) and a Revision of the Theropod Family Allosauridae." PhD diss., Columbia University.

Cichowolski, M., A. Ambrosio, and A. Concheyro. 2005. "Nautilids from the Upper Cretaceous of the James Ross Basin, Antarctic Peninsula." *Antarctic Science* 17(2):267–80.

Clark, B. E. 2015. "King of the River of Giants." *Aramcoworld*, September/October 2015. http://www.aramcoworld.com/issue/201505/king.of.the.river.of.giants.htm.

Clausing, G., and S. S. Renner. 2001. "Evolution of Growth Form in Epiphytic Dissochaeteae (Melastomataceae)." *Organisms Diversity & Evolution* 1:45–60.

Cobban, W. A. 1987. "The Upper Cretaceous (Cenomanian) ammonites Metengonoceras dumbli (Cragin) and M. acutum Hyatt: U.S." Denver, CO: Geological Survey Bulletin 1690.

Cobban, W. A., S. C. Hook, and W. J. Kennedy. 1989. *Upper Cretaceous rocks and ammonite faunas of southwestern New Mexico*. Memoir 45. Socorro: New Mexico Bureau of Mines & Mineral Resources.

Cobban, W. A., and W. J. Kennedy. 1990. "Evolution and Biogeography of the Cenomanian (Upper Cretaceous) Ammonite *Metoicoceras* Hyatt, 1903, with a Revision of *Metoicoceras praecox* Haas, 1949." In

Shorter Contributions to Paleontology and Stratigraphy: U.S. Geological Survey Bulletin 1934, edited by W. J. Sando. Denver, CO: Geological Survey Bulletin.

Cohen, C., dir. 2014. *NOVA*. Season 41, episode 20, "Bigger Than *T. rex*." Aired November 5, 2014, on PBS.

Colin, J-P., D. Néraudeau, A. Nel, and V. Perrichot. 2011. "Termite Coprolites (Insecta: Isoptera) from the Cretaceous of Western France: A Palaeoecological Insight." *Revue de Micropaléontologie* 54(3):129–39.

Collins, A. B., M. R. Heupel, R. E. Hueter, and P. J. Motta. 2007. "Hard Prey Specialists or Opportunistic Generalists? An Examination of the Diet of the Cownose Ray, *Rhinoptera bonasus*." *Marine & Freshwater Research* 58(1):135–44.

Compagno, L. J. V. 1984. *FAO Species Catalogue*. Vol. 4, part 1, *Sharks of the World: An Annotated and Illustrated Checklist of Species Known to Date*. Rome: FAO.

———. 2002. *Sharks of the World: An Annotated and Illustrated Catalogue of Shark Species Known to Date*. Vol. 2. Rome: FAO.

Contessi, M. 2013a. "First Report of Mammal-like Tracks from the Cretaceous of North Africa (Tunisia)." *Cretaceous Research* 42:48–54.

———. 2013b. "Late Albian/Cenomanian Dinosaur Tracks from Tunisia: New Constraints on the Paleobiogeography of the Mediterranean Platforms during Mid-Cretaceous." *Palaeogeography, Palaeoclimatology, Palaeoecology* 392:302–311.

Contessi, M., and F. Fanti. 2012a. "First Record of Bird Tracks in the Late Cretaceous (Cenomanian) of Tunisia." *PALAIOS* 27(27):455–64.

———. 2012b. "Vertebrate Tracksites in the Middle Jurassic-Upper Cretaceous of South Tunisia." *Ichnos* 19(4):211–27.

Coombs, S., C. B. Braun, and B. Donovan. 2001. "The Orienting Response of Lake Michigan Mottled Sculpin Is Mediated by Canal Neuromasts." *Journal of Experimental Biology* 204(2):337–48.

Cooper, S. L. A., and D. M. Martill. 2020. "A Diverse Assemblage of Pycnodont Fishes (Actinopterygii, Pycnodontomorpha) from the Mid-Cretaceous, Continental Kem Kem Beds of South-East Morocco." *Cretaceous Research* 112:104–456.

Cope, E. D. 1878. "Description of Fishes from the Cretaceous and Tertiary Deposits West of the Mississippi River." *B United States Geog Survey of the Territories*, 67–77.

Corbacho, J., S. Morrison, and M. Alonso. 2018. "First Mention of *Unusuropode castroi* Duarte & Santos, 1962 (Crustacea: Isopoda) in the Upper Cretaceous of Gara es Sbâa Lagerstätte, South-Eastern Morocco." *Earth Sciences* 7(6):288–92.

Costa, O. G. 1857. "Descrizione di pesci fossili del Libano." *Memorie della Reale Accademia delle Scienze di Napoli* 2:97–112.

Coughlin, B. L, and F. E. Fish. 2009. "Hippopotamus Underwater Locomotion: Reduced-Gravity Movements for a Massive Animal." *Journal of Mammalogy* 90(3):675–79.

Courville, P. 1992. "Les Vascoceratinae et les Pseudotissotiinae (Ammonitina) d'Ashaka (NE Nigeria): Relations avec leur environnement biose´dimentaire." *Bulletin des Centre de Recherches Exploration-Production Elf-Aquitaine* 16:407–57.

———. 1993. "Les formations marines et les faunes d'ammonites cénomaniennes et turoniennes (Crétacé supérieur) dans le Fosséde la Bénoué (Nigéria). Impacts des facteurs locaux et globaux sur les échanges fauniques à l'interface Téthys/ Atlantique Sud." Doctoral thesis, Universities of Dijon, Lyon I, Aix-Marseille I, Toulouse III.

Crane, P. R. 1988. "The Phylogenetic Position and Fossil History of the Magnoliaceae." Magnolias and Their Allies: Proceedings of an International Symposium, Royal Holloway, University of London.

Cruickshank, A. R. I. 1997. "A Lower Cretaceous Pliosauroid from South Africa." *Annals of the South African Museum* 105:207–26.

Cuny, G. 2012. "Freshwater Hybodont Sharks in Early Cretaceous Ecosystems: A Review." In *Bernissart Dinosaurs and Early Cretaceous Terrestrial Ecosystems*, edited by P. Godefoit, 519–532. Bloomington: Indiana University Press.

Cuny, G., M. Ouaja, D. Srarfi, L. Schmitz, E. Buffetaut, and M. J. Benton. 2004. "Fossil Sharks from the Early Cretaceous of Tunisia." *Revue de Paleobiologie, Geneve* 9:127–42.

Cuny, G., V. Suteethorn, E. Buffetaut, and M. Ouaja. 2007. "Hybodont Sharks from the Aptian-Albian of Tunisia and Thailand." *Bulletin de la societe D'Etude des sciences Naturelles D'Elbeuf*, 71–85

Currie, P.J., and K. Padian. 1997. *Encyclopedia of Dinosaurs*. Cambridge, MA: Academic Press.

Curry Rogers, C. 2006. "Titanosauria: A Phylogenetic Overview." In *The Sauropods: Evolution and Paleobiology*, edited by K. Curry Rogers and J. A. Wilson. Oakland: University of California Press.

Daber, R. 1968. "A *Weichselia-Stiehleria*-Matomaceaie Community within the Quedlinburg Estuary of Lower Cretaceous Age." *Botanical Journal of the Linnean Society* 61(384):75–85.

Daget, J., and M. Desoutter. 1983. "Essai de classification cladistique des Polypterides (Pisces,

Brachiopterygii)." *Bulletin du Museum National d'Histoire Naturelle de Paris* A, 661–74.
Dal Sasso, C., S. Maganuco, and A. Cioffi. 2009. "A Neurovascular Cavity within the Snout of the Predatory Dinosaur *Spinosaurus*." 1st International Congress on North African Vertebrate Palaeontology, Marrakech.
Darwish, M. H., and Y. Attia. 2007. "Plant Impressions from the Mangrove-Dinosaur Unit of the Upper Cretaceous Bahariya Formation of Egypt." *Taeckholmia* 27:105–25.
Darwish, M. H., A. Strougo, and W. El-Saadawi. 2000. "Fossil Plant Remains from Oligocene (?) of Farafra Oasis, Egypt." *Taeckholmia* 20:147–57.
Davesne, D., P. Gueriau., D. B. Dutheil, and L. Bertrand. 2018. "Exceptional Preservation of a Cretaceous Intestine Provides a Glimpse of the Early Ecological Diversity of Spiny-Rayed Fishes (Acanthomorpha, Teleostei)." *Scientific Reports* 8:1–12.
Daviero, V., B. Gomez, and M. Philippe. 2001. "Uncommon Branching Pattern within Conifers: *Frenelopsis turolensis*, a Spanish Early Cretaceous Cheirolepidiaceae." *Canadian Journal of Botany* 79(12):1400–1408.
Dawson, J. W. 1885. *Egypt and Syria: Their Physical Features in Relation to Bible History*. London: Forgotten Books.
Deaf, S. A. 2009. "Palynology, Palynofacies and Hydrocarbon Potential of the Cretaceous Rocks of Northern Egypt." PhD thesis, University of Southampton.
de Andrade, M. B., R. Edmonds, M. J. Benton, and R. Schouten. 2011. "A New Berriasian Species of *Goniopholis* (Mesoeucrocodylia, Neosuchia) from England, and a Review of the Genus." *Zoological Journal of the Linnean Society* 163(1):66–108.
de Lapparent, A-F. 1951. "Découverte de Dinosauriens, associés à une faune de Reptiles et de Poissons, dans le Crétacé inférieur de l'Extrême Sud tunisien." *Comptes Rendus de l'Académie des Sciences* 33:1430–32.
——. 1960. "Les dinosaurien du continental intercalair du Sahara central." *Memoirs de la Societe geologique de France (Nouvelle Serie)* BBA:1–57.
Seasky. 2016. "Viperfish." Accessed January 11, 2022. http://www.seasky.org/deep-sea/viperfish.html.
de Broin, F. de L. 2002. "*Elosuchus*, a New Genus of Crocodile from the Lower Cretaceous of the North of Africa." *Comptes Rendus Palevol* 1(5):275–85.
de Broin, F. de L., E. Buffetaut, J-C. Koeniguer, J-C. Rage, D. E. Russell, P. Taquet, C. Vergnaud-Grazzini, and S. Wenz. 1974. "La fauna de Vertébrés continentaux du gisement d'In Beceten (Sénonien du Niger)." *Comptes Rendus de l'Académie des Sciences* Série D: 469–472.
de Broin, F. de L., and P. Taquet. 1966. "Decouverte d'un Crocodilien nouveau dans le Cretace inferieur du Sahara." *Compte renduhebdomadaire des seances de l'Academie des Sciences* 262:2326–2329.
Delaney, R., H. Neave, Y. Fukuda, and W. K. Saalfeld. 2010. *Management Program for the Freshwater Crocodile* (Crocodylus johnstoni) *in the Northern Territory of Australia, 2010–2015*. Darwin: Northern Territory Department of Natural Resources, Environment, the Arts and Sport.
de Laubenfels, D. J. 1988. *Coniferales*. Series 1. Vol. 10 of *Flora Malesiana*. Dordrecht: Kluwer Academic.
Deleau, P. 1951. Les bassins Houillers du Sud oranais dans la région de Béchar-Habadla. *Bulletin du Service Géologique d'Algérie*, Livre I, Stratigraphie, 275p.
——. 1952. "La région de Colomb-Béchar." Monographie Régionale, XIXe Congrès de Géologie International, Algiers.
Delfaud, J., and K. Zellouf. 1995. "Existence, durant le Jurassique et le Crétacé inférieur, d'un Paléo-Niger coulant du Sud vers le Nord au Sahara occidental." In *Bassins Sédimentaires Africains: Géodynamique et Géologie Séquentielle, Biominéralisation, Sédimentation et Organismes*, edited by J. Lorenz, 93–104. Paris: CTHS.
Denny, M. W., and S. D. Gaines. 2007. "Crabs." In *Encyclopedia of Tidepools and Rocky Shores*. Berkeley: University of California Press.
Deparet, C., and J. Savornin. 1925. "Sur la decouverte d'une faune de vertebres albiens a Timimoun (Sahara occidental)." *Comptes Rendus, Academie du Sciences, Paris*, 1108–11.
Desio, A. 1928. Resultati Scientifici Missione Oasi di Giarabub. Rome: Roma Reale Societa Geografica Italian.
——. 1935. *Studi geologici sulla Cirenaica, sui deserto Libico, sulla Tripolitania e sui Fezzan orientali*. Vol. 1 of *Scientific Mission of the Royal Academy of Italy in Cufra*. Rome: Reale Accademia d'Italia.
——. 1936a. "Prime notizie presenza del Silurico fossilifero nel Fezzan." *Bollettino della Società Geologica Italiana* 55:116–120.
——. 1936b. "Riassunto sulla costituzione geologica del Fezzan." *Bollettino della Società Geologica Italiana* 55:319–56.
——. 1939. *Carta Geologique della Libya: Tripoli*. Tripoli: Annali del Museo Libico di Storia Naturale, v. 1, scale 1:3,000,000.
——. 1943. *L'esplorazione minerarie della Libia*. Vol. 10 of *Collezione scientifica e documentaria dell' Africa italiana*. Milan: 1st. Studi, Politica Internat.
DeSouza, O., and E. M. Cancello. 2011. "Termites and Ecosystem Function." *International*

Commission on Tropical Biology and Natural Resources 47(4):1–389.
Dhondt, A. V., N. Malchus, L. Boumaz, and E. Jaillard. 1999. "Cretaceous Oysters from North Africa: Origin and Distribution." *Bulletin de la Société Géologique de France* 170:67–76.
Dicker, S. E., and A. B. Elliott. 1970. "Water Uptake by the Crab-Eating Frog *Rana Cancrivora*, as Affected by Osmotic Gradients and by Neurohypophysial Hormones." *Journal of Physiology* 207(1):119–32.
Diedrich, C. G. 2012. "Stomach and Gastrointestinal Tract Contents in Late Cenomanian (Upper Cretaceous) Teleosts from the Black Shales of Germany and Analysis of Fish Mortality and Food Chains in the Upwelling-Influenced Pre–North Sea Basin of Europe." In *Vertebrate Coprolites*, edited by Hunt A. P., J. Milàn, S. G. Lucas, and J. A. Spielmann. *Bulletin of the New Mexico Museum of Natural History and Science*, no 57.
Diez, J. B., L. M. Sender, U. Villanueva-Amadoz, J. Ferrer, and C. Rubio. 2005. "New Data regarding *Weichselia reticulata*: Soral Clusters and the Spore Developmental Process." *Review of Palaeobotany and Palynology* 135:99–107.
Dixon, F. 1850. *Geology and Fossils of the Tertiary and Cretaceous Formations of Sussex*. London: Longman, Brown, Green and Longmans.
Dobruskina, I. A. 1997. "Turonian Plants from the Southern Negev, Israel." *Cretaceous Research* 18(1):87–107.
Dohlen, C. D. von, and N. A. Moran. 1995. "Molecular Phylogeny of the Homoptera: A Paraphyletic Taxon." *Journal of Molecular Evolution* 41:211–23.
Dolbeth, M., M. A. Pardal, A. I. Lillebø, U. Azeiteiro, and J. C. Marques. 2003. "Short and Long-Term Effects of Eutrophication on the Secondary Production of an Intertidal Macrobenthic Community." *Marine Biology* 143(6):1229–38.
d'Orbigny, A. 1840–42. "Paléontologie française—Terrains crétacé. 1. Cephalopodes." *Masson, Paris* 26:1–120.
Dubar, G. 1948. *Carte géologique provisoire du Haut Atlas de Midelt au 1/200.000. Notice explicative. Notes et Mémoires du Service Géologique du Maroc, 59 bis*. Laval: Impr. de Barnéoud frères.
Dumeril, A. M. C. 1806. *Zoologie analytique, ou methode naturelle de classification des animaux*. Paris.
Dunstan, A. J., P. D. Ward, and N. J. Marshall. 2011. "Vertical Distribution and Migration Patterns of *Nautilus pompilius*." *PLoS ONE* 6 (2):e16311.
Dutel, H., J. G. Maisey, D. R. Schwimmer, P. Janvier, M. Herbin, and G. Clément. 2012. "The Giant Cretaceous Coelacanth (Actinistia, Sarcopterygii) *Megalocoelacanthus dobiei* Schwimmer, Stewart & Williams, 1994, and Its Bearing on Latimerioidei Interrelationships." *PLoS ONE* 7 (11):e49911.
Dutel, H., E. Pennetier, and G. Pennetier. 2014. "A Giant Marine Coelacanth from the Jurassic of Normandy, France." *Journal of Vertebrate Paleontology* 34(5):1239–42.
Dutheil, D. B. 1999a. "The First Articulated Fossil Cladistian: *Serenoichthys kemkemensis*, Gen. et Sp. Nov., from the Cretaceous of Morocco." *Journal of Vertebrate Paleontology* 19(2):243–46.
———. 1999b. "An Overview of the Freshwater Fish Fauna from the Kem Kem Beds (Late Cretaceous: Cenomanian) of Southeastern Morocco." In *Mesozoic Fishes 2: Systematics and Fossil Record*, edited by G. Arratia and H-P. Schultze. Munich: Dr. Friedrich Pfeil.
———. 2000. "*Les Cladistia du Cénomanien continental du Sud-est marocain et les ichthyofaunes associées*." Unpublished EPHE Thesis, Ministère de l'Education Nationale, Ecole Pratique des Hautes Etudes Sciences de la Vie et de la Terre.
———. 2009. "Two New Short-Bodies Cladistia (Actinopterygii) from the Kem Kem Beds (Cenomanian) of Morocco." Abstract presented at the First International Congress on North African Vertebrate Palaeontology, Marrakech.
Dutheil, D. B., and P. M. Brito. 2009. "Articulated Cranium of *Onchopristis numidus* (Sclerorhynchidae, Elasmobranchii) from the Kem Kem Beds, Morocco." Abstract presented at the First International Congress on North African Vertebrate Palaeontology, Marrakech.
Dutta, D., K. Ambwani, and E. Estrada-Ruiz. 2011. "Late Cretaceous Palm Stem *Palmoxylon lametaei* Sp. Nov. from Bhisi Village, Maharashtra, India." *Revista Mexicana de Ciencias Geológicas* 28(1):1–9.
Earle, C. J. 2012. "*Agathis*." *The Gymnosperm Database*. https://www.conifers.org/ar/Agathis.php.
Ehrendorfer, F. 2013. *Woody Plants—Evolution and Distribution since the Tertiary: Proceedings of a Symposium Organized by Deutsche Akademie der Naturforscher LEOPOLDINA in Halle/Saale, German Democratic Republic*. Heidelberg: Springer.
El Atfy, H., A. Jasper, and D. Uhl. 2020. "A NewRrecord of *Paradoxopteris stromeri* Hirmer 1927 (Monilophyta, incertae sedis) from the Cenomanian of Sinai, Egypt." *Review of Palaeobotany & Palynology* 273:104–148.
El Beialy, S. Y., M. El-Soughier, S. A. Mohsen, and H. El Atfy. 2011. "Palynostratigraphy and Paleoenvironmental Significance of the Cretaceous Succession in the Gebel Rissu-1 Well, North Western Desert, Egypt." *Journal of African Earth Sciences* 59(2-3):215–226.

El Beialy, S. Y., M. J. Head, and H. S. El Atfy. 2010. "Palynology of the Mid-Cretaceous Malha and Galala Formations, Gebel El Minshera, North Sinai, Egypt." *PALAIOS* 25(7-8) 517–26.
Elewa, A. M. T., and O. Mohamed. 2014. "Migration Routes of the Aptian to Turonian Ostracod Assemblages from North Africa and the Middle East." *Paleontology Journal* 2014:1–7.
El Gezeery, M. N., and T. E. O'Connor. 1975. "Cretaceous Rock Units of Western Desert, Egypt." Abstract presented at the proceedings of the 13th Annual Meeting of the Geological Society of Egypt, Cairo.
Elgin, R. A., and E. Frey. 2011. "A New Azhdarchoid Pterosaur from the Cenomanian (Late Cretaceous) of Lebanon." *Swiss Journal of Geosciences* 104(1):21–33.
Elicki, O., and S. Gürsu. 2009. "First Record of *Pojetaia runnegari* Jell, 1980 and Fordilla Barrande, 1881 from the Middle East (Taurus Mountains, Turkey) and Critical Review of Cambrian Bivalves." *Paläontologische Zeitschrift* 83(2):267–91.
El-Khayal, A. A. 1985. "Occurrence of a Characteristic Wealden Fern (*Weichselia reticulata*) in the Wasia Formation, Central Saudi Arabia." *Scripta Geologica* 79:75–88.
El Qot, G. M., F. T. Fürsich, G. Abdel-Gawad, and W. Hannaa. 2009. "Taxonomy and Palaeoecology of Cenomanian-Turonian (Upper Cretaceous) Echinoids from Eastern Sinai, Egypt." *Beringeria* 40:55–98.
El-Saadawi, W., M. Kamal-El-Din, E. A. Wheeler, R. Osman, M. El-faramawi, and Z. El-noamani. 2014. "Early Miocene Woods of Egypt." *IAWA Journal* 35(1):35–50.
El-Shayeb, H., I. A. El-Hemaly, E. A. Aal, A. Saleh, A. Khashaba, H. Odah, and R. Mostafa. 2013. "Magnetization of Three Nubia Sandstone Formations from Central Western Desert of Egypt." *NRIAG Journal of Astronomy and Geophysics* 2(1):77–87.
El-Sisi, Z., M. Hassaouba, M. J. Odani, and J. C. Dolson. 2002. *Field Trip No. 8: The Geology of the Bahariya Oasis in the Western Desert of Egypt and Its Archeological Heritage*. Cairo: International Conference and Exhibition.
El-Soughier, M.I., A.S. Deaf, and M.S. Mahmoud. 2012. "Palynostratigraphy and Palaeoenvironmental Significance of the Cretaceous Palynomorphs in the Qattara Rim-1X Well, North Western Desert, Egypt." *Arabian Journal of Geosciences* 7(8):3051–3068.
Engel, M. S., R. C. McKellar, S. Gibb, and B. D. E. Chatterton. 2012. "A New Cenomanian-Turonian (Late Cretaceous) Insect Assemblage from Southeastern Morocco." *Cretaceous Research* 35:88–93
Ennih, N., and J-P. Liégeois. 2008. "The Boundaries of the West African Craton, with Special Reference to the Basement of the Moroccan Metacratonic Anti-Atlas Belt." *Geological Society, Special Publications* 297:1–17.
Erwin, T. L. 1997. "Biodiversity at Its Utmost: Tropical Forest Beetles." In *Biodiversity II*, edited by M. L. Reaka-Kudla, D. E. Wilson, and E. O. Wilson, 27–40. Washington, DC: Joseph Henry.
Escobar, H. 2015. "Author of 4-Legged-Snake Paper Defies Brazilian Fossil Laws." *Imagine Só!* (blog). *Ciencia*, July 24, 2015. http://ciencia.estadao.com.br/blogs/herton-escobar/author-of-4-legged-snake-paper-defies-brazilian-fossil-laws/.
Estes, R., K. de Queiroz, and J. Gauthier. 1988. "Phylogenetic Relationships within Squamata." In *Phylogenetic Relationships of the Lizard Families: Essays Commemorating Charles L. Camp*, edited by R. Estes and G. Pregill, 119–281. Stanford, CA: Stanford University Press.
Ettachfini, E. M., and B. Andreu. 2004. "Le Cénomanien et le Turonien de la Plate-forme Préafricaine du Maroc." *Cretaceous Research* 25(2):277–302.
Eugene, S. G., H. Tong, and P. A. Meylan. 2002. "*Galianemys*, a New Side-Necked Turtle (Pelomedusoides: Bothremydidae) from the Late Cretaceous of Morocco." *American Museum Novitates*, no. 3379, 1–20.
Evans, D. C., P. M. Barrett, K. S. Brink, and M. T. Carrano. 2014. "Osteology and Bone Microstructure of New Small Theropod Dinosaur Material from the Early Late Cretaceous of Morocco." *Gondwana Research* 27(3):1034–1041.
Evans, S. F., A. R. Milner, and C. Werner. 1996. "Sirenid Salamanders and a Gymnophionan Amphibian from the Cretaceous of the Sudan." *Palaeontology* 39(1):77–95.
Evans, T. A., T. Z. Dawes, P. R. Ward, and N. Lo. 2011. "Ants and Termites Increase Crop Yield in a Dry Climate." *Nature Communications* 2, article 262.
Everhart, M. 2013. "*Enchodus* sp.: The Sabre-Toothed Fish of the Cretaceous." *Oceans of Kansas*. Last modified 2013. http://oceansofkansas.com/enchodus.html.
———. 2014. "Sword-Eels of the Late Cretaceous." *Oceans of Kansas*. Last modified March 4, 2014. http://oceansofkansas.com/saurodon.html.
Evers, S. W. 2020. Personal communication, email to Jamale Ijouiher.
Evers, S. W., O. W. M. Rauhut, A. C. Milner, B. McFeeters, and R. Allain. 2015. "A Reappraisal of the Morphology and Systematic Position of the Theropod Dinosaur *Sigilmassasaurus* from the

'*Middle*' Cretaceous of Morocco." *PeerJ* 3:e1323. https://doi.org/10.7717/peerj.1323.
Fanti, F., A. Cau, L. Cantelli, M. Hassine, and M. Auditore. 2015. "New Information on *Tataouinea Hannibalis* from the Early Cretaceous of Tunisia and Implications for the Time and Mode of Rebbachisaurid Sauropod Evolution." *PLoS ONE* 10(4):e0123475.
Fanti, F., A. Cau, M. Hassine, and M. Contessi. 2013. "A New Sauropod Dinosaur from the Early Cretaceous of Tunisia with Extreme Avian-like Pneumatization." *Nature Communications* 4, article 2080.
Fanti, F., A. Cau, and M. Martinelli. 2014. "Integrating Palaeoecology and Morphology in Theropod Diversity Estimation: A Case from the Aptian-Albian of Tunisia." *Palaeogeography, Palaeoclimatology, Palaeoecology* 410:39–57.
FAO. 1994. *Mangrove Forest Management Guidelines*. Rome: Food and Agricultural Organisation.
Farrand, J., Jr. 1982. *The Audubon Society Encyclopedia of Animal Life*. New York: Clarkson N. Potter/ Paul Steiner.
Fastnacht, M. 2001. "First Record of *Coloborhynchus* (Pterosauria) from the Santana Formation (Lower Cretaceous) of the Chapada do Araripe, Brazil." *Paläontologische Zeitschrift* 75(23):23–36.
Fernández-Baldor, F. T., J. I. Canudo, P. Huerta, D. Montero, X. P. Suberbiola, and L. Salgado. 2011. "*Demandasaurus darwini*, a New Rebbachisaurid Sauropod from the Early Cretaceous of the Iberian Peninsula." *Acta Palaeontologica Polonica* 56:535–52.
Ferré, B., M. Benyoucef, D. Zaoui, M. Adaci, A. Piuz, S. Tchenar, C. Meister, K. Mebarki, and M. Bensalah. 2017. "Cenomanian-Turonian Roveacrinid Microfacies Assemblages (Crinoidea, Roveacrinida) from the Tinrhert Area (SE Algeria)." *Annales de Paléontologie* 102:225–35.
Field, R. D., and J. D. Reynolds. 2011. "Sea to Sky: Impacts of Residual Salmon-Derived Nutrients on Estuarine Breeding Bird Communities." *Proceedings of the Royal Society B: Biological Sciences* 278(1721):3081–88.
Fielitz, C. 2004. "The Phylogenetic Relationships of the †Enchodontidae (Teleostei: Aulopiformes)." In *Recent Advances in the Origin and Early Radiation of Vertebrates*, edited by G. Arratia, M. V. H. Wilson, and R. Cloutier, 619–634. Munich: Dr. Friedrich Pfeil.
Fielitz, C., and K. G. Rodríguez. 2008. "A New Species of *Ichthyotringa* from the El Doctor Formation (Cretaceous), Hidalgo. Mexico." In *Mesozoic Fishes 4—Homology and Phylogeny*, edited by G.Arratia, H-P. Schultze, and M. V. H. Wilson, 373–388. Munich: Dr. Friedrich Pfeil.
Figueiredo, F. J. (DE), V. Gallo, and P. M. Coelho. 2001. "First occurrence of *Rharbichthys* (Teleostei : Enchodontidae) in the Upper Cretaceous of Pelotas Basin (Atlântida Formation), Southern Brazil." *Boletim do Museu Nacional, Nova Série. Geologia (Brazil)*, no 61.
Figueredo, R. G., and W. Alexander. 2015. "A Review of the Genus *Araripesuchus* (Mesoeucrocodylia) from the Cretaceous of Gondwana." Abstract presented at the 75th Annual Meeting of the Society of Vertebrate Paleontology, Dallas.
Figueredo, R.G., B. M. Hormanseder, F. M. Dalla Vecchia, and A. W. Kellner. 2017. A New Crocodyliform Specimen from the Cretaceous of Morocco with *Hamadasuchus affinities* and the Morphological Variation within the Genus. Abstract. Society of Vertebrate Paleontology, 77th annual meeting, Calgary.
Filleul, A., and D.B. Dutheil. 2001. "*Spinocaudichthys oumtkoutensis*, a Freshwater Acanthomorph from the Cenomanian of Morocco." *Journal of Vertebrate. Paleontology* 21(4):774–80.
———. 2004. "A Peculiar Diplospondylous Actinopterygian Fish from the Cretaceous of Morocco." *Journal of Vertebrate Paleontology* 24(2):290–98.
Fischer, V., N. Bardet, R. B. J. Benson, M. S. Arkhangelsky, and M. Friedman. 2016. "Extinction of Fish-Shaped Marine Reptiles Associated with Reduced Evolutionary Rates and Global Environmental Volatility." *Nature Communications* 7: 1–11.
Floegel, S., and W. W. Hay. 2004. "The Hydrological Cycle on a Greenhouse Earth and Its Implications; Different from Today." *Geological Society of America Abstracts with Programs* 36:196–97.
Fontaine W. M. 1889. The Potomac or Younger Mesozoic Flora. Vol. 15 of *Monographs of the United States Geological Survey*. Washington: Government Printing Office.
Foote, M. 1999. "Morphological Diversity in the Evolutionary Radiation of Paleozoic and Post-Paleozoic Crinoids." *Paleobiology* 25(2):1–116.
Ford, L. S., and D. C. Cannatella. 1993. "The Major Clades of Frogs." *Herpetological Monographs* 7:94–117.
Forest, J., and M. de Saint Laurent. 1975. "Présence dans la faune actuelle d'un représentant du groupe mésozoïque des Glyphéides: Neoglyphea inopinata gen. nov., sp. nov. (Crustacea Decapoda Glypheidae)." Comptes rendus hebdomadaires des Séances de l'Académie des Sciences, Paris D (281): 155–58.
———. 1989. *Résultats des Campagnes MUSORSTOM*. Vol 5. Paris: Mémoires du Muséum National d'Histoire Naturelle.
Forey, P. L. 1973. "A Revision of the Elopiform Fishes, fossil and Recent." Supplement, *Bulletin*

of the British Museum (Natural History), Geology 10:1–222.

———. 1997. "A Cretaceous Notopterid (Pisces: Osteoglossomorpha) from Morocco." *South African Journal of Science* 93:564–69.

———. 1998. *History of the Coelacanth Fishes*. London: Chapman and Hall.

Forey, P. L., and L. Cavin. 2007. "A New Species of *Cladocyclus* (Teleostei: Ichthyodectiformes) from the Cenomanian of Morocco." *Palaeontologia Electronica* 10(3).

Forey, P. L., and L. Grande. 1998. "An African Twin to the Brazilian *Calamopleurus* (Actinopterygiia: Amiidae)." *Zoological Journal of the Linnean Society* 123:179–95.

Forey, P. L., A. López-Arbarello, and N. MacLeod. 2011. "A New Species of *Lepidotes* (Actinopterygii: Semiontiformes) from the Cenomanian (Upper Cretaceous) of Morocco." *Palaeontologia Electronica* 14(1):1–12.

Fortier, D., D. Perea, and C. Schultz. 2011. "Redescription and Phylogenetic Relationships of *Meridiosaurus vallisparadisi*, a Pholidosaurid from the Late Jurassic of Uruguay." *Zoological Journal of the Linnean Society* 163:257–72.

Foster, J. R. 2003. "Paleoecological Analysis of the Vertebrate Fauna of the Morrison Formation (Upper Jurassic), Rocky Mountain Region, U.S.A." *Bulletin of the New Mexico Museum of Natural History and Science* 23.

———. 2007. *Jurassic West: The Dinosaurs of the Morrison Formation and Their World*. Bloomington: Indiana University Press.

Fragoso, L. G. C., P. Brito, and Y. Yabumoto. 2018. "*Axelrodichthys araripensis* Maisey, 1986 Revisited." *Historical Biology* 31 (10):1350–1372.

Freund, F., and M. Raab. 1969. "Lower Turonian Ammonites from Israel." *Special Papers in Palaeontology* 4:1–83.

Frey, R. W., and S. G. Pemberton. 1987. "The Psilonichnus Ichnocoenose, and Its Relationship to Adjacent Marine and Nonmarine Ichnocoenoses along the Georgia Coast." *Bulletin of Canadian Petroleum Geology* 35:333–57.

Freymann, B. P., R. Buitenwerf, O. De Souza, and H. Olff. 2008. "The Importance of Termites (Isoptera) for the Recycling of Herbivore Dung in Tropical Ecosystems: A Review." *European Journal of Entomology* 105(2):165–73.

Frías-Quintana, C. A., G. Márquez-Couturier, C. A. Alvarez-González, D. Tovar-Ramírez, H. Nolasco-Soria, M. A. Galaviz-Espinosa, R. Martínez-García, S. Camarillo-Coop, R. Martínez-Yañez, and E. Gisbert. 2015. "Development of Digestive Tract and Enzyme Activities during the Early Ontogeny of the Tropical Gar *Atractosteus tropicus*." *Fish Physiology and Biochemistry* 41(5):1075–91.

Friis, E. M., P. R. Crane, and K. R. Pedersen. 2011. *Early Flowers and Angiosperm Evolution*. Cambridge: Cambridge University Press.

Friis, E. M., J. A. Doyle, P. K. Endress, and Q. Leng. 2003. "*Archaefructus*: Angiosperm Precursor or Specialized Early Angiosperm?" *Trends in Plant Science* 8:369–73.

Frost, B. W. 1996. "Phytoplankton Bloom on Iron Rations." *Nature* 383:474–76.

Fry, B. G., N. Vidal, J. A. Norman, F. J. Vonk, H. Scheib, S. F. R. Ramjan, S. Kuruppu, K. Fung, S. B. Hedges, M. K. Richardson, W. C. Hodgson, V. Ignjatovic, R. Summerhayes, and E. Kochva. 2006. "Early Evolution of the Venom System in Lizards and Snakes." *Nature* 4(39):584–88.

Fry, B. G., N. Vidal, L. van der Weerd, E. Kochva, and C. Renjifo. 2009. "Evolution and Diversification of the Toxicofera Reptile Venom System." *Journal of Proteomics* 72(2):127–36.

Fukuda, Y., G. Webb, C. Manolis, R. Delaney, M. Letnic, G. Lindner, P. Whitehead. 2011. "Recovery of Saltwater Crocodiles following Unregulated Hunting in Tidal Rivers of the Northern Territory, Australia." *Journal of Wildlife Management* 75(6):1253–66.

Gaffney, E. S., H. Tong, and P. A. Meylan. 2002. "*Galianemys*: A New Side-Necked Turtle (Pelomedusoides: Bothremydidae) from the Late Cretaceous of Morocco." *American Museum Novitates* 3379:1–20.

Gaffney, E. S., H. Tong, and P. A. Meylan. 2006. "Evolution of the Side-Necked Turtles: The Families Bothremydidae, Euraxemydidae, and Araripemydidae." *Bulletin of the American Museum of Natural History* 300, 1–318.

Gaffney, E. S., P. A. Meylan, R. C. Wood, E. L. Simons, and D. de A. Campos. 2011. "Evolution of the Side-Necked Turtles: The Family Podocnemididae." *Bulletin of the American Museum of Natural History* 350, 1–98.

Gaffney, G. 2001. "*Hamadachelys escuilliei*." Digital Morphology. January 15, 2001. http://digimorph.org/specimens/Hamadachelys_escuilliei/.

Gaire, A. "Water Storage In *Terminalia Tomentosa*." Tunza Eco Generation (blog), Accessed January 18, 2022. https://tunza.eco-generation.org/ambassadorReportView.jsp?viewID=51299.

Gale, A. S., P. Bengtson, and W. J. Kennedy. 2005. "Ammonites at the Cenomanian–Turonian Boundary in the Sergipe Basin, Brazil." *Bulletin of the Geological Society of Denmark* 52:167–91.

Gale, A. S., A. B. Smith, N. E. A. Monks, J. A. Young, A. Howard, D. S. Wray and J. M. Huggett. 2000.

"Marine Biodiversity through the Late Cenomanian-Early Turonian: Palaeoceanographic Controls and Sequence Stratigraphic Biases." *Journal of the Geological Society* 157(4):745–7.

Gallina, P. A., S. Apesteguía, A. Haluza, and J. I. Canale. 2014. "A Diplodocid Sauropod Survivor from the Early Cretaceous of South America." *PLoS ONE* 9 (5):e97128.

Gallo, V., M. S. S. de Carvalho, and H. R. S. Santos. 2010. "New Occurrence of Mawsoniidae (Sarcopterygii, Actinistia) in the Morro do Chaves Formation, Lower Cretaceous of the Sergipe- Alagoas Basin, Northeastern Brazil." *Bulletin of the Museu Paraense Emílio Goeldi Natural Sciences* 5:195–205.

Garassino A., A. De Angeli, and G. Pasini. 2007. "New Decapod Assemblage from the Upper Cretaceous (Cenomanian-Turonian) of Gara Sbaa, Southeastern Morocco." 3rd Symposium on Mesozoic and Cenozoic Decapod Crustaceans. Museo di Storia Naturale di Milano, May 23–25, 2007." *Memorie della Società italiana di Scienze Naturali e del Museo civico di Storia naturale di Milano* 149:45–47.

———. 2008a. "New Decapod Assemblage from the Upper Cretaceous (Cenomanian) of Southeastern Marocco." *Atti della Società italiana di Scienze naturalie del Museo civico di Storia naturale di Milano* 149:37–67.

———. 2008b. "A New Porcellanid genus (Crustacea, Decapoda) to accommodate the Late Cretaceous *Paragalathea africana* Garassino, De Angeli & Pasini, 2008 from Southeast Morocco (R.H.B. Fraaije. et al. (eds.))." *Proceedings of the 5th Symposium on Mesozoic and Cenozoic Decapod Crustaceans*, Krakow, Poland, 2013: A Tribute to Pál Mihály Müller. Scripta Geologica, 147.

———. 2011. "A New Species of Ghost Shrimp (Decapoda, Thalassinidea, Callianassidae) from the Late Cretaceous (Cenomanian) of Agadir (W Morocco)." *Atti della Società italiana di Scienze naturalie del Museo civico di Storia naturale di Milano* 152:45–55.

Garassino, A., and G. Pasini. 2018. "Amazighopsidae, a New Family of Decapod Macruran Astacideans from the late Cretaceous (Cenomanian-Turonian) of Gara Sbaa, Southeastern Morocco." *Natural History Sciences* 5(1):11–18.

Garassino, A., G. Pasini, and D. Dutheil. 2006. "*Cretapenaeus berberus* N. Gen., N. Sp. (Crustacea, Decapoda, Penaeidae) from the Late Cretaceous (Cenomanian) of Southeastern Morocco." *Atti della Società italiana di scienze naturali e del museo civico di storia naturale di Milano* 149:117–24.

Gaudant, M. 1978. "Contribution à une révision des poissons crétacés du Jbel Tselfat (Rides prérifaines, Maroc)." *Notes du Service géologique du Maroc* 39:79–124.

Gayet, M. 1981. "Contribuition à l'etude anatomique et systématique de l'ichthyofaune Cénomanien du Portugal, deuxième partie: les ostariophysaires." *Communicações dos Serviços Geológicos de Portugal* 71:173–90.

Gebhardt, H. 1999. "Cenomanian to Coniacian Biogeography and Migration of North and West African Ostracods." *Cretaceous Research* 20(2):215–29.

Gee, C. T. 2011. "Dietary Options for the Sauropod Dinosaurs from an Integrated Botanical and Paleobotanical Perspective." In *Biology of the Sauropod Dinosaurs: Understanding the Life of Giants*, edited by N. Klein, K. Remes, C. T. Gee, and P. M. Sander, 34–56. Bloomington: Indiana University Press.

Gertsch, B., T. Adatte, G. Keller, A. A. A. M. Tantawy, Z. Berner, H. P. Mort, and D. Fleitmann. 2010b. "Middle and Late Cenomanian Oceanic Anoxic Events in Shallow and Deeper Shelf Environments of NW Morocco." *Sedimentology* 57(6):1430–1462.

Gertsch, B., G. Keller, T. Adatte, Z. Berner, A. S. Kassab, A. A. A. Tantawy, A. M. El-Sabbagh, and D. Stueben. 2010a. "Cenomanian-Turonian Transition in a Shallow Water Sequence of the Sinai, Egypt." *International Journal of Earth Sciences* 99(1):165–182.

Ghenim, A. F., M. Benyoucef, G. El Qot, M. Adaci, and M. Bensalah. 2019. "Upper Cenomanian Bivalves from the Guir Basin (Southwestern Algeria): Order Veneroida." *Annales de Paléontologie* 105(1):21–38.

Ghorab, M.A.1961. "Abnormal Stratigraphic Features in Ras Gharib Oilfield." Abstract presented at the 3rd Arab Petroleum Congress, Alexandria, Egypt, v. 2.

Gili, E., J. Masse, and P. W. Skelton. 1995. "Rudists as Gregarious Sediment-Dwellers, Not Reef-Builders, on Cretaceous Carbonate Platforms." *Palaeogeography, Palaeoclimatology, Palaeoecology* 118(3-4):245–67.

Gill, G. A., and F. Chikhi. 1991. "Remarks on New Occurrences of *Aspidiscus*, a Cenomanian Scleractinian Coral, in the Persian Gulf and in Algeria." *Lethaia* 24(3):349–50.

Gillikin, D. P., and C. D. Schubart. 2004. "Ecology and Systematics of Mangrove Crabs of the Genus Perisesarma (Crustacea: Brachyura: Sesarmidae) from East Africa." *Zoological Journal of the Linnean Society* 141(3):435–45.

Gimsa, J., R. Sleigh, and U. Gimsa. 2015. "The Riddle of *Spinosaurus aegyptiacus*' Dorsal Sail." *Geological Magazine* 153(3):544–47

Glut, D. F. 1982. *The New Dinosaur Dictionary*. Secaucus, NJ: Citadel.
Goff, M. L. 1993. "Estimation of Postmortem Interval Using Arthropod Development and Successional Patterns." *Forensic Science Review* 5(2):81–94.
Golovneva, L.B. 2008. "A New Platanaceous Genus *Tasymia* (Angiosperms) from the Turonian of Siberia." *Paleontological Journal* 42:192–202.
Golovneva, L. B., P. Alekseev, E. Bugdaeva, and E. Volynets. 2018. "An Angiosperm Dominated Herbaceous Community from the Early Middle Albian of Primorye, Far East of Russia." *Fossil Imprint* 74(1-2):165–78.
Gomez, B., V. Daviero, and M. Philippe. 2001. "Uncommon Branching Pattern within Conifers: *Frenelopsis turolensis*, a Spanish Early Cretaceous Cheirolepidiaceae." *Canadian Journal of Botany* 79(12):1400–1408.
Gomez, B., C. Martín-Closas, G. Barale, and F. Thévenard. 2000. "A New Species of *Nehvizdya* (Ginkgoales) from the Lower Cretaceous of the Iberian Ranges." *Review of Palaeobotany and Palynology* 111(1-2):49–70.
Gomez, B., T. Gillot, V. Daviero-Gomez, C. Coiffard, P. Spagna, and J. Yans. 2012. "Mesofossil Plant Remains from the Barremian of Hautrage (Mons Basin, Belgium), with Taphonomy, Palaeoecology, and Palaeoenvironment Insights." In *Bernissart Dinosaurs and Early Cretaceous Terrestrial Ecosystems*, edited by P. Godefoit, 97–112. Bloomington: Indiana University Press.
Gomez, B., C. Martín-Closas, B. Georges, and N. S. de Porta. 2002. "*Frenelopsis* (Coniferales: Cheirolepidiaceae) and Related Male Organ Genera from the Lower Cretaceous of Spain." *Palaeontology* 45(5):997–1036.
Goodyear, C. P. 1967. "Feeding Habits of Three Species of Gars, *Lepisosteus*, along the Mississippi Gulf Coast." *Transactions of the American Fisheries Society* 96:297–300.
Gordon, M. S., J. B. Graham, and T. Wang. 2004. "Revisiting the Vertebrate Invasion of the Land." *Physiological and Biochemical Zoology* 75:697–99.
Gordon, W. A. 1973. "Marine Life and Ocean Surface Currents in the Cretaceous." *Journal of Geology* 81(3):269–84.
Gorscak, E. 2017. "An Emerging Model on the Paleobiogeographic Role(s) of Late Cretaceous Africa: New Titanosaurian Sauropod Dinosaur Signal Distinct Northern and Southern African Regions." Abstract presented at the Romer Prize Session, 77th Annual Meeting of the Society of Vertebrate Paleontology, Calgary.
Goudarzi, G. H. 1970. "Geology and Mineral Resources of Libya: A Reconnaissance." Geological survey professional paper no. 660. Washington, DC: Government Printing Office.
Goulding, M., R. B. Barthem, and R. Duenas. 2003. *The Smithsonian Atlas of the Amazon*. Washington, DC: Smithsonian Books.
Graham, J. 1997. *Air-Breathing Fishes: Evolution, Diversity, and Adaptation*. Cambridge, MA: Elsevier Academic Press.
Grande, L. 2010. "An Empirical Synthetic Pattern Study of Gars (Lepisosteiformes) and Closely Related Species, Based Mostly on Skeletal Anatomy: The Resurrection of Holostei." American Society of Ichthyologists and Herpetologists, Special Publication 6, supplement, *Copeia* 10: 2a.
Grande, L. 1982. "A revision of the Fossil Genus *Diplomystus*, with Comments on the Interrelationships of Clupeomorph Fishes." *American Museum Novitates* 2728:1–34.
Grande, L., and W. E. Bemis. 1998. "A comprehensive phylogenetic study of amiid fishes (Amiidae) based on comparative skeletal anatomy. An empirical search for interconnected patterns of natural history." Supplement, *Journal of Vertebrate Palaeontology* 18, memoir 4: l–690.
Grandstaff, B. S. 2006. "Giant Fishes from the Bahariya Formation, Bahariya Oasis, Western Desert, Egypt." Dissertation, University of Pennsylvania. ProQuest, paper AAI3225464.
Grandstaff, B. S. 2011. Personal communications, email to Jamale Ijouiher.
———. 2018. Personal communications, email to Jamale Ijouiher.
Grandstaff, B. S., J. B. Smith, M. C. Lamanna, K. J. Lacovara, and M. S. Abdel-Ghan. 2012. "*Bawitius*, Gen. Nov., a Giant Polypterid (Osteichthyes, Actinopterygii) from the Upper Cretaceous Bahariya Formation of Egypt." *Journal of Vertebrate Paleontology* 32(1):17–26.
Grandstaff, B. S., A. Yousry, and J. B. Smith. 2002. "New Specimens of *Mawsonia* (Actinistia, Coelacanthiformes) from the Cenomanian (Late Cretaceous) of Bahariya Oasis, Western Desert, Egypt." Abstract. *Journal of Vertebrate Paleontology* 22:32.
Gratton, C., and M. J. Vander Zanden. 2009. "Flux of Aquatic Insect Productivity to Land: Comparison of Lentic and Lotic Ecosystems." *Ecology* 90(10):2689–99.
Greenwood, P. H., and M. V. Wilson. 1998. "Bonytongues and Their Allies." In *Encyclopaedia of Fishes*, edited by J. R. Paxton and W. N. Eschmeyer, 2nd ed, 80–84. Fog City Press.
Gregory, J. W. 1899. "*Polytremacis* and the Ancestry of the Helioporidae." *Proceedings of the Royal Society of London* 46:291–305.

Grigg, G., and D. Kirshner. 2015. *Biology and Evolution of Crocodylians*. Ithaca, NY: Cornell University Press.
Guinot, D., A. De Angeli, and A. Garassino. 2008. "Marocarcinidae, a New eubrachyuran Family, and *Marocarcinus pasinii* N. Gen., N. Sp. from the Upper Cretaceous (Cenomanian-Turonian) of Gara Sbaa, Southeastern Morocco (Crustacea, Decapoda, Brachyura)." *Atti della Società italiana di Scienze naturalie del Museo civico di Storia naturale in Milano* 149:25–36.
Gullan, P. J. 1999. "Why the Taxon Homoptera Does Not Exist." *Entomologica* 33:101–4.
Haas, G. 1979. "On a New Snakelike Reptile from the Lower Cenomanian of Ein Jabrud, near Jerusalem." *Bulletin du Muséum national d'Histoire Naturelle de Paris* 1:51–64.
Haddoumi, H., A. Charrière, A. Bernard, and M. Pierre-Olivier. 2008. "Les dépôts continentaux du Jurassique moyen au Crétacé inférieur dans le Haut Atlas oriental (Maroc): paléoenvironnements successifs et signification paléogéographique." *Notebooks on Geology* 6:1–29.
Hallett, D., and D. Clark-Lowes. 2016. *Petroleum Geology of Libya*. 2nd ed. Amsterdam: Elsevier.
Hanski, I. 1999. *Metapopulation Ecology*. Oxford: Oxford University Press.
Hans-Volker, K. 2010. "Turtle Shell Remains (Testudines: Bothremydidae) from the Cenomanian of Morocco." *Studia geologica salmanticensia* 46(1):47–54.
Hartman, C. 2014. "*Spinosaurus* Fishiness Part Deux." *Skeletaldrawing.com*. September 13, 2014. http://www.skeletaldrawing.com/home/there-may-be-more-fishiness-in-spinosaurus9132014.
Hartshorne, P. M. 1989. "Facies Architecture of a Lower Cretaceous Coral-Rudist Patch Reef, Arizona." *Cretaceous Research* 10(4):311–36.
Hassler, A., J. E. Martin, R. Amiot, T. Tacail, F. Arnaud Godet, R. Allain, and V. Balter. 2018. "Calcium Isotopes Offer Clues on Resource Partitioning among Cretaceous Predatory Dinosaurs." *Proceedings of the Royal Society B: Biological Sciences* 285:1–8.
Hassouba, A. B., A. Mokhtar, and Z. E. Sisi. 1988. *Sedimentary Environment and Petrology of the Bahariya and Kharita Reservoirs, Abu Gharadig Gas Project 7*. Maadi: Gulf of Suez Petroleum Company.
Haug, E. 1905. *Paléontologie. Documents Scientifiques de la mission saharienne (mission Foureau-Lamy)*. Paris: Publications de la Société de Géographie.
Headden, J.A. 2014. "The Outlaw *Spinosaurus*." *Bite the Stuff* (blog), September 12, 2014. https://qilong.wordpress.com/2014/09/12/the-outlaw-spino-saurus/.
———. 2018. "*Sigilmassasaurus* Hump and Humpless." *Bite the Stuff* (blog), October 15, 2018. https://qilong.wordpress.com/sigilmassasaurus-hump-and-humpless/.
Heckeberg, N. S., and O. W. M. Rauhut. 2020. "Histology of Spinosaurid Dinosaur Teeth from the Albian-Cenomanian of Morocco: Implications for Tooth Replacement and Ecology." *Palaeontologia Electronica* 23(3):1–18.
Hedrick, M. S., and D. R. Jones. 1999. "Control of Gill Ventilation and Air-Breathing in the Bowfin *Amia calva*." *Journal of Experimental Biology* 202(1):87–94.
Helfield, J., and R. Naiman. 2006. "Keystone Interactions: Salmon and Bear in Riparian Forests of Alaska." *Ecosystems* 9(2):167–80.
Henderson, D. M. 2018. "A Buoyancy, Balance and Stability Challenge to the Hypothesis of a Semi-aquatic *Spinosaurus* Stromer, 1915 (Dinosauria: Theropoda)." *PeerJ* 6:e5409.
Henderson, D. M., and R. Nicholls. 2015. "Balance and Strength: Estimating the Maximum Prey Lifting Potential of the Large Predatory Dinosaur *Carcharodontosaurus saharicus*." *Journal of Vertebrate Paleontology* 298(8):108–15.
Hendrick, C., O. Mateus, and E. Buffetaut. 2016. "Morphofunctional Analysis of the Quadrate of Spinosauridae (Dinosauria: Theropoda) and the Presence of *Spinosaurus* and a Second Spinosaurine Taxon in the Cenomanian of North Africa." *PLoS ONE* 11 (1):e0144695.
Herkat, M. 2003. "Palaeogeographic Context of Cenomanian and Turonian Carbonate Platforms in the Eastern Atlasic Domain." In *North African Cretaceous Carbonate Platform Systems*, edited by E. Gili, M. N. El Negra, and P. W. Skelton, 143–160. NATO Science Series 4. Heidelberg: Springer.
Hermsen, E. J., M. A. Gandolfo, K. C. Nixon, and W. L. Crepet. 2003. "*Divisestylus* genus Novus (Affinity Iteaceae), a Fossil Saxifrage from the Late Cretaceous of New Jersey, USA." *American Journal of Botany* 90(9):1373–88.
Hilton, E. J. 2003. "Comparative Osteology and Phylogenetic Systematics of Fossil and Living Bony-Tongue Fishes (Actinopterygii, Teleostei, Osteoglossomorpha)." *Zoological Journal of the Linnean Society* 137(1):1–100.
Hinsley, S. T. 2010. "Notes on Fossil Wood." *Malvacea.info*. Accessed April 25, 2019. http://www.malvaceae.info/Fossil/Wood.html.
Hluštík, A. 1987. "Frenelopsidaceae Fam. Nov., a Group of Highly Specialized *Classopollis*-Producing Conifers." *Acta Palaeobotanica* 27(2):3–20.
Hoffstetter, R. 1955. *Squamates de type moderne—Traité de Paléontologie*, 5:606–662.

———. 1959. "Un serpent terrestre dans le Crétacé inférieur du Sahara." *Bull. Soc. geol. France* 7 (1): 897–902.
Hogarth, P. 2007. *The Biology of Mangroves and Seagrasses*. Biology of Habitats. Oxford: Oxford University Press.
Holgado, B., and R. V. Pêgas. 2020. "A Taxonomic and Phylogenetic Review of the Anhanguerid Pterosaur Group Coloborhynchinae and the New Clade Tropeognathinae." *Acta Palaeontologica Polonica* 65(4):743–61.
Holliday, C. M., and N. M. Gardner. 2012. "A New Eusuchian Crocodyliform with Novel Cranial Integument and Its Significance for the Origin and Evolution of Crocodylia." *PLoS ONE* 7 (1):e30471.
Holloway, W. L., K. M. Claeson, H. M. Sallam, S. El-Sayed, M. Kora, J. J.W. Sertich, and P. M. O'Connor. 2017. "A New Species of the Neopterygian Fish *Enchodus* from the Duwi Formation, Campanian, Late Cretaceous, Western Desert, Central Egypt." *Acta Palaeontologica Polonica* 62(3):603–11.
Holtz, T. R., Jr. 2012. "The Facts and Fictions of Terra Nova." *Jonja.net*. Accessed April 25, 2019. http://jonja.net/forums/viewtopic.php?f=3&t=13115.
———. 2016. Personal communications, email to Jamale Ijouiher.
Holtz, T. R., Jr., and V. L. Rey. 2011. *Dinosaurs: The Most Complete, Up-to-Date Encyclopaedia for Dinosaur Lovers of All Ages*. New York: Random House.
Holwerda, F. M., V. D. Díaz, A. Blanco, R. Montie, and J. W. F. Reumer. 2018. "Could Late Cretaceous Sauropod Tooth Morphotypes Provide Supporting Evidence for Faunal Connections between North Africa and Southern Europe?" *PeerJ Preprints* 6:e27286v1. https://doi.org/10.7287/peerj.preprints.27286v1.
Hone, D. W. E, and T. R. Holtz Jr. 2017. "A Century of Spinosaurs, a Review and Revision of the Spinosauridae with Comments on Their Ecology." *Acta Geologica Sinica* 91(3):1120–32.
———. 2019. "Comment On: Aquatic Adaptation in the Skull of Carnivorous Dinosaurs (Theropoda: Spinosauridae) and the Evolution of Aquatic Habits in Spinosaurids." *Cretaceous Research* 93:104–52.
———. 2021. "Evaluating the Ecology of *Spinosaurus*: Shoreline Generalist or Aquatic Pursuit Specialist?" *Palaeontologia Electronica* 24(3):1–28.
Hossain, M. D., and A. A. Nuruddin. 2016. "Soil and Mangrove: A Review." *Journal of Environmental Science and Technology* 9(2):198–207.
Hoyoux, C., M. Zbinden, S. Samadi, F. Gaill, and P. Compère. 2012. "Diet and Gut Microorganisms of Squat Lobsters *Munidopsis* Associated with Natural Woods and Mesh-Enclosed Substrates in the Deep South Pacific." *Marine Biology Research* 8(1):28–47.
Hsiang, A. Y., D. J. Field, T. H. Webster, A. D. B. Behlke, M. B. Davis, R. A. Racicot, and J. A. Gauthier. 2015. "The Origin of Snakes: Revealing the Ecology, Behaviour, and Evolutionary History of Early Snakes Using Genomics, Phenomics, and the Fossil Record." *BMC Evolutionary Biology* 15(87):2–22.
Huber, B. T., R. M. Leckie, R. D. Norris, T. J. Bralower, and E. CoBabe. 1999. "Foraminiferal Assemblage and Stable Isotopic Change across the Cenomanian-Turonian Boundary in the Subtropical North Atlantic." *Journal of Foraminiferal Research* 29(4):392–417.
Huber, B. T., R. D. Norris, and K. G. MacLeod. 2002. "Deep-Sea Paleotemperature Record of Extreme Warmth during the Cretaceous." *Geology* 30(2):123–26.
Huber, M. 2010. *Compendium of Bivalves: A Full-Colour Guide to 3,300 of the World's Marine Bivalves.* A Status on Bivalvia after 250 Years of Research. Hackenheim: ConchBooks.
Huene, F. von. 1948. "Short Review of the Lower Tetrapods." In *Robert Broom Commemorative Volume, Royal Society of South Africa Special Publications*, edited by A. L. Du Toit, 65–106. Cape Town: Royal Society of South Africa.
Hughes, N. F. 2010. *Palaeobiology of Angiosperm Origins: Problems of Mesozoic Seed-Plant Evolution*. Cambridge Earth Sciences. Cambridge: Cambridge University Press.
Hummel, J., and M. Clauss. 2011. "Sauropod Feeding and Digestive Physiology." In *Biology of the Sauropod Dinosaurs: Understanding the Life of Giants*, edited by N. Klein, K. Remes, C. T. Gee, and P. M. Sander, 11–32. Bloomington: Indiana University Press.
Hummel, J., C. T Gee, K-H. Südekum, P. M. Sander, G. Nogge, and M. Clauss. 2008. "In Vitro Digestibility of Fern and Gymnosperm Foliage: Implications for Sauropod Feeding Ecology and Diet Selection." *Proceedings of the Royal Society B: Biological Sciences* 275:1015–20.
Hunt, E. S. E. 2018. "A New Crocodylomorph from the Early Late Cretaceous Kem Kem Beds of Morocco." *Palaeontological Association Newsletter*, no. 99, 91–93.
Hussakof, L. 1916. "A New Pycnodont Fish, *Coelodus syriacus*, from the Cretaceous of Syria." *Bulletin American Museum of Natural History* 35:135–37.
Hutt, S., and P. Newbery. 2004. *A New Look at Baryonyx walkeri (Charig and Milner, 1986) Based upon a Recent Fossil Find from the Wealden*. SVPCA, Platform Presentation, Leicester.

Ibrahim, B. U., J. Auta, and J. K. Balogun. 2009. "An Assessment of the Physicochemical Parameters of Kontagora Reservoir, Niger Stata, Nigeria." *Bayero Journal of Pure and Applied Sciences* 2(1):64–69.
Ibrahim, N., S. Maganuco, C. D. Sasso, M. Fabbri, M. Auditore, G. Bindellini, D. M. Martill, S. Zouhri, D. A. Mattarelli, D. M. Unwin, J. Wiemann, D. Bonadonna, A. Amane, J. Jakubczak, U. Joger, G. V. Lauder, and S. E. Pierce. 2020b. "Tail-Propelled Aquatic Locomotion in a Theropod Dinosaur." *Nature* 518:1–4.
Ibrahim, N., C. D. Sasso, S. Maganuco, M. Fabbri, D. M. Martill, E. Gorscak, and M. C. Lamanna. 2016. "Evidence of a Derived Titanosaurian (Dinosauria, Sauropoda) in the Kem Kem Beds of Morocco, with Comments on Sauropod Paleoecology in the Cretaceous of Africa." In *Cretaceous Period: Biotic Diversity and Biogeography*, edited by A. Khosla and S. G. Lucas, 71:149–59. Albuquerque: New Mexico Museum of Natural History and Science.
Ibrahim, N., P. C. Sereno, C. D. Sasso, S. Maganuco, M. Fabbri, D. M. Martill, S. Zouhri, N. Myhrvold, and D. A. Iurino. 2014b. "Semiaquatic Adaptations in a Giant Predatory Dinosaur." *Science* 345(6204):1613–16.
Ibrahim, N., P. C. Sereno, D. J. Varricchio, D. M. Martill, D. B. Dutheil, D. M. Unwin, L. Baidder, H. C. E. Larsson, S. Zouhri, and A. Kaoukaya. 2020a. "Geology and Paleontology of the Upper Cretaceous Kem Kem Group of Eastern Morocco." *ZooKeys* 928:1–216.
Ibrahim, N., D. M. Unwin, D. M. Martill, ,L. Baidder, and S. Zouhri. 2010. "A New Pterosaur (Pterodactyloidea: Azhdarchidae) from the Upper Cretaceous of Morocco." *PLoS ONE* 5 (5):e10875.
Ibrahim, N., D. J. Varricchio, P. C. Sereno, J. A. Wilson, D. B. Dutheil, D. M. Martill, L. Baidder, and S. Zouhri. 2014a. "Dinosaur Footprints and Other Ichnofauna from the Cretaceous Kem Kem Beds of Morocco." *PLoS ONE* 9 (3):e90751.
Ifrim, C. 2013. "Paleobiology and Paleoecology of the Early Turonian (Late Cretaceous) Ammonite *Pseudaspidoceras flexuosum*." *PALAIOS* 28(1):9–22.
Ijouiher, J. 2016a. "A Reconstruction of the Palaeoecology and Environmental Dynamics of the Bahariya Formation of Egypt." *PeerJ Preprints* 4: e2470v1. https://peerj.com/preprints/2470/.
Ijouiher, J. 2016b. "Faunal Differentiation in the Cenomanian of North Africa." Poster presented at the 2016 meeting of the Progressive Palaeontology Association, Oxford.
Iles, W. J. D. 2014. "Reconstructing the Age and Historical Biogeography of the Ancient Flowering-Plant Family Hydatellaceae (Nymphaeales)." *BMC Evolutionary Biology* 14(102):1–10.
International Commission on Stratigraphy. 2018. "International Chronostratigraphic Chart." *International Commission on Stratigraphy*. Accessed April 25, 2019. https://stratigraphy.org/ICSchart/ChronostratChart2021-10.pdf.
International Geological Congress. 1952. *Carte geologique du nord-ouest de 1' Afrique, [Sheet] 2, Algerie-Tunisie: 19th international geological congress, 1952, scale 1:2,000,000*. International Geological Congress, Algiers.
Ismail, A. A., and S. I. Soliman. 1997. "Cenomanian-Santonian Foraminifera and Ostracodes from Horus Well-1, North Western Desert, Egypt." *Micropaleontology* 43(2):165–83.
Issawi, B., M. H. Francis, E. A. A. Youssef, and R. A. Osman. 2009. *The Phanerozoic Geology of Egypt: A Geodynamic Approach*. 2nd ed. Special Publication 81. Cairo: Ministry of Petroleum, Egyptian Mineral Resources Authority.
Jablonski, D., and W. G. Chaloner. 1994. "Extinctions in the Fossil Record and Discussion." *Philosophical Transactions of the Royal Society of London B* 344:11–17.
Jacobs, D. K., and J. A. Chamberlain. 1996. "Buoyancy and Hydrodynamics in Ammonoids." In *Ammonoid Paleobiology: Topics in Paleobiology*, edited by N. Landman, K. Tanabe, and R. A. Davis, 169–224. Heidelberg: Springer.
Jacobs, L. L., O. Mateus, M. J. Polcyn, A. S. Schulp, C. R. Scotese, A. Goswami, K. M. Ferguson, J. A. Robbins, D. P. Vineyard, A. B. Neto. 2009. "Cretaceous Paleogeography, Paleoclimatology, and Amniote Biogeography of the Low and Mid-latitude South Atlantic Ocean." *Bulletin de la Société Géologique de France* 180:333–41.
Jacobs, M. L., D. M. Martill, N. Ibrahim, and N. Longrich. 2018. "A New Species of *Coloborhynchus* (Pterosauria, Ornithocheiridae) from the Mid-Cretaceous of North Africa." *Cretaceous Research* 95:77–88.
Jacobs, M. L., D. M. Martill, D. M. Unwin, N. Ibrahim, S. Zouhri, and N. R. Longrich. 2020. "New Toothed Pterosaurs (Pterosauria: Ornithocheiridae) from the Middle Cretaceous Kem Kem Beds of Morocco and Implications for Pterosaur Palaeobiogeography and Diversity." *Cretaceous Research* 110:104–413.
Jacobsen, I. P., J. W. Johnson, and M. B. Bennett. 2009. "Diet and Reproduction in the Australian Butterfly Ray *Gymnura australis* from Northern and North-eastern Australia." *Journal of Fish Biology* 75(10):2475–89.

Johnson, C. 2002. "The Rise and Fall of Rudist Reefs." *American Scientist* 90. 148–153.

Johnson, G. D., and C. Patterson. 1993. "Percomorph Phylogeny: A Survey of Acanthomorphs and a New Proposal." *Bulletin of Marine Science* 52:554–626.

Jones, J. H., S. R. Manchester, and D. L. Dilcher. 1988. "*Dryophyllum debey EX saporta*, Juglandaceous Not Fagaceous." *Review of Palaeobotany and Palynology* 56:205–11.

Jonet, S. 1981. "Contribution à l'étude des Vertébrés du Crétacé portugais et spécialement du Cénomanien de l'Estremadure." *Comunicações dos Serviços Geólogicos de Portugal* 67:191–306.

Jordan, D. S. 1898. "Description of a Species of Fish (*Mitsukurina owstoni*) from Japan, the Type of a Distinct Family of lamnoid Sharks." *Proceedings of the California Academy of Sciences Zoology* 3:199–204.

Jordi, H. A., and F. Lonfat. 1963. "Stratigraphic Subdivision and Problems in Upper Cretaceous-Lower Tertiary Deposits in Northwestern Libya: Petroleum Exploration Society." *Libya Saharan Symposium, First, Tripoli 1963, Revue de l'Institut Francais du Petrole* 18:1428–36.

Kaplan, M. 2010. "Sea Stars Suck Up Carbon." *Nature*, January 7, 2010. https://doi.org/10.1038/news.2009.1041.

Karanth, K. P. 2006. "Out-of-India, Gondwanan Origin of Some Tropical Asian Biota." *Current Science* 90(6): 789–92.

Kartonegoro, A., J. F. Veldkamp, P. Hovenkamp, and P. van Welzen. 2018. "A Revision of *Dissochaeta* (Melastomataceae, Dissochaeteae)." *PhytoKeys* 107:1–178.

Kathiresan, K., and B. L. Bingham. 2001. "Biology of Mangroves and Mangrove Ecosystems." *Advances in Marine Biology* 40:81–251.

Kauffman, E. G. 1995. "Global Change Leading to Biodiversity Crisis in a Greenhouse World: The Cenomanian-Turonian (Cretaceous) Mass Extinction." In *The Effects of Past Global Change on Life Studies in Geophysics*, edited by S. M. Stanley and T. Usselmann, 47–71. Washington, DC: National Academies Press.

Kear, B. P., and P. M. Barrett. 2011. "Reassessment of the Early Cretaceous (Barremian) Pliosauroid *Leptocleidus superstes* Andrews, 1922 and Other Plesiosaur Remains from the Nonmarine Wealden Succession of Southern England." *Zoological Journal of the Linnean Society* 161(3):663–91.

Keller, G., and A. Pardo. 2004. "Age and Paleoenvironment of the Cenomanian-Turonian Global Stratotype Section and Point at Pueblo, Colorado." *Marine Micropaleontology* 51(1–2) 95–128.

Kellner, A. W. A., S. A. K. Azevedo, E. B. Machado, L. B. de Carvalho, and D. D. R. Henriques. 2011. "A New Dinosaur (Theropoda, Spinosauridae) from the Cretaceous (Cenomanian) Alcântara Formation, Cajual Island, Brazil." *Anais da Academia Brasileira de Ciências* 83:99–108.

Kellner, A. W. A., A. M. S. Mello, and T. Ford. 2007. "A Survey of Pterosaurs from Africa with the Description of a New Specimen from Morocco." Vol. 1 of *Paleontologia: Cenários de Vida*: 257–267.

Kemp, A. 1977. "The Pattern of Tooth-Plate Formation in the Australian Lungfish, *Neoceratodus forsteri* (Krefft)." *Zoological Journal of the Linnean Society, London* 60(3):223–58.

———. 1996. "*Sagenodus* (*Proceratodus*) *carlinvillensis* (Romer & Smith, 1934) (Osteichthyes: Dipnoi), Short Ridge Anomaly and Classification of Dipnoans." *Journal of Vertebrate Paleontology* 16(1):16–19.

Kennedy, W. J., and W. A. Cobban. 1976. "Aspects of Ammonite Biology, Biogeography, and Biostratigraphy." *Special Papers in Palaeontology* 17:1–94.

Kennedy, W. J., and P. Juignet. 1981. "Upper Cenomanian Ammonites from the Environs of Saumur, and the Provenance of the Types of Ammonites *vibrayeanus* and Ammonites *geslinianum*." *Cretaceous Research* 2(1):19–49.

———. 1994. "A Revision of the Ammonite Faunas of the Type Cenomanian 6: Acanthoceratinae (*Calycoceras* (*Proeucalycoceras*), *Eucalycoceras*, *Pseudocalycoceras*, *Neocardioceras*), Euomphaloceratinae, Mammitinae and Vascoceratidae." *Cretaceous Research* 15(4):469–501.

Kennedy, W. J., P. Juignet, and J. Girard. 2003. "Uppermost Cenomanian Ammonites from Eure, Haute-Normandie, Northwest France." *Acta Geologica Polonica* 53(1):1–18.

Kennedy, W. J., and H. C. Klinger. 2010. "Cretaceous Faunas from Zululand and Natal, South Africa: The Ammonite Subfamily Acanthoceratinae de Grossouvre, 1894." *African Natural History* 6:1–76.

Kennedy, W. J., I. Walaszczyk, A. Gale, K. Dembicz, T. Praszkier. 2013. "Lower and Middle Cenomanian Ammonites from the Morondava Basin, Madagascar." *Acta Geologica Polonica* 63(4):625–55.

Kent, B. W. 1994. *Fossil Sharks of the Chesapeake Region*. Maryland: Egan Rees Boyer Inc.

Kerr, A. C. 1998. "Oceanic Plateau Formation: A Cause of Mass Extinction and Black Shale Deposition around the Cenomanian-Turonian Boundary." *Journal of the Geological Society* 155(4):619–26.

Khaled, K. A. 1999. "Cretaceous Source Rocks at the Abu Gharadig Oil- and Gasfield, Northern Western Desert, Egypt." *Journal of Petroleum Geology* 22:377–95.

Khalifa, M. A., M. M. Askalany, and M. M. Seleim. 2003. "Lithofacies and Depositional Environments of the Cenomanian Galala Formation and Lower Turonian Abu Qada Formation, Central Western Sinai, Egypt." *Annals Geological Survey of Egypt* 26:217–34.

Khalloufi, B., D. B. Dutheil, and P. M. Brito. 2018. A New "Triple Armoured" Clupeomorph (Actinopterygii) from the Continental Kem Kem Beds (Cenomanian, Southeastern Morocco). 5th International Paleontological Congress, Paris.

Khalloufi, B., D. B. Dutheil, P. M. Brito, T. Mora, and R. Z. Bagils. 2017. "Mesozoic Clupeomorphs of North Africa: Diversity and Phylogeny." *Research & Knowledge* 3(2):46–49.

Khalloufi, B., D. Ouarhache, and H. Lelièvre. 2010. "New Paleontological and Geological Data about Jbel Tselfat (Late Cretaceous of Morocco)." *Historical Biology* 22:57–70.

Kilian, C. 1931. "Des principaux complexes continentaux du Sahara." *Compterendus Sommaire de la Société Géologique de France* 9:109–11.

Kim, B. K. 1969. "A Study of Several Sole Marks in the Haman Formation." *Journal of the Geological Society of Korea* 5:243–58.

Kin, A., and B. Błażejowski. 2014. "The Horseshoe Crab of the Genus *Limulus*: Living Fossil or Stabilomorph?" *PLoS ONE* 9 (10):e108036.

Kiritchkova, A. I. 1962. "Rod *Cladophlebis* v nizhnemezozoyskih otlozheniyah vostochnovo Urala" [The genus *Cladophlebis* in Mesozoic layers of North Ural]. *Trudy VNIGRI* 196:495–44.

Kirkland, J. I. 1996. "Paleontology of the Greenhorn Cyclothem (Cretaceous: Late Cenomanian to Middle Turonian) at Black Mesa, Northeastern Arizona." *Bulletin of the New Mexico Museum of Natural History and Science* 9.

Kirkland, J. I., and M. C. Aguillón-Martínez. 2002. "*Schizorhiza*: A Unique Sawfish Paradigm from the Difunta Group, Coahuila, Mexico." *Revista Mexicana de Ciencias Geológicas* 19(1):16–24.

Klein, C. G., N. R. Longrich, N. Ibrahim, S. Zouhri, and D. M. Martill. 2016. "A New Basal Snake from the Mid-Cretaceous of Morocco." *Cretaceous Research* 72:134–41.

Knight, J. L., D. J. Cicimurri, and R. W. Prudy. 2007. "New Western Hemisphere Occurrences of *Schizorhiza* Weiler, 1930 and Eotorpedo White, 1934 (Chondrichthyes, Batomorphii)." *Paludicola* 6(3):87–93.

Knüppe, J. 2014. Personal communications, email to Jamale Ijouiher.

———. 2015. Personal communications, email to Jamale Ijouiher.

———. 2016. Personal communications, email to Jamale Ijouiher.

Kora, M., H. Khalil, and M. Sobhy. 2001. "Stratigraphy and Microfacies of Some Cenomanian-Turonian Successions in the Gulf of Suez Region, Egypt." *Egyptian Journal of Geology* 45/1A:413–39.

Kost, M.A. 2002. *Natural Community Abstract for Rich Conifer Swamp*. Lansing: Michigan Natural Features Inventory.

Krassilov, V. A. 1969. "Approach to Classification of Mesozoic 'Ginkgoalean' Plants from Siberia." *Palaeobotanist* 18:12–19.

———. 1978. "Late Cretaceous Gymnosperms from Sakhalin and the Terminal Cretaceous Event." *Palaeontology* 21:893–905.

———. 1979. "Cretaceous Flora of Sakhalin, Moscow." Moscow: Nauka, 138–144.

Krassilov, V. A., and F. Bacchia. 2000. "Cenomanian Florule of Nammoura, Lebanon." *Cretaceous Research* 21(6):785–99.

———. 2013. "New Cenomanian Florule and a Leaf Mine from Southeastern Morocco: Palaeoecological and Climatological Inferences." *Cretaceous Research* 72:218–26.

Krassilov, V. A., and A. P. Rasnitsyn. 2008. *Plant and Arthropod Interactions in the Early Angiosperm History*. Sofia: Pensoft, Brill.

Krassilov, V. A., and S. Shuklina. 2008. "Arthropod Trace Diversity on Fossil Leaves from the Mid-Cretaceous of Negev, Israel." *Alavesia* 2:239–45.

Kräusel, R. 1939. "Ergebnisse der Forschungsreisen Prof. E. Stromer's in den Wüsten Ägyptens. IV. Die fossilen Floren Ägyptens." Vol. 47 of *Abhandlungen der Bayerischen Akademie der Wissenschaften, Mathematisch-naturwissenschaftliche Abteilung, München.*

Kriwer, J., E. V. Nunn, and S. Klug. 2009. "Neoselachians (Chondrichthyes, Elasmobranchii) from the Lower and Lower Upper Cretaceous of Northeastern Spain." *Zoological Journal of the Linnean Society* 155(2):316–7.

Kriwet, J., F. J. Poyatoariza, and S. Wenz. 1999. "A Revision of the Pycnodontid Fish *Coelodus subdiscus* Wenz 1989, from the Early Cretaceous of Montsec (Lleida, Spain)." *Treballs del Museu de Geologia de Barcelona* 8:33–65.

Kroner, A. 1977. "The Precambrian Geotectonic Evolution of Africa: Plate Accretion vs. Plate Destruction." *Precambrian Research* 4(2):163–213.

Kruta, I., N. Landman, I. Rouget, F. Cecca, and P. Tafforeau. 2011. "The Role of Ammonites in the Mesozoic Marine Food Web Revealed by Jaw Preservation." *Science* 331(6013):70–72.

Krutzsch, W. 1961. "Uber funde von '*ephedroidem*' Pollen im deutschen Tertiar." *Geology* 10:15–53.

Kružić, P., and A. Požar-Domac. 2003. "Banks of the Coral *Cladocora caespitosa* (Anthozoa, Scleractinia) in the Adriatic Sea." *Coral Reefs* 22(4):536.
Kuhn, O. 1936. *Crocodilia*. Fossilium Catalogus I: Animalia. 75:1–144. Berlin: Gravenhage.
Kuhnt, W., A. Holbourn, A. Gale, E-H, Chellai, and W. J. Kennedy. 2009. "Cenomanian Sequence Stratigraphy and Sea-Level Fluctuations in the Tarfaya Basin (SW Morocco)." *GSA Bulletin* 121(11-12):1695–1710.
Kvacek, J., and A. Herman. 2004. "The Campanian Grünbach Flora of Lower Austria: Palaeoecological interpretations." Vol. 106 of *Annalen des Naturhistorischen Museums. Vienna: Serie A für Mineralogie und Petrographie, Geologie und Paläontologie, Anthropologie und Prähistorie*.
Kvacek, Z. 1983. "Cuticular Studies in Angiosperms of the Bohemian Cenomanian." *Acta Palaeontologica Poionica* 28(1-2):159–70.
Labandeira, C. 2002. "Paleobiology of Predators, Parasitoids, and Parasites: Accommodation and Death in the Fossil Record of Terrestrial Invertebrates." *Paleontological Society Papers* 8:211–49.
Lacovara, K. J., J. R. Smith, J. B. Smith, and M. C. Lamanna. 2003. "The Ten Thousand Islands Coast of Florida: A Modern Analog to Low-Energy Mangrove Coasts of Cretaceous Epeiric Seas." In *Proceedings of the 5th International Conference on Coastal Sediments, Clearwater Beach, Florida*, edited by R.A. Davis.
Lacovara, K. J., J. R. Smith, J. B. Smith, M. C. Lamanna, and P. Dodson. 2002. "Evidence of Semidiurnal Tides along the African Coast of the Cretaceous Tethys Seaway: Bahariya Oasis, Egypt." *Geological Society of America Abstracts with Programs* 34:32.
Lake, J. S. 1978. *Australian Freshwater Fishes: An Illustrated Field Guide*. Melbourne: Nelson Field Guides.
Lakin, R. J., and N. R. Longrich. 2019. "Juvenile *Spinosaurs* (Theropoda: Spinosauridae) from the Middle Cretaceous of Morocco and Implications for Spinosaur Ecology." *Cretaceous Research* 93:129-142.
Lamanna, M. C., and Y. Hasegawa. 2014. "New Titanosauriform Sauropod Dinosaur Material from the Cenomanian of Morocco: Implications for Paleoecology and Sauropod Diversity in the Late Cretaceous of North Africa." *Bulletin of Gunma Museum of Natural History* 18:1–19.
Lamanna, M. C., J. B. Smith, Y. S. Attia, and P. Dodson. 2004. "From Dinosaurs to Dyrosaurids (Crocodyliformes): Removal of the Post-Cenomanian (Late Cretaceous) Record of Ornithischia from Africa." *Journal of Vertebrate Paleontology* 24(3):764–68.
Lamsdell, J. C., J. N. Tashman, G. Pasini, and A. Garassino. 2019. "A New Limulid (Chelicerata, Xiphosurida) from the Upper Cretaceous (Cenomanian–Turonian) of Gara Sbaa, Southeast Morocco." *Cretaceous Research* 106:104–230.
Landemaine, O. 1991. "Sélaciens nouveaux du Crétacé supérieur du Sud-Ouest de la France. Quelques apports à la systématique des élasmobranches." *Société Amicale des Géologues Amateurs* 1:1–45.
Lane, J. A. 2010. "The Morphology and Relationships of the Hybodont shark *Tribodus limae* with a Phylogenetic Analysis of Hybodont sharks (Chondrichthyes, Hybodontiformes)." PhD diss., University of New York, 2010.
Lane, J. A., and J. G. Maisey. 2012. "The Visceral Skeleton and Jaw Suspension in the Durophagous Hybodontid Shark *Tribodus limae* from the Lower Cretaceous of Brazil." *Journal of Paleontology* 86(5):886–905.
Läng, E. 2014. Personal communications, email to Jamale Ijouiher.
Läng, E., L. Boudad, L. Maio, E. Samankassou, J. Tabouelle, H. Tonge, and L. Cavina. 2013. "Unbalanced Food Web in a Late Cretaceous Dinosaur Assemblage." *Palaeogeography, Palaeoclimatology, Palaeoecology* 381–382:26–32.
Lang, J. C., W. D. Hartman, and L. S. Land. 1975. "Sclerosponges: Primary Framework Constructors on the Jamaican Fore-reef." *Journal of Marine Research* 33:223–31.
Lanza, B., S. Vanni, and A. Nistri. 1998. *Encyclopaedia of Reptiles and Amphibians*. San Diego: San Diego Academic Press.
Large, M. F., and J. E. Braggins. 2004. *Tree Ferns*. Portland: Timber Press.
Larghi, C. 2004. "Brachyuran Decapod Crustacea from the Upper Cretaceous of Lebanon." *Journal of Paleontology* 78(3):528–41.
Larsson, H. C. E. 2001. "Endocranial Anatomy of *Carcharodontosaurus saharicus* (Theropoda: Allosauroidea) and Its Implications for Theropod Brain Evolution." In *Mesozoic Vertebrate Life*, edited by D. H. Tank, K. Carpenter, and M. W. Skrepnick, 19–33. Bloomington: Indiana University Press.
Larsson, H. C. E., P. C. Sereno, and D. C. Evans. 2015. "New Giant Late Cretaceous Crocodyliform with Feeding Adaptations Convergent on Spinosaurids." Abstract presented at the 75th Annual Meeting of the Society of Vertebrate Paleontology, Dallas.
Larsson, H. C. E., and H-D Sues. 2007. "Cranial Osteology and Phylogenetic Relationships of

Hamadasuchus rebouli (Crocodyliformes: Mesoeucrocodylia) from the Cretaceous of Morocco." *Zoological Journal of the Linnean Society*, 149(4):533–67.

Lavocat, R. 1948. "Découverte de Crétacé à vertébrés dans le soubassement de l'Hammada du Guir. Sud Marocain." *Comptes rendus de l'Académie des Sciences de Paris* 226:1291–92.

———. 1954a. "Reconnaissance géologique dans les hammadas des confins Algéro-Marocains du sud." *Notes et Mémoires du Service Géologique du Maroc* 116:1–122.

———. 1954b. "Sur les dinosauriens du Continental Intercalaire des Kem-Kem de la Daoura." *Comptes Rendus 19th Intenational Geological Congress, 1952* 1: 65–68.

———. 1955a. "Decouverte d'un Crocodilien du genre *Thoracosaurus* dans le Cretace Superiuer d'Afrique." *Bulletin du Museum National d'Historie Naturelle* 2:338–40.

———. 1955b. *Titres et travaux scientifiques de M. Rene Lavocat*. Romorantin, France: Imprimerie Centrale, Loir-et-Cher.

Lebedel, V., C. Lézin, B. Andreu, El M. Ettachfini, and D. Groshenyd. 2015. "The Upper Cenomanian-Lower Turonian of the Preafrican Trough, Morocco: Platform Configuration and Palaeoenvironmental Conditions." *Sedimentary Geology* 245–46:1–16.

Leckie, R. M., T. J. Bralower, and R. Cashman. 2002. "Oceanic Anoxic Events and Plankton Evolution: Biotic Response to Tectonic Forcing during the Mid-Cretaceous." *Paleoceanography and Paleoclimatology* 17(3): 13-1–13-29.

Lecointre, G., and H. Le Guyader. 2007. *The Tree of Life: A Phylogenetic Classification*. Cambridge, MA: Harvard University Press.

Lee, M. S. Y., and M. W. Caldwell. 1998. "Anatomy and Relationships of *Pachyrhachis*, a Primitive Snake with Hindlimbs." *Philosophical Transactions of the Royal Society of London: Biological Sciences* 353:1521–52.

Lee, S. Y. 1997. "Potential Trophic Importance of the Faecal Material of the Mangrove Sesarmine Crab *Sesarma messa*." *Marine Ecology Progress* 159:275–84.

Lehman, T. M. 2001. "Late Cretaceous Dinosaur Provinciality." In *Mesozoic Vertebrate Life*, edited by D. H. Tanke and K. Carpenter, 310–328. Bloomington: Indiana University Press.

Lejal-Nicol, A., and W. Dominik. 1990. "Sur la paleoflore a Weichseliaceae et a angiosperms du Cenomanien de la region de Bahariya (Egypte du Sud-Oest)." *Berliner geowiss*, 120:957–91.

Le Loeuff, J., E. Läng, L. Cavin, and E. Buffetaut. 2012. "Between Tendaguru and Bahariya: On the Age of the Early Cretaceous Dinosaur Sites from the Continental Interclaire and Other African Formation." *Journal of Stratigraphy* 36(2):1–18.

Lelubre, M. 1948. "Le paleozoique du Fezzan sud-oriental." *Bulletin de la Société géologique de France* 18:79–81.

———. 1952. "Aperçu sur la geologie du Fezzan." *Algeria Service de la Carte Géologique, Travaux Recents Des Collaborateur* 3:109–48.

Lézin, C., B. Andreu, M. El Ettachfini, M. J. Wallez, V. Lebedel, and C. Meister. 2012. "The Upper Cenomanian–Lower Turonian of the Preafrican Trough, Morocco." *Sedimentary Geology*, vol 245-246:1–16.

Li, X., N. C. Duke, Y. Yang, L. Huang, Y. Zhu, Z. Zhang, R. Zhou, C. Zhong, Y. Huang, and S. Shi. 2016. "Re-evaluation of Phylogenetic Relationships among Species of the Mangrove Genus *Avicennia* from Indo-West Pacific Based on Multilocus Analyses." *PLoS ONE* 11 (10):e0164453.

Lieberman, B. S. 2005. "Geobiology and Paleobiogeography: Tracking the Coevolution of the Earth and Its Biota." *Palaeogeography, Palaeoclimatology, Palaeoecology* 219(1-3):23–33.

Liégeois, J-P., M. G. Abdelsalam, N. Ennih, and A. Ouabadi. 2013. "Metacraton: Nature, Genesis and Behaviour." *Gondwana Research* 23(1):220–37.

Lio, G., F. Agnolin, A. Cau, and S. Maganuco. 2012. "Crocodyliform Affinities for *Kemkemia auditorei* Cau and Maganuco, 2009, from the Late Cretaceous of Morocco." *Atti della Società Italiana di Scienze Naturali e del Museo di Storia Naturale di Milano* 159:119–26.

Liston, J., and A. Long. 2015. "Macbeth's Nature Paper: The Remains of the Oldest Dinosaur Egg and Embryo, and the Obstructive Potential of Fossil Protection Legislation." Abstract presented at the 13th Annual Meeting of the European Association of Vertebrate Palaeontologists, Opole.

Liston, J., M. Newbrey, T. Challands, and C. Adams. 2013. "Growth, Age and Size of the Jurassic Pachycormid *Leedsichthys problematicus* (Osteichthyes: Actinopterygii)." In *Mesozoic Fishes 5: Global Diversity and Evolution*, edited by G. Arratia, H. Schultze, and M. Wilson, 145–175. Munich: Dr. Friedrich Pfeil.

Little, S. A., R. A. Stockey, and B. Penner. 2009. "Anatomy and Development of Fruits of Lauraceae from the Middle Eocene Princeton Chert." *American Journal of Botany* 93(3):637–51.

Lohmann, H. 1996. "Das phylogenetische System der Anisoptera (Odonata)." *Deutsche Entomologische Zeitschrift* 106:209–66.

López-Arbarello, A. 2012. "Phylogenetic Interrelationships of Ginglymodian Fishes (Actinopterygii: Neopterygii)." *PLoS ONE* 7 (7):e39370.
Luger, P., and M. Gröschke. 1989. "Late Cretaceous Ammonites from the Wadi Quena Area in the Egyptian Eastern Desert." *Palaeontology* 32(2):355–407.
Lugo, A., A. E. M. Sell, and C. Snedaker. 1976. "Mangrove Ecosystem Analysis." *System Analysis and Simulation in Ecology* 5:113–46.
Luiz, M. A., V. L. Reynaldo, A. H. Devol, J. E. Richey, and B. R. Forsberg. 1989. "Suspended Sediment Load in the Amazon Basin: An Overview." *GeoJournal* 19:381–38.
Lüning, S., S. Kolonic, E. M. Belhadj, Z. Belhadj, L. Cota, G. Barić, and T. Wagner. 2004. "Integrated Depositional Model for the Cenomanian-Turonian Organic-Rich Strata in North Africa." *Earth-Science Reviews* 64(1-2):51–117.
Luque, J. 2015. "The Oldest Higher True Crabs (Crustacea: Decapoda: Eubrachyura): Insights from the Early Cretaceous of the Americas." *Palaeontology* 58(2):251–63.
Lythgoe, J., and G. Lythgoe. 1991. *Fishes of the Sea: The North Atlantic and Mediterranean*. London: Blandford.
Lyon, M. A. 2001. "Research Interests." *UPENN Paleobiology*. Accessed April 25, 2019. http://www.sas.upenn.edu/~mlyon/research.html.
Lyon, M. A., R. K. Johnson, D. J. Nichols, K. J. Lacovara, and J. B. Smith. 2001. "Late Cretaceous Equatorial Coastal Vegetation: New Megaflora Associated with Dinosaur Finds in the Bahariya Oasis, Egypt." *Geological Society of America Abstracts with Programs*, no. 82-0:1–2.
Lyons, D. A., C. Arvanitidis, A. J. Blight, E. Chatzinikolaou, T. Guy-Haim, J. Kotta, H. Orav-Kotta, A. M Queirós, G. Rilov, P. J. Somerfield, and T. P. Crowe. 2014. "Macroalgal Blooms Alter Community Structure and Primary Productivity in Marine Ecosystems." *Global Change Biology* 20(9):2712–24.
Mabberley, D. J. 1997. *The Plant-Book*. Cambridge: Cambridge University Press.
Mädel-Angeliewa, E., and W. R. Müller-Stoll. 1973. "Kritische Studien über fossile Combretaceen-Hölzer: über Hölzer von Typus *Terminalioxylon* G. Schönfeld mit einer Revision der bisher zu *Evodioxylon* Chiarugi gestellten Arten." *Palaeontographica, Abteilung B* 142 (4–6): 117–36.
Mader, B. J., and A. W. A. Kellner. 1999. "A New Anhanguerid Pterosaur from the Cretaceous of Morocco." *Boletim do Museu Nacional, nova série Geologia* 45:1–11.
Maganuco, S., and C. Dal Sasso. 2018. "The Smallest Biggest Theropod Dinosaur: A Tiny Pedal Ungual of a Juvenile Spinosaurus from the Cretaceous of Morocco." *PeerJ* 6:e4785. https://doi.org/10.7717/peerj.4785.
Magnusson, W. E., and A. P. Lima. 1991. "The Ecology of a Cryptic Predator, *Paleosuchus trigonatus*, in a Tropical Rainforest." *Journal of Herpetology* 25(1):41–48.
Mahler, L. 2005. "Record of Abelisauridae (Dinosauria: Theropoda) from the Cenomanian of Morocco." *Journal of Vertebrate Paleontology* 25:236–39.
Mahmoud, M. S., and A. M. M. Moawad. 2002. "Cretaceous Palynology of the Sanhur-1X Borehole, Northwestern Egypt." *Revista Espãnola de Micropaleontologia* 38:129–44.
Maisch, M. W., and J. Lehmann. 2000. "*Tselfatia formosa* Arambourg, 1943 (Teleostei) from the Upper Cretaceous of Lower Saxony (Northern Germany)." *Neues Jahrbuch fur Geologie und Palaontologie Monatshefte* 2000:499–512.
Maisey, J. G. 1986. "Coelacanths from the Lower Cretaceous of Brazil." *American Museum Novitates* (2866):1–30.
———. 1991. *Santana Fossils: An Illustrated Atlas*. New Jersey: TFH.
Makovicky, P. J., S. Apesteguía, and F. L. Agnolín. 2005. "The Earliest Dromaeosaurid Theropod from South America." *Nature* 437(7061):1007–11.
Malek, J. 1987. *In the Shadow of the Pyramids: Egypt during the Old Kingdom*. Norman: University of Oklahoma Press.
Manchester, S. R., M. A. Akhmetiev, and T. M. Kodrul. 2002. "Leaves and Fruits of *Celtis aspera* (Newberry) Comb. Nov. (Celtidaceae) from the Paleocene of North America and Eastern Asia." *International Journal of Plant Sciences* 163(5):725–36.
Mannion, P. D., and P. M. Barrett. 2013. "Additions to the Sauropod Dinosaur Fauna of the Cenomanian (Early Late Cretaceous) Kem Kem Beds of Morocco: Palaeobiogeographical Implications of the Mid-Cretaceous African Sauropod Fossil Record." *Cretaceous Research* 45:49–59.
Mannion, P. D., P. Upchurch, R. N. Barnes, and O. Mateus. 2013. "Osteology of the Late Jurassic Portuguese Sauropod Dinosaur *Lusotitan atalaiensis* (Macronaria) and the Evolutionary History of Basal Titanosauriforms." *Zoological Journal of the Linnean Society* 168(1):98–206.
Mantell, G. A. 1822. *The Fossils of the South Downs, or Illustrations of the Geology of Sussex*. London: Lupton Relfe.
Marchant, S., and P. J. Higgins. 1990. *Handbook of Australian, New Zealand and Antarctic Birds*. Vol.

1, *Ratites to Ducks*. Melbourne, Victoria: Oxford University.
MarineBio Conservation Society. 2016. "Trophic Structure." *MarineBio*. Last modified February 16, 2019. http://marinebio.org/oceans/trophic-structure/.
Marsh, M.E. 2003. "Regulation of $CaCO_3$ Formation in Coccolithophores." *Comparative Biochemistry and Physiology Part B: Biochemistry and Molecular Biology* 136(4): 743–754.
Martill, D. M., G. Bechly, and R. F. Loveridge. 2007. *The Crato Fossil Beds of Brazil: Window into an Ancient World*. Cambridge: Cambridge University Press.
Martill, D. M., and P. M. Brito. 2000. "First Record of *Calamopleurus* (Actinopterygii: Halecomorphi: Amiidae) from the Crato Formation (Lower Cretaceous) of North-East Brazil." *ORYCTOS* 3:3–8.
Martill, D. M., and N. Ibrahim. 2012. "Aberrant Rostral Teeth of the Sawfish *Onchopristis numidus* from the Kem Kem Beds (Early Late Cretaceous) of Morocco and a Reappraisal of *Onchopristis* in New Zealand." *Journal of African Earth Sciences* 64:71–76.
———. 2015. "An Unusual Modification of the Jaws in *cf. Alanqa*, a Mid-Cretaceous Azhdarchid Pterosaur from the Kem Kem Beds of Morocco." *Cretaceous Research* 53:59–67.
Martill, M. D., N. Ibrahim, P. M. Brito, L. Baider, S. Zhourid, R. Loveridge, D. Naish, and R. Hinga. 2011. "A New Plattenkalk Konservat Lagerstätte in the Upper Cretaceous of Gara Sbaa, South-eastern Morocco." *Cretaceous Research* 32(4):433–46.
Martill, D. M., R. Smith, D. M. Unwin, A. Kao, J. McPhee, and N. Ibrahim. 2020. "A New Tapejarid (Pterosauria, Azhdarchoidea) from the Mid-Cretaceous Kem Kem Beds of Takmout, Southern Morocco." *Cretaceous Research* 112:104–424.
Martill, D. M., D. M. Unwin, N. Ibrahim, and N. Longrich. 2018. "A New Edentulous Pterosaur from the Cretaceous Kem Kem Beds of South Eastern Morocco." *Cretaceous Research* 84:1–12.
Martin, J. E., and F. L. de Broin. 2016. "A Miniature Notosuchian with Multicuspid Teeth from the Cretaceous of Morocco." *Journal of Vertebrate Paleontology* 36(6):1–17.
Martin, J. H., and S. E. Fitzwater. 1988. "Iron Deficiency Limits Phytoplankton Growth in the Northeast Pacific Subarctic." *Nature* 331:341–43.
Martin, M. 1981. "Les Ceratodontiformes (Dipnoi) de Gadoufaoua (Aptien supérieur, Niger)." *Bulletin du Muséum national d'Histoire naturelle, Paris, 4e sér., sect. C* 3 (3): 267–83.
———. 1982a. "Nouvelles données sur la phylogénie et la systématique des Dipneustes postpaléozoïques." *Comptes Rendus de l'Académie des Sciences, Paris* 2:611–14.
———. 1982b. "Nouvelles données sur la phylogénie et la systématique des Dipneustes postpaléozoïques, conséquences stratigraphiques et paléogéographiques." In *Phylogénie et paléobiogéographie*, Livre jubilaire en l'honneur de Robert Hoffstetter, Geobios Mémoire special, Villeurbanne: 53–64.
———. 1984a. "Deux Lepidosirenidae (Dipnoi) crétacés du Sahara, *Protopterus humei* (Priem) et *Protopterus protopteroides* (Tabaste)." *Paläontologische Zeitschrift* 58:265–77.
———. 1984b. "Révision des arganodontidés et des néocératodontidés (Dipnoi, Ceratodontiformes) du Crétacé africain." *Neues Jahrbuch für Geologie und Paläontologie, Abhandlungen* 169:225–60.
Mateus, O., L. L. Jacobs, A. S. Schulp, M. J. Polcyn, T. S. Tavares, A. B. Neto, M. L. Morais, and M. T Antunes. 2011. "*Angolatitan adamastor,* a New Sauropod Dinosaur and the First Record from Angola." *Anais da Academia Brasileira de Ciências* 83(1):221–33.
McCabe, P. J., and J. T. Parrish. 1992. "Controls on the Distribution and Quality of Cretaceous Coals." Boulder, CO, Geological Society of America, Special Paper 267.
McCallum, M. L. 2007. "Amphibian Decline or Extinction? Current Declines Dwarf Background Extinction Rate." *Journal of Herpetology* 41:483–91.
McClain, M. E., and R. J. Naiman. 2008. "Andean Influences on the Biogeochemistry and Ecology of the Amazon River." *BioScience* 58(4):325–38.
McFeeters, B. D. 2020. "New Mid-cervical Vertebral Morphotype of Spinosauridae from the Kem Kem Group of Morocco." *Vertebrate Anatomy Morphology Palaeontology* 8:182–93.
McFeeters, B. D., M. J. Ryan, S. Hinic-Frlog, and C. Schröder-Adams. 2013. "A Reevaluation of *Sigilmassasaurus brevicollis* (Dinosauria) from the Cretaceous of Morocco." *Canadian Journal of Earth Sciences* 50(6):636–49.
McGowan, A. J., and G. J. Dyke. 2009. "A Surfeit of Theropods in the Moroccan Late Cretaceous? Comparing Diversity Estimates from Field Data and Fossil Shops." *Geology* 37(9):843–46.
McNulty, C. L., and B. H. Slaughter. 1968. "Teeth of Cretaceous Batoid Genus, *Ptychotrygon*." Abstract. *AAPG Bulletin* 52:541.
———. 1972. "The Cretaceous Selachian Genus, *Ptychotrygon* Jaekel 1894." *Eclogae Geologicae Helvetiae* 65(3):647–56.
McPhee, J., N. Ibrahim, A. Kao, D. M. Unwin, R. Smith, and D. M. Martill. 2020. "A New Chaoyangopterid (Pterosauria: Pterodactyloidea) from the

Cretaceous Kem Kem Beds of Southern Morocco." *Cretaceous Research* 110:104–410.
Medeiros, M. A. 2006. "Large Theropod Teeth from the Eocenomanian of Northeastern Brazil and the Occurrence of Spinosauridae." *Revista bras. paleont* 9(3):333–38.
Medeiros, M. A., R. M. Lindoso, I. D. Mendes, and I. S. Carvalho. 2014. "The Cretaceous (Cenomanian) Continental Record of the Laje do Coringa Flagstone (Alcântara Formation), Northeastern South America." *Journal of South American Earth Sciences* 53:50–58.
Meister, C. 1989. "Ammonites from the Upper Cretaceous of Ashaka, Nigeria: Taxonomic Analysis, Ontogenetic, Biostratigraphic and Scalable." *Bulletin des Centres de Recherche Exploration-Production Elf-Aquitaine* 13:1–84.
Meister, C., and H. Abdallah. 2005. "Précision sur les successions d'ammonites du Cénomanien-Turonien dans la région de Gafsa, Tunisie du centre-sud." *Revue de Paléobiologie, Genève* 24:111–99.
———. 2012. "Les ammonites du Cénomanien-Turonien de la région de Kasserine, Tunisie centrale." *Revue de Paléobiologie, Genève* 31:425–81.
Meister, C., K. Alzouma, J. Lang, and B. Mathey. 1992. "Les ammonites du Niger (Afrique occidentale) et la transgression transsaharienne au cours du Cénomaniene/Turonien." *Geobios* 25:55–100.
Meister, C., and M. Rhalmi. 2002. "Quelques ammonites du Cénomaniene Turonien de la région d'Errachidia-Boudnid-Erfoud (partie méridionale du Haut Atlas Central, Maroc)." *Revue de Paléobiologie* 21:759–79.
Megerisi, M., and V. D. Mamgain. 1980a. "Al Khowaymat Formation: An Enigma in the Stratigraphy of Northeastern Libya." In *The Geology of Libya*, edited by M. J. Salem and M. T. Busrewil, 73–88. London: Academic Press.
———. 1980b. "The Upper Cretaceous-Tertiary Formations of Northern Libya." In *The Geology of Libya*, edited by M. J. Salem and M. T. Busrewil, 67–72. London: Academic Press.
Mekawy, M. S. 2010. "Factors Affecting the Behavior and Traits of Some Cenomanian Oysters from the Sinai, Egypt." *Egyptian Journal of Paleontology* 10:107–21.
Menchikoff, N. 1936. "Etudes géologiques sur les confins algéro-marocains du Sud." *Bulletin de la Société Géologique de France, Paris, 5e* 6 (6): 131–48.
Mendes, M. M., J. L. Dinis, B. Gomez, and J. Pais. 2010. "Reassessment of the Cheirolepidiaceous Conifer *Frenelopsis teixeirae* Alvin et Pais from the Early Cretaceous (Hauterivian) of Portugal and Palaeoenvironmental Considerations." *Review of Palaeobotany and Palynology* 161:30–42.
Meng, J., Y. Hu, C. Li, and Y. Wang. 2006. "The Mammal Fauna in the Early Cretaceous Jehol Biota: Implications for Diversity and Biology of Mesozoic Mammals." *Geological Journal* 41(3-4):439–63.
Meng, X. 2008. "A New Species of *Sinopterus* from Jehol Biota and Reconstruction of Stratigraphic Sequence of the Jiufotang Formation." PhD diss., Chinese Academy of Sciences.
Meunier, F. J., R-P. Eustache, D. Dutheil, and L. Cavin. 2016. "Histology of Ganoid Scales from the Early Late Cretaceous of the Kem Kem Beds, SE Morocco: Systematic and Evolutionary Implications." *Cybium* 40(2):121–32.
Meunier, L. M. V., and H. C. Larsson. 2015. "Redescription and Phylogenetic Affinities of *Elosuchus cherifiensis* (Crocodyliformes)." Abstract presented at the 75th Annual Meeting of the Society of Vertebrate Palaeontology, Dallas.
———. 2017. "Revision and Phylogenetic Affinities of *Elosuchus* (Crocodyliformes)." *Zoological Journal of the Linnean Society* 179(1):169–200.
Meyer, K. M., and L. R. Kump. 2008. "Oceanic Euxinia in Earth History: Causes and Consequences." *Annual Review of Earth and Planetary Science* 36:251–88.
Michael, S. W. 1993. *Reef Sharks and Rays of the World: A Guide to Their Identification, Behavior, and Ecology*. Monterey, CA: Sea Challengers.
Micheli, F., F. Gherardi, and M. Vannini. 1991. "Feeding and Burrowing Ecology of Two East African Mangrove Crabs." *Marine Biology* 111:247–54.
Milner, A. C. 2003. "Fish-Eating Theropods: A Short Review of the Systematics, Biology and Palaeobiogeography of *Spinosaurs*." Proceedings of the II International Conference on Dinosaur Paleontology and Its Environment, Burgos:129–38.
Mkhitaryan, T. G., and A. O. Averianov. 2011. "New Material and Phylogenetic Position of *Aidachar paludalis* Nesov, 1981 (Actinopterygii, Ichthyodectiformes) from the Late Cretaceous of Uzbekistan." *Proceedings of the Zoological Institute RAS* 315:181–92.
Mohr, B. A. R., M. E. C. Bernardes-de-Oliveira, and D. W. Taylor. 2008. "*Pluricarpellatia*, a Nymphaealean Angiosperm from the Lower Cretaceous of Northern Gondwana, Crato Formation, Brazil." *Taxon* 57(4):1147–58.
Moiseeva, M. G. 2011. "New Species of the Genus *Macclintockia* (Angiosperms) from the Campanian of the Ugol'naya Bay (Northeastern Russia)." *Paleontological Journal* 45(2):207–23.
———. 2012. "*Barykovia*, a New Genus of Angiosperms from the Campanian of Northeastern Russia." *Review of Palaeobotany and Palynology* 178:1–12.

Monbiot, G. 2013. "How Wolves Can Alter the Course of Rivers." TEDGlobal Conference, Edinburgh.
Monnet, C. 2009. "The Cenomanian-Turonian Boundary Mass Extinction (Late Cretaceous): New Insights from Ammonoid Biodiversity Patterns of Europe, Tunisia and the Western Interior (North America)." *Palaeogeography, Palaeoclimatology, Palaeoecology* 282(1-4):88–104.
Morri, C., A. Peirano, and N. C. Bianchi. 2001. "Is the Mediterranean Coral *Cladocora caespitosa* an Indicator of Climatic Change?" *Limnology and Oceanography* 22:139–44.
Morrison, C. L., A. W. Harvey, S. Lavery, K. Tieu, Y. Huang, and C. W. Cunningham. 2001. "Mitochondrial Gene Rearrangements Confirm the Parallel Evolution of the Crab-like Form." *Proceedings of the Royal Society B: Biological Sciences* 269 (1489):345–50.
Morrissey, J. F., and S. H. Gruber. 1993a. "Habitat Selection by Juvenile Lemon Sharks, *Negaprion brevirostris*." *Environmental Biology of Fishes* 38:311–19.
———. 1993b. "Home Range of Juvenile Lemon Sharks, *Negaprion brevirostris*." *Copeia*, vol. 1993 (2):425–434.
Mortimer, M. 2014. "No Giant Egyptian *Deltadromeus* After-All." *Theropod Database* (blog). Accessed April 25, 2019. https://theropoddatabase.blogspot.com/2014/09/no-giant-egyptian-deltadromeus.html.
———. 2015. "Carnosauria." *Theropod Database* (blog). Accessed April 25, 2019. http://theropoddatabase.com/Carnosauria.htm.
———. 2017. "Ornithoscelida Tested: Adding Taxa and Checking Characters." *Theropod Database* (blog), March 28, 2017. http://theropoddatabase.blogspot.co.uk/2017/03/ornithoscelida-tested-adding-taxa-and.html.
Motani, R., X-H. Chen, D-Y. Jiang, L. Cheng, A. Tintori, and O. Rieppel. 2015. "Lunge Feeding in Early Marine Reptiles and Fast Evolution of Marine Tetrapod Feeding Guilds." *Scientific Reports* 5 (8900):1–8.
Motta, M. J., A. M. A. Ronaldo, S. Rozadilla, F. E. Agnolín, N. R. Chimento, F. B. Egli, and F. E. Novas. 2016. "New Theropod Fauna from the Upper Cretaceous (Huincul Formation) of Northwestern Patagonia, Argentina." *Bulletin of the New Mexico Museum of Natural History and Science*, no. 71:123–253.
Moustafa, T. F., and G. A. Lashin. 2012. "Aptian-Turonian Palynomorphs from El-Waha-1 Well, Southwestern Part of the Western Desert, Egypt." *Journal of Applied Sciences Research* 8(4):1870–77.
Muller-Feuga, R. 1954. "Contribution a l'etude de la geologie, de la petrographic et des resources hydrauliques et minerales du Fezzan: Tunisia." Direction Trav. Publics, Annales Mines Geologic, no. 12:354.
Murch, A. 2015. "Butterfly Ray." *Elasmodiver*. Accessed April 25, 2019. http://www.elasmodiver.com/SpinyButterflyRay.htm.
Murdy, E. O., R. S. Birdsong, and J. A. Musick. 1997. *Fishes of Chesapeake Bay*. Washington, DC: Smithsonian Institution Press.
Murray, A. M. 2000. "The Palaeozoic, Mesozoic and Early Cenozoic Fishes of Africa." *Fish and Fisheries* 1(2):111–145.
Murray, A. M., and M. V. H. Wilson. 2009. "A New Late Cretaceous Macrosemiid Fish (Neopterygii, Halecostomi) from Morocco, with Temporal and Geographical Range Extensions for the Family." *Palaeontology* 52(2):429–440.
———. 2011. "A New Species of Sorbinichthys (Teleostei: Clupeomorpha: Ellimmichthyiformes) from the Late Cretaceous of Morocco." *Canadian Journal of Earth Sciences* 48(1):1–9.
———. 2013. "Two New Paraclupeid Fishes (Clupeomorpha: Ellimmichthyiformes) from the Upper Cretaceous of Morocco." In *Mesozoic Fishes 5: Global Diversity and Evolution*, edited by G. Arratia, H-P. Schultze, and M. V. H. Wilson, 267–290. Munich: Dr. Friedrich Pfeil.
———. 2014. "Four New Basal Acanthomorph Fishes from the Late Cretaceous of Morocco." *Journal of Vertebrate Paleontology* 34(1):34–48.
Murray, A. M., M. V. H. Wilson, and B. D. E. Chatterton. 2007. "A Late Cretaceous Actinopterygian Fauna from Morocco." Supplement, *Journal of Vertebrate Paleontology* 27(3):122A.
Murray, A. M., M. V. H. Wilson, S. Gibb, and B. D. E. Chatterton. 2013. "Additions to the Late Cretaceous (Cenomanian/Turonian) Actinopterygian Fauna from the Agoult Locality, Akrabou Formation, Morocco, and Comments on the Palaeoenvironment." In *Mesozoic Fishes 5: Global Diversity and Evolution*, edited by G. Arratia, H-P. Schultze, and M. V. H. Wilson, 525–548. Munich: Dr. Friedrich Pfeil.
Myers, T. S., N. J. Tabor, and L. L. Jacobs. 2011. "Late Jurassic Paleoclimate of Central Africa." *Palaeogeography, Palaeoclimatology, Palaeoecology* 311:111–25.
Myrow, P. M. 1995. "*Thalassinoides* and the Enigma of Early Paleozoic Open-Framework Burrow Systems." *PALAIOS* 10(1):58–74.
Nagelkerken, I., S. J. M. Blaber, S. Bouillon, P. Greene, M. Haywood, L. G. Kirtong, J-O. Meyneckeh, J. Pawliki, H. M. Penrose, A. Sasekumar, and P. J. Somerfield. 2008. "The Habitat Function

of Mangroves for Terrestrial and Marine Fauna: A Review." *Aquatic Botany* 89(2):155–85.

Nagm, E. 2009. "Integrated Stratigraphy, Palaeontology and Facies Analysis of the Cenomaniane/Turonian (Upper Cretaceous) Galala and Maghra El Hadida Formations of the Western Wadi Araba, Eastern Desert, Egypt." PhD thesis, Würzburg University. http://www.opus-bayern.de/uni-wuerzburg/volltexte/2009/3988/.

Nagm, E., and M. Wilmsen. 2012. "Late Cenomanian-Turonian (Cretaceous) Ammonites from Wadi Qena, Central Eastern Desert, Egypt: Taxonomy, Biostratigraphy and Palaeobiogeographic Implications." *Acta Geologica Polonic* 62(1):63–89.

Nagm, E., M. Wilmsen, M. F. Aly, and A-G. Hewaidy. 2010. "Upper Cenomanian-Turonian (Upper Cretaceous) Ammonoids from the Western Wadi Araba, Eastern Desert, Egypt." *Cretaceous Research* 3(5)1:473–499.

Naish, D. 2020. "Stop Saying That There Are Too Many Sauropod Dinosaurs, Part 6." *Tetrapod Zoology Podcast* (blog), May 5, 2020. http://tetzoo.com/blog/2020/5/5/stop-saying-that-there-are-too-many-sauropod-dinosaurs-part-6.

Nash, D. 2012. "Planet Predator 2." *Antediluvian Salad* (blog). Accessed April 25, 2019. https://antediluviansalad.blogspot.com/2012/09/planet-predator-ii-kem-kem.html.

——. 2015a. "*Carcharodontosaurs* Deadlifts Less Than World's Strongest Man!?!" *Antediluvian Salad* (blog), May 19, 2015. http://antediluviansalad.blogspot.co.uk/2015/05/carcharodontosaurs-deadlifts-less-than.html.

——. 2015b. "*Masiakasaurus* Proposed as a Fossorial Animal Hunting Specialist." *Antediluvian Salad* (blog), September 7, 2015. http://antediluviansalad.blogspot.co.uk/2015/09/masiakasaurus-proposed-as-fossorial.html.

——. 2015c. "*Spinosaurus* Unauthorized I: Hippos Are Not Really Fat and Can't Swim." *Antediluvian Salad* (blog), October 12, 2015. http://antediluviansalad.blogspot.co.uk/2015/10/spinosaurus-unauthorized-i-hippos-are.html.

——. 2015d. "*Spinosaurus* Unauthorized II: Spino Identity Crisis & Island-Hopping Hippos." *Antediluvian Salad* (blog), November 2, 2015. http://antediluviansalad.blogspot.co.uk/2015/11/spinosaurus-unauthorized-ii-spino.html.

National Research Council. 2011. "Overview of Climate Changes and Illustrative Impacts." In *Climate Stabilization Targets: Emissions, Concentrations, and Impacts over Decades to Millennia*. Washington, DC: National Academies Press.

Neilson, J.D., and R. I. Perry. 1990. "Diel Vertical Migrations of Marine Fishes: An Obligate or Facultative Process?" *Advances in Marine Biology* 26:115–168.

Néraudeau, D., and P. Courville. 1997. "Cenomanian and Turonian Echinoids from Nigeria." *Géobios* 30(6):835–847.

Néraudeau, D., and B. Mathey. 2000. "Biogeography and Diversity of South Atlantic Cretaceous Echinoids: Implications for Circulation Patterns." *Palaeogeography, Palaeoclimatology, Palaeoecology* 156(1-2):71–88.

Nessov, L. A., V. I. Zhegallo, and A. O. Averianov. 1998. "A New Locality of Late Cretaceous Snakes, Mammals and Other Vertebrates in Africa (Western Libya)." *Annales de Paléontologie* 84(3-4):265–274.

Newman, M. E. J. 1997. "A Model of Mass Extinction." *Journal of Theoretical Biology* 189(3):235–252.

New Scientist. 2008. "Submarine Eruption Bled Earth's Oceans of Oxygen." *New Scientist*, July 16, 2008. https://www.newscientist.com/article/mg19926655.300-submarine-eruption-bled-earths-oceans-of-oxygen/.

Nguenang, G. S., A. T. Mbaveng, A. G. Fankam, H. T. Manekeng, P. Nayim, B. E. N. Wamba, and V. Kuete. 2018. "*Tristemma hirtum* & Five Other Cameroonian Edible Plants with Weak or No Antibacterial Effects Modulate the Activities of Antibiotics against Gram-Negative Multidrug-Resistant Phenotypes." *Scientific World Journal* 25: 1–12.

Nicholl, C.S.C., E. S. .E. Hunt, D. Ouarhache, and P. D. Mannion. 2021. "A Second Peirosaurid Crocodyliform from the Mid Cretaceous Kem Kem Group of Morocco and the Diversity of Gondwanan Notosuchians outside South America." *Royal Society, Open Science* 8(10): 211254:1–31.

Nicol, E. A. T. 1932. "The Feeding Habits of the Galatheidea." *Journal of the Marine Biological Association of the United Kingdom* 18(1):87–106.

Nienhuis, S., A. Palmer, and C. Harley. 2010. "Elevated CO_2 Affects Shell Dissolution Rate but Not Calcification Rate in a Marine Snail." *Proceedings of the Royal Society B: Biological Science* 277(1693):2553–2558.

Nopcsa, F. 1925. "*Die Symoliophis Reste - In Ergbenisse der Forschungsreisen Prof. E. Stromers in den Wüsten Ägyptens, II*." Abhand lungen der Bayerischen Akademie der Wissenschaften, Mathematisch- naturwissenschaftliche Abteilung, Munich 30, 1–27.

——. 1926. "Neue Beobachtungen an *Stomatosuchus*." *Centralblatt fur Mineralogie, Geologie und Palaontologie* B:212–15.

Norell, M. A., J. M. Clark, T. A. Hamilton, P. J. Makovicky, R. Barsbold, and T. Timothy. 2006. "A New Dromaeosaurid Theropod from Ukhaa Tolgod

(Omnogov, Mongolia)." *American Museum Novitates*, no. 3545:1–51.
Norton, P. 1967. *Rock Stratigraphic Nomenclature of the Western Desert, Egypt*. General Petroleum Corporation of Egypt, Cairo.
Noske, R.A. 1996. "Abundance, Zonation and Foraging Ecology of Birds in Mangrove of Darwin Harbour, Northern Territory." *Wildlife Research* 23(4):443–474.
Nothdurft, W. E., and J. B. Smith. 2002. *The Lost Dinosaurs of Egypt*. New York: Random House.
Noto, C. R., and A. Grossman. 2010. "Broad-Scale Patterns of Late Jurassic Dinosaur Paleoecology." *PLoS ONE* 5 (9):e12553.
Novas, F. E. 2009. *The Age of Dinosaurs in South America*. Bloomington: Indiana University Press.
Novas, F. E., F. Dalla Vecchia, and D. Pais. 2005. "Theropod Pedal Unguals from the Late Cretaceous (Cenomanian) of Morocco, Africa." *Revista del Museo Argentino de Ciencias Naturales nueva series*, no. 7(2):167–75.
Novas, F. E., S. de Valais, P. Vickers-Rich, and T. Rich. 2005. "A Large Cretaceous Theropod from Patagonia, Argentina, and the Evolution of Carcharodontosaurids." *Naturwissenschaften* 92(5):226–230.
Nursall, J. R. 1996. "The Phylogeny of Pycnodont Fishes." In *Mesozoic Fishes: Systematics and Paleoecology*, edited by G. Arratia and G. Viohl, 125–152. Munich: Dr. Friedrich Pfeil.
O'Brien, M., R. Crossley, and K. Karlson. 2006. *The Shorebird Guide*. Boston, MA: Houghton Mifflin.
O'Connor, P., J. Sertich, H. M. Sallam, and E. Seiffert. 2010. "Reconnaissance Paleontology in the Late Cretaceous of Dakhla and Kharga Oases, Western Desert, Egypt." Abstract presented at the 70th Anniversary Meeting of the Society of Vertebrate Paleontology, Pittsburgh.
Oliver, R. 1982. *Whales and Giants of the Sea*. London: Treasure House.
Olshevsky, G. 1991. "A Revision of the Parainfraclass Archosauria Cope, 1869, Excluding the Advanced Crocodylia." *Mesozoic Meanderings* 2:1–196.
Ong, J. E. 1993. "Mangroves: A Carbon Source and Sink." *Chemosphere* 27(6):1097–107.
Ortega-Hernandez, J., J. Tremewan, and S. J. Braddy. 2010. "Euthycarcinoids." *Geology Today* 26(5):195–198.
Ouaja, M., M. Philippeb, G. Barale, S. Ferry, and M. B. Youssef. 2004. "Mise en évidence d'une flore oxfordienne dans le Sud-Est de la Tunisie: intérêts stratigraphique et paléoécologique." *Geobios* 37(1):89–97.
Ouroumova, O., K. Shimada, and J. I. Kirkland. 2016. "Fossil Marine Vertebrates from the Blue Hill Shale Member (Middle Turonian) of the Upper Cretaceous Carlile Shale in Northeastern Nebraska." *Transactions of the Kansas Academy of Science* 119(2):211–221.
Owen, R. 2012. *The Rise and Fall of Arab Presidents for Life*. Cambridge, MA: Harvard University Press.
Paladino, F. V., J. R. Spotila, and P. Dodson. 1997. "A Blueprint for Giants: Modeling the Physiology of Large Dinosaurs." In *The Complete Dinosaur*, edited by J. O. Farlow and M. K. Brett-Surman, 491–504. Bloomington: Indiana University Press.
Pandey, D. K., F. T. Fürsich, G. Gameil, and W. S. Ayoub-Hannaa. 2011. "*Aspidiscus cristatus* (Lamarck) from the Cenomanian Sediments of Wadi Quseib, East Sinai, Egypt." *Journal of the Palaeontological Society of India* 56(1):29–37.
Patterson, C. 1970. "A Clupeomorph Fish from the Gault (Lower Cretaceous)." *Zoological Journal of the Linnean Society* 49(3):161–82.
Paterson, N., dir. 2011. *Planet Dinosaur*. Season 1, episode 1, "Lost World." Aired October 12, 2011, on BBC ONE.
Paul, G. S. 1988a. "The Brachiosaur Giants of the Morrison and Tendaguru with a Description of a New Subgenus, *Giraffatitan*, and a Comparison of the World's Largest Dinosaurs." *Hunteria* 2(3):1–14.
———. 1988b. *Predatory Dinosaurs of the World*. New York: Simon & Schuster.
———. 2000. *The Scientific American Book of Dinosaurs*. New York: St. Martin's.
Paulina-Carabajal, A., and R. A. Coria. 2015. "An Unusual Frontal Theropod from the Upper Cretaceous of North Patagonia." *Alcheringa* 39(4):514–518.
Pauly, D., and J. Ingles. 1986. "The Relationship between Shrimp Yields and Intertidal Vegetation (Mangrove) Areas: A Reassessment." In *IOC/FAO Workshop on Recruitment in Tropical Coastal Demersal Communities*, Ciudad del Carmen, edited by A. Yanez-Arancibia and D. Pauly, 277–283.
Pei, R., M. Pittman, P. A.Goloboff, T. A. Dececchi, M. B. Habib, T. G. Kaye, H. C. E. Larsson, M. A. Norell, S. L.Brusatte, and X. Xu. 2020. "Potential for Powered Flight Neared by Most Close Avialan Relatives, but Few Crossed Its Thresholds." *Current Biology* 30(20):1–14.
Penman, D. E. 2007. "Biostratigraphy of Santonian-Campanian Genus *Baculites* in the Western Interior of North America: Implications for Evolutionary Timing and Migration." BA paper, Carleton College, Northfield, Minnesota.
Persons, W. S., P. J. Currie, and G. M. Erickson. 2020. "An Older and Exceptionally Large Adult Specimen of *Tyrannosaurus rex*." Special issue, *Anatomical Record* 303(4):656–672.

Persson, P. O. 1963. "A Revision of the Classification of the Plesiosauria with a Synopsis of the Stratigraphical and Geographical Distribution of the Group." *Lunds Universitets Årsskrift* 59:1–59.

Petti, F. M., M. Bernardi, P. Ferretti, R. Tomasoni, and M. Avanzini. 2011. "Dinosaur Tracks in a Marginal Marine Environment: The Coste dell'Anglone Ichnosite (Early Jurassic, Trento Platform, NE Italy)." *Italian Journal of Geoscience* 130(1):27–41.

Peyer, B. 1925. "Die Ceratodus-Funde. II. Wirbeltier-Reste der BaharõÃje-Stufe (unterstes Cenoman). Ergebnisse der Forschungsreisen Prof. E. Stromers in den WuÈsten AÈ gyptens." *Abhandlungen der Bayerischen Akademie der Wissenschaften, Mathematisch-naturwissenschaftliche*:5–32.

Philippe, M., G. Cuny, M. Bamford, E. Jaillard, G. Barale, B. Gomez, M. Ouaja, F. Thévenard, M. Thiébaut, and P. V. Sengbusch. 2003. "The Palaeoxylological Record of *Metapodocarpoxylon libanoticum* (Edwards) Dupéron-Laudoueneix et Pons and the Gondwana Late Jurassic-Early Cretaceous Continental Biogeography." *Journal of Biogeography* 30:389–400.

Pittet, F., L. Cavin, and F. J. Poyato-Ariza. 2010. "A New Ostariophysan Fish from the Early Late Cretaceous (Cenomanian) of SE Morocco, with a Discussion of Its Phylogenetic Relationships." In *A Comprehensive Review of Gonorynchiforms and of Ostariophysan Relationships*, edited by T. Grande, F. J. Poyato-Ariza, and R. Diogo, 332–355. Boca Raton, FL: CRC Press.

Piveteau, J. 1936. "Une forme ancestrale des amphibiens anoures dans le Trias inférieur de Madagascar." *Comptes Rendus hebdomadaires des séances de l'Académie des Sciences* 202:1607–1608.

Plant List. 2013. "*Tristemma intermedium*." *Theplantlist.org*. Accessed July 24, 2020. http://www.theplantlist.org/tpl1.1/record/ifn-80847.

Poinar, G. O. 2009. "Description of an Early Cretaceous Termite (Isoptera: Kalotermitidae) and Its Associated Intestinal Protozoa, with Comments on Their Co-evolution." *Parasites & Vectors* 2:1–25.

Pol, D., and O. W. M. Rauhut. 2012. "A Middle Jurassic Abelisaurid from Patagonia and the Early Diversification of Theropod Dinosaurs." *Proceedings of the Royal Society B: Biological Sciences* 279(1741):3170–3175.

Pole, M. 2014. "*Cladophlebis*: New Zealand's Mesozoic Weed." *Mikepole*. June 22, 2014. https://mikepole.wordpress.com/2014/06/22/cladophlebis-new-zealands-mesozoic-weed/.

Poore, G. C. B., S. T. Ahyong, and J. Taylor. 2011. *The Biology of Squat Lobsters*. Canberra: CSIRO.

Porchetti, S. D., U. Nicosia, A. Biava, and S. Maganuco. 2011. "New Abelisaurid Material from the Upper Cretaceous (Cenomanian) of Morocco." *Rivista Italiana di Paleontologia e Stratigrafia* 117(3):463–472.

Poulsen, C. J., C. R. Tabor, and J. D. White. 2015. "Long-Term Climate Forcing by Atmospheric Oxygen Concentrations." *Science* 348(6240):1238–1241.

Powell, J. H. 1989. "Stratigraphy and Sedimentation of the Phanerozoic Rocks in Central and South Jordan. Pt. B: Kurnub, Ajlun and Belqa Groups." *National Resources Authority, Geological Bulletin*, no. 11:1–130.

Poyato-Ariza, F. J. 2013. "*Sylvienodus*: A New Replacement Genus for the Cretaceous Pycnodontiform Fish." *Pycnodus" laveirensis*." Comptes Rendus Palevol 12(2):91–100.

Poyato-Ariza, F. J., and S. Wenz. 2002. "A New Insight into Pycnodontiform Fishes." *Geodiversitas* 24(1):139–248.

Prasad, V., C. A. E. Strömberg, H. Alimohammadian, and A. Sahn. 2005. "Dinosaur Coprolites and the Early Evolution of Grasses and Grazers." *Science* 310(5751):1177–1180.

Prince, H. C. 1997. *Wetlands of the American Midwest: A Historical Geography of Changing Attitudes*. Chicago: University of Chicago Press.

Quammen, D. 1998. "Planet of Weeds." *Harper's Magazine*, October:57–70.

Rage J-C., and H. Cappetta. 2002. "Vertebrates from the Cenomanian, and the Geological Age of the Draa Ubari Fauna (Libya)." *Annales de Paléontologie* 88(2):79–84.

Rage, J-C., and D. Dutheil. 2008. "Amphibians and Squamates from the Cretaceous (Cenomanian) of Morocco." *Palaeontographica Abteilung A* 285(1–3): 1–22.

Rage, J-C., and F. Escuillie. 2003. "The Cenomanian: Stage of Hindlimbed Snakes." *Notebooks on Geology* 24:1–11.

Rage, J-C., R. Vullo, and D. Neraudeau. 2016. "The Mid-Cretaceous Snake *Simoliophis rochebrunei* Sauvage, 1880 Squamata: Ophidia) from Its Type Area (Charentes, Southwestern France): Redescription, Distribution, and Palaeoecology." *Cretaceous Research* 58:234–53.

Rauhut, O. W. M. 1995. "Zur systematischen Stellung der afrikanischen Theropoden *Carcharodontosaurus* Stromer 1931 und *Bahariasaurus* Stromer 1934." *Berliner Geowissenschaftliche Abhandlungen* E:357–75.

Rauhut, O. W. M., and M. T. Carrano. 2016. "The Theropod Dinosaur *Elaphrosaurus bambergi* Janensch, 1920, from the Late Jurassic of Tendaguru,

Tanzania." *Zoological Journal of the Linnean Society* 178(3):546–610.
Rauhut, O. W. M., and A. Lopez-Arbarello. 2009. "Considerations on the Age of the Tiouaren Formation (Iullemmeden Basin, Niger, Africa): Implications for Gondwanan Mesozoic Terrestrial Vertebrate Faunas." *Palaeogeography, Palaeoclimatology, Palaeoecology* 271(3-4):259–267.
Rauhut, O. W. M., and C. Werner. 1995. "First Record of the Family Dromaeosauridae (Dinosauria: Theropoda) in the Cretaceous of Gondwana (Wadi Milk Formation, Northern Sudan)." *Palaontologische Zeitschriften* 69(3):475–489.
Rees, J., and C. J. Underwood. 2002. "The Status of the Shark Genus *Lissodus* Brough, 1935, and the Position of Nominal *Lissodus* Species within the Hybodontoidea (slachii)." *Journal of Vertebrate Paleontology* 22(3):471–479.
Rego, A. I., and D. C. Evans. 2017. "A New Species of Notosuchian Crocodyliform from the Late Cretaceous of Morocco." Abstract presented at the 77th Annual Meeting of the Society of Vertebrate Paleontology, Calgary.
Rehn, A. C. 2003. "Phylogenetic Analysis of Higher-Level Relationships of Odonata." *Systematic Entomology* 28(2):181–240.
Reimchen, T. E. 2001. "Salmon Nutrients, Nitrogen Isotopes and Coastal Forests." *Ecoforestry* 2001 fall edition:13–16.
Reineck, H. E., and F. Wunderlich. 1968. "Classification and Origin of Flaser and Lenticular Bedding." *Sedimentology* 11(1-2):99–104.
Reitner, J. 1992. "Coralline Spongien. Der Versuch einer phylogenetisch-taxonomischen Analyse." *Berliner Geowissenschaftliche Abhandlungen Reihe E (Paläobiologie)*, 1–352. Berlin: Fachbereich Geowiss.
Renema, W., D. R. Bellwood, J. C. Braga, K. Bromfield, R. Hall, K. G. Johnson, P. Lunt, C. P. Meyer, L. B. McMonagle, R. J. Morley, A. O'dea, J. A. Todd, F. P. Wesselingh, M. E. J. Wilson, and J. M. Pandolfi. 2008. "Hopping Hotspots: Global Shifts in Marine Biodiversity." *Science* 321(5889):654–657.
Reyment, R. A. 1980. "Paleo-oceanology and Paleobiogeography of the Cretaceous South Atlantic Ocean." *Oceanologica Acta* 3:127–34.
Richer de Forges, B. 2006. "Découverte en mer du Corail d'une deuxième espèce de glyphéide (Crustacea, Decapoda, Glypheoidea)." *Zoosystema* 28(1):17–29.
Richter, M., and M. Smith. 1995. "A Microstructural Study of the Ganoine Tissue of Selected Lower Vertebrates." *Zoological Journal of the Linnean Society* 114(2):173–212.
Richter, U., A. Mudroch, and G. L. Buckley. 2012. "Isolated Theropod Teeth from the Kem Kem Beds (Early Cenomanian) near Taouz, Morocco." *Paläontologische Zeitschrift* 87(2):291–309.
Rieppel, O. 2002. "Feeding Mechanisms in Triassic Stem-Group Sauropterygians: The Anatomy of a Successful Invasion of Mesozoic Seas." *Zoological Journal of the Linnean Society* 135(1):33–63.
Rieppel, O., and J. J. Head. 2004. "New Specimens of the Fossil Snake Genus *Eupodophis* Rage & Escuillie, from the Cenomanian (Late Cretaceous) of Lebanon." *Memorie della Societa Italiana di Scienze Naturali e del Museo Civico di Storia Naturale di Milano* 32:3–24.
Riff, D., A. W. A. Kellner, B. Mader, and D. Russell. 2004. "On the Occurrence of an Avian Vertebra in Cretaceous Strata of Morocco, Africa." *Anais da Academia Brasileira de Ciências* 74(2):367–368.
Ripple, W. J., and R. L. Beschta. 2012. "Trophic Cascades in Yellowstone: The First 15 Years after Wolf Reintroduction." *Biological Conservation* 145(1):205–213.
Robertson, A. I., and P. A. Daniel. 1989. "The Influence of Crabs on Litter Processing in High Intertidal Mangrove Forests in Tropical Australia." *Oecologia* 78(2):191–198.
Robertson Research International and Associated Research Consultants. 1982. *Petroleum Potential Evaluation, Western Desert*. Vol. 8. Cairo: Associated Research Consultants, in association with Scott Pickford and Associates Limited and Energy Resource Consultants Limited, for the Egyptian General Petroleum Corporation.
Robin, C., E. Ostanciaux, F. Guillocheau, L. Husson, and G. Trotin. 2013. "Cenomanian Sea Level High: A Global Signal Modified by Long Wavelength Deformations of Mantellic Origin." Conference Papers. EGU General Assembly 2013, Vienna.
Robins, C. R., and G. C. Ray. 1986. *A Field Guide to Atlantic Coast Fishes of North America*. Boston, MA: Houghton Mifflin.
Rocci, G. 1965. "Essai d'interpretation de mesures geochronologiques. La structure de l'Ouest africain." *Sciences de la terra* 3–8:463–479.
Rodrigues, T., and A. W. A. Kellner. 2008. "Review of the Pterodactyloid Pterosaur *Coloborhynchus*." In *Flugsaurier: Pterosaur Papers in Honor of Peter Wellnhofer*, edited by D. W. E. Hone and E. Buffetaut, 28. Zitteliana B, Munich.
Rodrigues, T., A. W. A. Kellner, B. J. Mader, and D. A. Russell. 2011. "New Pterosaur Specimens from the Kem Kem Beds (Upper Cretaceous, Cenomanian) of Morocco." *Rivista Italiana di Paleontologiae Stratigrafia* 117(1):149–160.

Roger, J. 1946. "Les invertébrés des couches à poissons du Crétacé supérieur du Liban." *Mémoires de la Société géologique de France*, no. 51, 1–92.
Romer, A. S. 1956. *Osteology of the Reptiles*. Malabar, FL: Krieger Publishing.
Romer, A. S., and T. S. Parsons. 1977. *The Vertebrate Body*. Philadelphia, PA: Holt-Saunders International.
Roney, R. O. 2013. Paleobiogeographical Variation of Cretaceous *Mecaster batnensis* and *Mecaster fourneli* (Echinoidea: Spatangoida). Master's thesis, University of Tennessee.
Rosen, D. E. 1985. "An Essay on Euteleostean Classification." *American Museum Novitates*, no. 2827, 1–57.
Rosen, D. E., P. L. Forey, B. G. Gardiner, and C. Patterson. 1981. "Lungfishes, Tetrapods, Palaeontolgy and Plesiomorphy." *Bulletin of the American Museum of Natural History* 167:159–276.
Rothwell, G. 2020. Personal communications, email to Jamale Ijouiher.
Rothwell, G., and R. Stockey. 2018. "Penetrating the Perplexing Upper Cretaceous *Parataxodium plexus* from the Price Creek Formation of North Slope Alaska." Abstract presented at Botany Conference, Rochester, MN.
Russell, D. A. 1995. "China and the Lost Worlds of the Dinosaurian Era." *Historical Biology* 10(1):3–12.
———. 1996. "Isolated Dinosaur Bones from the Middle Cretaceous of the Tafilalt, Morocco." *Bulletin du Musém National d'Histoire Naturelle, Paris* 4 (18): 349–402.
———. 2009. *Islands in the Cosmos: The Evolution of Life on Land*. Bloomington: Indiana University Press.
Russell, D. A., and M. A. Paesler. 2003." Environments of Mid-Cretaceous Saharan Dinosaurs." *Cretaceous Research* 24(5):569–88.
Sachs, S. 2014. "Research Interests." *Art by Joschua Knüppe*. Facebook. Accessed April 25, 2019. https://www.facebook.com/pages/Art-by-Joschua-Kn%C3%BCppe/732426953481529?ref_type=bookmark.
Said, R. 1962. *The Geology of Egypt*. Amsterdam: Elsevier Publishing Company.
———. 1971. "Explanatory Notes to Accompany the Geological Map of Egypt." Paper no. 56, Geological Survey of Egypt, Cairo.
———. 1990. "Cretaceous Paleogeographic Maps." In *The Geology of Egypt*, edited by R. Said, 439–449. Amsterdam: Elsevier Publishing Company.
Salaj, J. 1979. *Geological Map of Libya, Explanatory Booklet*. Sheet Al Qaryrat Al Gharbiya, NH 33–5, scale I:250 000. Tripoli: Industrial Research Center.
Salem, B. S., M. C. Lamanna, P. O'Connor, G. M. El-Qot, F. Shaker, W. A. Thabet, S. El-Sayed, and H. M. Sallam. 2021. "First Definitive Record of Abelisauridae from Bahariya Formation, Bahariya Oasis: Western Desert of Egypt Increases Diversity of Large-Bodied Theropods in the Middle Cretaceous of North Eastern Africa." Talk given at the 81st Annual Meeting of the Society of Vertebrate Paleontology, Virtual Meeting.
Sales, M. A. F., M. B. Lacerda, B. L. D. Horn, I. A. P. de Oliveira, and C. L. Schultz. 2016. "The '?' of the Matter: Testing the Relationship between Paleoenvironments and Three Theropod Clades." *PLoS ONE* 11 (2):e0147031.
Sales, M. A. F, and C. L. Schultz. 2017. "Spinosaur Taxonomy and Evolution of Craniodental Features: Evidence from Brazil." *PLoS ONE* 12 (11):e0187070.
Sallam, H. M., B. S. Salem, W. A. Thabet, S. El-Sayed, and M. C. Lamanna. 2021. "The First Pterosaur from the Upper Cretaceous (Lower Cenomanian) Bahariya Formation, Bahariya Oasis, Western Desert of Egypt." Poster presentation at the 81st Annual Meeting of the Society of Vertebrate Paleontology, Virtual Meeting.
Sampson, S. D., and L. M. Witmer. 2007. "Craniofacial Anatomy of *Majungasaurus crenatissimus* (Theropoda: Abelisauridae) from the Late Cretaceous of Madagascar." In *Majungasaurus crenatissimus (Theropoda: Abelisauridae) from the Late Cretaceous of Madagascar*, edited by S. F. D. Sampson and D. W. Krause, memoir 8. Society of Vertebrate Paleontology, McLean, VA .
Sánchez-Hernández, B., and M. J. Benton. 2014. "Filling the Ceratosaur Gap: A New Ceratosaurian Theropod from the Early Cretaceous of Spain." *Acta Palaeontologica Polonica* 59(3):581–600.
Sandin, S. A., J. E. Smith, E. E. DeMartini, E A. Dinsdale, S. D. Donner, A. M. Friedlander, T. Konotchick, M. Malay, J. E. Maragos, D. Obura, O. Pantos, G. Paulay, M. Richie, F. Rohwer, R. E. Schroeder, S. Walsh, J. B. C. Jackson, N. Knowlton, and E. Sala. 2008. "Baselines and Degradation of Coral Reefs in the Northern Line Islands." *PLoS ONE* 3 (2):e1548.
Sasekumar, A. 1974. "Distribution of Macrofauna on a Malayan Mangrove Shore." *Journal of Animal Ecology* 43(1):51–69.
Sauvage, H. E. 1880. "Sur l'existence d'un reptile du type ophidien dans les couches à Ostrea columba des Charentes." *Comptes rendus hebdomadaires des séances de l'Académie des Sciences* 91:671–672.
Scanlon, J. D., M. S. Y. Lee, M. W. Caldwell, and R. Shiner. 1999. "The Palaeoecology of the Primitive Snake *Pachyrhachis*." *Historical Biology* 13:127–52.

Schaal, S. 1984. "Oberkretazische Osteichthyes (Knochenfische) aus dem Bereich von Bahariya und Kharga, Aegypten, und ihre Aussagen zur Palaekologie und Stratigraphie." *Berliner geowissenschaftliche Abhandlungen. Reihe A. Berlin/West*:1–79.

Scheyer, T. M., O. A. Aguilera, M. Delfino, D. C. Fortier, A. A. Carlini, R. Sánchez, J. D. Carrillo-Briceño, L. Quiroz, and M. R. Sánchez-Villagra. 2013. "Crocodylian Diversity Peak and Extinction in the Late Cenozoic of the Northern Neotropics." *Nature Communications* 4:1–9.

Schlanger, S. O., M. A. Arthur, H. C. Jenkyns, and P. A. Scholle. 1987. "The Cenomanian-Turonian Oceanic Anoxic Event I: Stratigraphy and Distribution of Organic Carbon-Rich Beds and the Marine d13C Excursion." Special publications no. 26, Geological Society, London.

Schlumberger, S. A. 1995. Well Evaluation Conference—Egypt. Houston, TX:1–87.

Schöbel, J. 1975. "Ammoniten der Familie Vascoceratidae aus dem Unterturon des Damergou-Gebietes, Republique du Niger." *Special Publications of the Palaeontological Institution of the University of Uppsala*, no. 3, 125–34.

Schram, F. R. 1970. "Isopod from the Pennsylvanian of Illinois." *Science* 169:854–855.

Schram, S. R., and S. T. Ahyong. 2002. "The Higher Affinities of *Neoglyphea inopinata* in Particular and the Glypheoidea (Decapoda, Reptantia) in General." *Crustaceana* 75(3-4):629–35.

Schrank, E. 1990. "Palynology of the Clastic Cretaceous Sediments between Dongola and Wadi Muqaddam, Northern Sudan." *Berliner Geowissenschaftliche Abhandlungen—Reihe* A 120:149–68.

Schulze, F., Z. Lewy, J. Kuss, and A. Gharaibeh. 2003. "Cenomanian/Turonian Carbonate Platform Deposits in West Central Jordan." *International Journal of Earth Sciences* 92(4):641–660.

Schweitzer, C. E., H. Karasawa, J. Luque, and R. Feldmann. 2016. "Phylogeny and Classification of Necrocarcinoidea Förster, 1968 (Brachyura: Raninoida) with Description of Two New Genera." *Journal of Crustacean Biology* 36(3):338–372.

Schweitzer, C. E., K. J. Lacovara, J. B. Smith, M. C. Lamanna, M. A. Lyon, and Y. Attia. 2003. "Mangrove-Dwelling Crabs (Decapoda: Brachyura: Necrocarcinidae) Associated with Dinosaurs from the Upper Cretaceous (Cenomanian) of Egypt." *Journal of Paleontology* 77(5):888–894.

Schwimmer, D. R., J. D. Stewart, and G. D. Williams. 1997. "Scavenging by Sharks of the Genus *Squalicorax* in the Late Cretaceous of North America." *PALAIOS* 12:71–83.

Scott, R. F. [1912] 1946. *Scott's Last Expedition: The Personal Journals of Captain R. F. Scott on His Journey to the South Pole*. London: J. Murray.

Seebacher, F. 2001. "A New Method to Calculate Allometric Length-Mass Relationships of Dinosaurs." *Journal of Vertebrate Paleontology* 21(1):51–60.

Sendera, L. M., U. Villanueva-Amadoz, D. Pons, J. B. Diez, M. García-Ávila, and J. Ferrer. 2015. "New Reconstruction of *Weichselia reticulata* (Stokes et Webb) Fontaine in Ward Emend. Alvin, 1971 Based on Fertile Remains from the Middle Albian of Spain." *Historical Biology* 27(3-4):460–468.

Senter, P. 2010. "Vestigial Skeletal Structures in Dinosaurs." *Journal of Zoology* 280(1):60–71.

Sereno, P. C. 2010. "Noasaurid (Theropoda: Abelisauroidea) Skeleton from Africa Shows Derived Skeletal Proportions and Functions." Abstract presented at the 70th Anniversary Meeting of the Society of Vertebrate Paleontology. Pittsburgh.

Sereno, P. C., A. L. Beck, D. B. Dutheil, B. Gado, H. C. E. Larsson, G. H. Lyon, J. D. Marcot, O. W. M. Rauhut, R. W. Sadleir, C. A. Sidor, D. J. Varricchio, G. P. Wilson and J. A. Wilson. 1998. "A Long-Snouted Predatory Dinosaur from Africa and the Evolution of Spinosaurids." *Science* 282(5392):1298–1302.

Sereno, P. C., A. L. Beck, D. B. Dutheil, H. C. E. Larsson, G. H. Lyon, B. Moussa, R. W. Sadleir, C. A. Sidor, D. J. Varricchio, G. P. Wilson and J. A. Wilson. 1999. "Cretaceous Sauropods from the Sahara and the Uneven Rate of Skeletal Evolution among Dinosaurs." *Science* 286(5443):1342–1347.

Sereno, P. C., D. B. Dutheil, M. Iarochene, H. C. E. Larsson, G. H. Lyon, P. M. Magwene, C. A. Sidor, D. J. Varricchio and J. A. Wilson. 1996. "Predatory Dinosaurs from the Sahara and Late Cretaceous Faunal Differentiation." *Science* 272(5264):986-991.

Sereno, P. C., F. E. Fish, and N. Myhrvold. 2015. "Swimming Function in the Cretaceous Giant *Spinosaurus aegyptiacus* Based on the Kinematics of Undulatory Swimming in the American Alligator." Abstract presented at the 75th Annual Meeting of the Society of Vertebrate Paleontology, Dallas.

Sereno, P. C., and H. C. E. Larsson. 2009. "Cretaceous Crocodyliforms from the Sahara." *ZooKeys* 28:1–143.

Sereno, P. C., J. A. Wilson, L. M. Witmer, J. A. Whitlock, A. Maga, O. Ide, and T. A. Rowe. 2007. "Structural Extremes in a Cretaceous Dinosaur." *PLoS ONE* 2 (11):e1230.

Shahin, A., and S. El Baz. 2014. "Paleoenvironmental Changes of the Cenomanian-Early Turonian Shallow Marine Carbonate Platform Succession in West Central Sinai, Egypt." *Revue de Paléobiologie, Genève* 33(2):561–581.

Shaw, E. 2016. "When Crocodiles Go to Sea." *Earth Touch News Network*. https://www.earthtouchnews.com/natural-world/animal-behaviour/when-crocodiles-go-to-sea/.

Shears-Ozeki, C. 2017. "Bored Bones from the Terrestrial Middle Cretaceous Kem-Kem Beds of Southeast Morocco." *PeerJ Preprints* 5:e3200v2. https://doi.org/10.7287/peerj.preprints.3200v2.

Sheaves, M. 2005. "Nature and Consequences of Biological Connectivity in Mangrove Systems." *Marine Ecology Progress* 302:293–305.

Shen, L., X-Y. Chen, X. Zhang, Y-Y. Li, C-X. Fu, and Y-X. Qiu. 2005. "Genetic Variation of Ginkgo biloba L. (Ginkgoaceae) Based on cpDNA PCR-RFLPs: Inference of Glacial Refugia." *Heredity* 94:396–401.

Shibamoto, T., and F. L. Bjeldanes. 2009. *Introduction to Food Toxicology*. 2nd ed. Cambridge, MA: Elsevier Academic Press.

Shimada, K. 2006. "Marine Vertebrates from the Blue Hill Shale Member of the Carlile Shale (Upper Cretaceous: Middle Turonian) in Kansas." *Transactions of the Kansas Academy of Science* 119(2):211–221.

———. 2007. "Skeletal and Dental Anatomy of Lamniform Shark *Cretalamna appendiculata*, from the Upper Cretaceous Niobrara Chalk of Kansas." *Journal of Vertebrate Paleontology* 27(3):584–602.

Shimada, K., and D. J. Cicimurri. 2005. "Skeletal Anatomy of the Late Cretaceous Shark, *Squalicorax* (Neoselachii: Anacoracidae)." *Paläontologische Zeitschrift* 79:241–61.

Shimada, K., M. Everhart, B. Reilly, and C. Rigsby. 2011. "First Associated Specimen of the Late Cretaceous Shark, *Cretodus* (Elasmobranchii: Lamniforms)." Poster presentation at the 71st Annual Meeting of the Society of Vertebrate Paleontology, Las Vegas.

Shyamaprassad, R. 2004. "Water Storage in *Terminalia tomentosa*." *Current Science* 87(4):416–417.

Silverman, D. P. 2003. *Ancient Egypt*. Oxford: Oxford University Press.

Singer, T. 2015. "Jurapark on the Trail of New Dinosaurs from Morocco." *Jurapark*. Accessed April 25, 2019. http://jurapark.pl/jurapark-na-tropie-nowych-dinozaurow-z-maroka/.

Sims, D. W. 2000. "Filter-Feeding and Cruising Swimming Speeds of Basking Sharks Compared with Optimal Models: They Filter-Feed Slower Than Predicted for Their Size." *Journal of Experimental Marine Biology & Ecology* 249(1):65–76.

Sinton, C. W., and R. A. Duncan. 1997. "Potential Links between Ocean Plateau Volcanism and Global Ocean Anoxia at the Cenomanian-Turonian Boundary." *Economic Geology* 92(7-8):836–842.

Slaughter, B. H., and J. T. Thurmond. 1974. "A Lower Cenomanian (Cretaceous) Ichthyofauna from the Bahariya Formation of Egypt." *Annals of the Geological Survey of Egypt* 4:25–40.

Smith, C. L. 1997. *National Audubon Society Field Guide to Tropical Marine Fishes of the Caribbean, the Gulf of Mexico, Florida, the Bahamas, and Bermuda*. New York: Alfred A. Knopf.

Smith, J. B., B. S. Grandstaff, and M. S. Abdel-Ghani. 2006. "Microstructure of Polypterid Scales (Osteichthyes: Actinopterygii: Polypteridae) from the Upper Cretaceous Bahariya Formation, Bahariya Oasis, Egypt." *Journal of Paleontology* 80(6):1179–1185.

Smith, J. B., M. C. Lamanna, Y. Attiya, and K. J. Lacovara. 2001a. "On a Small Predatory Dinosaur from the Late Cretaceous of Egypt." *Geological Society of America Abstracts with Programs*, 33(6): 389.

Smith, J. B., M. C. Lamanna, H. Mayr, and K. J. Lacovara. 2006. "New Information Regarding the Holotype of *Spinosaurus aegyptiacus*, Stromer, 1915." *Journal of Paleontology* 80(2):400–406.

Smith, J. B., M. C. Lamanna, K. J. Lacovara, P. Dodson, J. R. Smith, J. C. Pool, R. Giegengack and Yousry Attis. 2001b. "A Giant Sauropod Dinosaur from an Upper Cretaceous Mangrove Deposit in Egypt." *Science* 292(5522):1704–1706.

Smith, J. W., and J. V. Merriner. 1985. "Food Habits and Feeding Behaviour of the Cownose Ray, *Rhinoptera bonasus*, in Lower Chesapeake Bay." *Estuaries* 8(3):305–310.

Smith, M. M., A. Riley, G. J. Fraser, C. Underwood, M. Welten, J. Kriwet, C. Pfaff, and Z. Johanson. 2015. "Early Development of Rostrum Saw-Teeth in a Fossil Ray Tests Classical Theories of the Evolution of Vertebrate Dentitions." *Proceedings of the Royal Society of London B: Biological Science* 282(1816):1–8.

Smith, R. 2020. "A New and Unique Kem Kem Beds (Mid Cretaceous) Locality near the Oasis of Tarda, South Eastern Morocco." *Palaeontological Association Newsletter, N* 103:24.

Smith, R. E., D. M. Martill, A. Kao, S. Zouhri, and N. R. Longrich. 2020. "A Long-Billed, Possible Probe-Feeding Pterosaur (Pterodactyloidea: ?Azhdarchoidea) from the Mid-Cretaceous of Morocco, North Africa." *Cretaceous Research* 118:104–643.

Smyth, R. S. H., N. Ibrahim, A. Kao, and D. M. Martill. 2020. "Abelisauroid Cervical Vertebrae from the Cretaceous Kem Kem Beds of Southern Morocco and a Review of Kem Kem Abelisauroids." *Cretaceous Research* 108:104–330.

Smyth, R. S. H., N. Ibrahim, and D. M. Martill. 2020. "*Sigilmassasaurus* Is *Spinosaurus*: A Reappraisal of African Spinosaurines." *Cretaceous Research* 114:104–520.

Sokolov, M. 1965. "Teeth Evolution of Some Genera of Cretaceous Sharks and Reconstruction of Their Dentition." *Moskovkoe Obshchestvo Ispytatelie Prirody, Biulleten Otodel Geologicheskii* 40:133–34.

Squires, D. F. 1958. "The Cretaceous and Tertiary Corals of New Zealand." *New Zealand Geological Survey Paleontological Bulletin*, no. 29, 1–107.

Stafford-Deitsch, J. 1996. *Mangrove: The Forgotten Habitat*. London: Immel.

Stanley, S. M. 1974. "What Has Happened to the Articulate Brachiopods?" *Geological Society of America Abstracts with Programs* 6(7): 966–967.

———. 2008. "Predation Defeats Competition on the Seafloor." *Paleobiology* 34:1–21.

Stauffer, J. 2007. *Fishes of West Virginia*. Philadelphia, PA: Academy of Natural Sciences.

Sternes, P. C., and K. Shimada. 2019. "Paleobiology of the Late Cretaceous Sclerorhynchid Sawfish, *Ischyrhiza mira* (Elasmobranchi: Rajiformes), from North America Based on New Anatomical Data." *Historical Biology* 31(10):1323–1340.

Stevens, J., and P. R. Last. 1998. *Encyclopaedia of Fishes*. Edited by J. R. Paxton and W. N. Eschmeyer. San Diego: Academic.

Stevens, K.A. 2002. "DinoMorph: Parametric Modeling of Skeletal Structures." *Senckenbergiana lethaea* 82(1):23-34.

Stevens, K. A., and J. M. Parrish. 1999. "Neck Posture and Feeding Habits of Two Jurassic Sauropod Dinosaurs." *Science* 284:798-800.

Stewart, A. C. 1898. *Fossil Plants*. Vol. 1. Cambridge: Cambridge University Press.

Stilwell, J. D., and R. A. Henderson. 2002. "Description and Paleobiogeographic Significance of a Rare Cenomanian Molluscan Faunule from Bathurst Island, Northern Australia." *Journal of Paleontology* 76(3):447–471.

Stock, T. 2012. "It's a Huge Ornithomimus! It's a Giant *Effigia*! No, It's a *Deltadromeus*!" *Mesozoic Archives* (blog), April 3, 2012. http://tyranno-teen.blogspot.co.uk/2012/04/its-huge-ornithomimimus-its-giant.html.

Stromer, E. 1914. "Ergebnisse der Forschungsreisen Prof. E. Stromers in den Wüsten Ägyptens. II. Wirbeltier- Reste der Baharîje-Stufe (unterstes Cenoman). 1. Einleitung und 2. *Libycosuchus*." *Abhandlungen der Königlich Bayerischen Akademie der Wissenschaften, Mathematisch-Physikalische Klasse* 27:1–16.

———. 1915. "Ergebnisse der Forschungsreisen Prof. E. Stromer in den Wüsten Agyptens. II. Wirbeltier- Resteder Bahar.je-Stufe (unterstes Cenoman). 3. Das Original des Theropoden *Spinosaurus aegyptiacus* nov. gen., nov. spec." *Abhandlungen der Königlich Bayerischen Akademie der Wissenschaften, Mathematisch-Physikalische* 28:1–32.

Stromer, E. 1917. "Ergebnisse der Forschungreisen Prof. E. Stromers in den Wüsten Ägyptens. II. Wirbeltier-Reste der Baharîje-Stufe (unterstes Cenoman). 4. Die Säge des Pristiden *Onchopristis numidus* Haug sp. und über die Sägen der Sägehaie." *Abhandlungen der Königlich Bayerischen Akademie der Wissenschaften, Mathematisch-physikalische* 18:1–28.

———. 1925. "Ergebnisse der Forschungsreisen Prof. E. Stromers in den Wüsten Ägyptens. II. Wirbeltier- Reste der Baharije-Stufe (unterstes Cenoman). 7. *Stomatosuchus inermis* Stromer, ein schwach bezahnter Krokodilier und 8. Ein Skelettrest des Pristiden *Onchopristis numidus* Huag sp." *Abhandlungen der königlichen Bayerischen Akademie der Wissenschaften, Mathematisch Physikalische Klasse* 30:1–22.

———. 1927. "Ergebnisse der Forschungsreisen Prof. E. Stromers in den Wüsten Ägypten. II. Wirbeltier-Reste der Baharije-Stufe (Unterstes Cenoman). 9. Die Plagiostomen mit einem Anhang über Käno-undmesozoische Rückenflossenstacheln von Elasmobranchiern." *Abhandlungen der Bayerischen Akademie der Wissenschaften. Mathematisch- naturwissenschaftliche Abteilung, Neue Funde* 31:1–64.

———. 1931. "Ergebnisse der Forschungsreisen Prof. E. Stromers in den Wüsten Ägyptens. II. Wirbeltierreste der Baharje-Stufe (unterstes Cenoman). 10. Ein Skelett- Rest von *Carcharodontosaurus* nov. gen." *Abhandlungen der Bayerischen Akademie der Wissenschaften, athematisch-Naturwissenschaftliche Abteilung, Neue Folge* 9:1–23.

———. 1933. "Ergebnisse der Forschungsreisen Prof. E. Stromers in den Wüsten Ägyptens. II. Wirbeltierreste der Baharîje-Stufe (unterstes Cenoman). 12. Die procölen Crocodilia." *Abhandlungen der Bayerischen Akademie der Wissenschaften Mathematisch- naturwissenschaftliche Abteilung, Neue Folge* 15:1–31.

———. 1934a. "Ergebnisse der Forschungsreisen Prof. E. Stromers in den Wüsten Ägyptens. II. Wirbeltier-Reste der Baharije-Stufe (unterstes Cenoman). 13. Dinosauria." *Abhandlungen der Bayerischen Akademie der Wissenschaften Mathematisch- Naturwissenschaftliche Abteilung, Neue Folge* 22:1–79.

———. 1934b. "Erbebnisse der Forschungsreisen Prof. E. Stromers in den Wusten Agyptens. II. Wirbeltierreste der Baharije-Stufe, Unterestes Cenoman (14:Testudinata.)" *Abhandlungen der Bayerischen Akademieder Wissenschaften Mathematisch- n aturwissenschafliche Abtelunge Neue Folge, heft* 25:4–26.

———. 1935. "Ergebnisse der Forschungsreisen Prof. E. Stromers in den Wusten Agyptens: II.

Wirbeltierreste der Baharije-Stufe (unterstes Cenoman), 15. Plesiosauria." *Abhandlungen der Bayerischen Akademieder Wissenschaften Mathematisch-naturwissenschafliche, Heft.* 26:1–55.

———. 1936. "Ergebnisse der Forschungsreisen Prof. E. Stromers in den Wüsten Ägyptens. VII. Baharije-Kessel und-Stufe mit deren Fauna und Flora. Eine ergänzende Zusammenfassung." *Abhandlungen der Bayerischen Akademie der Wissenschaften, Mathematisch- naturwissenschaftliche Abteilung, Neue Folge* 33:1–102.

Stromer, E., and W. Weiler. 1930. "Ergebnisse der Forschungsreisen Prof. E. Stromers in den Wüsten Ägyptens. VI. Beschreibung von Wirbeltier-Resten aus dem nubischen Sandsteine Oberägyptens und aus ägyptischen Phosphaten nebst Bemerkungen über die Geologie der Umgegend von Mahamîd in Oberägypten." *Abhandlungen der Bayerischen Akademie der Wissenschaften Mathematisch-naturwissenschaftliche Abteilung, Neue Folge* 7:1–42.

Sues, H. D., E. Frey, D. M. Martill, and D. M. Scott. 2001. "*Irritator challengeri*, a Spinosaurid (Dinosauria: Theropoda) from the Lower Cretaceous of Brazil." *Journal of Vertebrate Paleontology* 22(3):535–547.

Sun, G., and D. L. Dilcher. 2002. "Early Angiosperms from the Lower Cretaceous of Jixi, Eastern Heilongjiang, China." *Review of Palaeobotany and Palynology* 121(2):91–112.

Tabaste, N. 1963. "Études de restes de poissons du Crétacé saharien, in Mélanges ichthyologiques dédiés à la mémoire d'Achille Valenciennes (1794–1865)." *Mémoires de l'Institut français d'Afrique noire* 68:475–485.

Tahoun, S. S., W. A. Makled, and T. F. Mostafa. 2013. "Stratigraphic Distribution of the Palynomorphs and the Particulate Organic Matter in Subsurface Lower/Middle Cretaceous Deposits, Western Desert of Egypt: Palynological and Geochemical Approach." *Egyptian Journal of Petroleum* 22(3):435–449.

Takashima, R., H. Nishi, B. T. Huber, and R. M. Leckie. 2006. "Greenhouse World and the Mesozoic Ocean." *Oceanography* 19(4):82–92.

Takemoto, H., Y. Kawamoto, and T. Furuichi. 2015. "How Did Bonobos Come to Range South of the Congo River? Reconsideration of the Divergence of *Pan paniscus* from Other Pan Populations." *Evolutionary Anthropology* 24(5):170–184.

Tan, C. G. S., and P. K. L. Ng. 1994. "An Annotated Checklist of Mangrove Brachyuran Crabs from Malaysia and Singapore." *Hydrobiologia* 285:75–84.

Tank, D. C. J. M Eastman, M. W. Pennell, P. S. Soltis, D. E Soltis, C. E Hinchliff, J. W Brown, E. B Sessa, and L. J Harmon. 2015. "Nested Radiations and the Pulse of Angiosperm Diversification: Increased Diversification Rates Often Follow Whole Genome Duplications." *New Phytologist* 207(2):454–467.

Tanner, L. H., and M. A. Khalifa. 2009. "Origin of Ferricretes in Fluvial-Marine Deposits of the Lower Cenomanian Bahariya Formation, Bahariya Oasis, Western Desert, Egypt." *Journal of African Earth Sciences* 56(1-4):179–189.

Taquet, P. 1976. *Géologie et paléontologie du gisement de Gadoufaoua (Aptien du Niger)*. Cahiers de Paléontologie. Paris: Centre National de la Recherche Scientifique.

Tarailo, D. A., D. Hester, and C. A. Brochu. 2017. "Oceanic Dispersal Rates within Crocodylia and Their Significance to the Development of Salt Tolerance in Crocodyloids." Abstract presented at the 77th Annual Meeting of the Society of Vertebrate Paleontology, Calgary.

Tarduno, J. A., D. B. Brinkman, P. R. Renner, D. Cottrell, H. Scherland, and P. Castillo. 1998. "Evidence for Extreme Climatic Warmth from the Late Cretaceous Arctic Vertebrates." *Science* 18(5397):2241–2243.

Taverne, L. 1977. "Ostéologie de *Clupavus maroccanus* (Crétacé supérieur du Maroc) et considérations sur la position systématique det les relations des Clupavidae au sein de l'ordre des Clupéiformes sensu stricto (Pisces, Teleostei)." *Géobios* 10:697–722.

———. 1987. "Ostéologie de *Cyranichthys ornatissimus* nov. gen. du Cénomanien du Zaïre et de *Rhynchodercetis yovanovitchi* du Cénomanien de l'Afrique du Nord. Les relations intergénériques et la position systématique de la famille néocrétacique marine des Dercetidae (Pisces, Teleostei)." Musée Royal de l'Afrique Centrale, Tervuren, Département de Géologie et Minéralogie, Leuvensesteenweg.

———. 1991. "Révision du genre *Protostomias*, Téléostéen Stomiiforme crétacique de la Mésogée eurafricaine." *Biologisch Jaarboek Dodonaea* 59:57–76.

———. 1993. "Révision de *Kermichthys daguini* (Arambourg, 1954) nov. gen., Téléosté en Salmoniforme du Crétacé de la Mésogée eurafricaine." *Biologisch Jaarboek Dodonaea* 60:76–95.

———. 1995. "Description de l'appareil de Weber du téléostéen crétacé marin *Clupavus maroccanus* et ses implications phylogénétiques." *Belgian Journal of Zoology* 125:267–282.

———. 1996. "Révision de *Tingitanichthys heterodon* (Arambourg, 1954) nov. gen. (Teleostei, Pachyrhizodontoidei) du Crétacé supérieur marin du Maroc." *Biologisch Jaarboek Dodonaea* 63:133–151.

———. 2000a. "Nouvelles données osteologique et-phylogénétiques sur *Palaeonotopterus greenwoodi*, notopteridé (Teleostei, Osteoglossomorpha) du Cénomanianinferieur continental (Crétacé) du Maroc." *Stuttgarter Beitétra¨ge zur Naturkunde B (Geologie und Palaontologie)* 293:1–24.
———. 2000b. "*Tselfatia formosa*, téléostéen marin du Crétacé (Pisces, Actinopterygii), et la position systématique des Tselfatiiformes ou Bananogmiiformes." *Geodiversitas* 22:5–22.
———. 2006. "Révision d'*Ichthyotringa africana*, poisson marin (Teleostei, Aulopiformes) du Crétacé supérieur de la Mésogée eurafricaine. Considérations sur les relations phylogénétiques du genre *Ichthyotringa*." *Belgian Journal of Zoology* 136:31–41.
———. 2008. "Considerations about the Late Cretaceous Genus Chirocentrites and Erection of the New Genus *Heckelichthys* (Teleostei, Ichthyodectiformes): A New Visit inside the Ichthyodectid Phylogeny." *Bulletin de l'Institut Royal des Sciences naturelles de Belgique* 78:209–28.
———. 2011. "Les poissons crétacés de Nardò. 33°. *Lecceclupea ehiravaensis* gen. et sp. nov. (Teleostei, Clupeidae)." *Bollettino del Museo Civico di Storia Naturale di Verona*, 35, 2011 *Geologia Paleontologia Preistoria* 35:3–17.
Taverne, L., and M. Gayet. 2005. "Phylogenetical Relationships and Paleozoogeography of the Marine Cretaceous Tselfatiiformes (Teleostei, Clupeocephala)." *Cybium* 29(1):65–87.
Taverne L., M. Layeb, Y. Layeb-Tounsi, and J. Gaudant. 2015. "*Paranursallia spinosa* N. Gen., N. Sp., a New Upper Cretaceous Pycnodontiform Fish from the Eurafrican Mesogea." *Geodiversitas* 37(2):215–227.
Tawadros, E. 2001. *Geology of Egypt and Libya*. Milton Park Oxfordshire: Taylor & Francis.
———. 2011. *Geology of North Africa*. Boca Raton, FL: CRC Press.
Taylor, M. 2008a. "Aetogate: What Professional Palaeontologists Are Saying." *Miketaylor.org*. Last modified June 6, 2008. http://www.miketaylor.org.uk/dino/nm/comments.html.
———. 2008b. "Off-Topic: We're Going to Need a Bigger Ethics Committee." *Sv-pow*. July 10, 2008. http://svpow.com/2008/07/10/off-topic-were-going-to-need-a-bigger-ethics-committee/.
Taylor, M. L., and J. H. Williams. 2009. "Consequences of Pollination Syndrome Evolution for Post-pollination Biology in an Ancient Angiosperm Family." *International Journal of Plant Sciences* 170(5):584–598.
Taylor, M. P. 2018. "*Xenoposeidon* Is the Earliest Known Rebbachisaurid Sauropod Dinosaur." *PeerJ Preprints* 5:e3415v1. https://doi.org/10.7287/peerj.preprints.3415v1.
Taylor, M. P., M. J. Wedel, and D. Naish. 2009. "Head and Neck Posture in Sauropod Dinosaurs Inferred from Extant Animals." *Acta Palaeontologica Polonica* 54(2):213–20.
Teichert, C. 1967. "Major Features of Cephalopod Evolution." In *Essays in Stratigraphy and Paleontology: Special Publication*, edited by C. Teichert and E. L. Yochelson, 162–210. Lawrence: University of Kansas Press.
Teodoridis, V., Z. Kvacek, M. Sami, and E. Martinetto. 2015. "Palaeoenvironmental analysis of the Messinian macrofossil floras of Tossignano and Monte Tondo (Vena del Gesso Basin, Romagna Apennines, northern Italy)." *Acta Musei Nationalis Pragae, Series B—Historia Naturalis* 71:249–92.
Therrien, F., and D. M. Henderson. 2007. "My Theropod Is Bigger Than Yours . . . or Not: Estimating Body Size from Skull Length in Theropods." *Journal of Vertebrate Paleontology* 27(1):108–115.
Therrien, F., D. M. Henderson, and C. B. Ruff. 2005. "Bite Me: Biomechanical Models of Theropod Mandibles and Implications for Feeding Behaviour." In *The Carnivorous Dinosaurs*, edited by K. Carpenter, 179–237. Bloomington: Indiana University Press.
Thomas, H. 2017a. "Pterosaurs of the Kem Kem Beds." *Those with Pycnofibres* (blog), November 14, 2017. https://zhejiangopterus.wordpress.com/2017/11/14/pterosaurs-of-the-kem-kem-beds/.
Thomas, H. 2017b. Personal communications, email to Jamale Ijouiher.
Thomas, N., and S. Nigam. 2018. "Twentieth-Century Climate Change over Africa: Seasonal Hydroclimate Trends and Sahara Desert Expansion." *Journal of Climate* 31(9):33-49.
Thomson, L. A. J. 2006. "Species Profiles for Pacific Island Agroforestry, ver. 1.2. *Agathis macrophylla* (Pacific kauri)." *Traditionaltree.org*. Accessed April 25, 2019. http://www.traditionaltree.org.
Titus, A. L., and M. A. Loewen. 2013. *At the Top of the Grand Staircase: The Late Cretaceous of Southern Utah*. Bloomington: Indiana University Press.
Tjørve, M. C., G. E. Garcia-Peña, and T. Székely. 2009. "Chick Growth Rates in Charadriiformes: Comparative Analyses of Breeding Climate, Development Mode and Parental Care." *Journal of Avian Biology* 40:553–558.
Tjørve, M. C., and E. Tjørve. 2010. "Shapes and Functions of Bird-Growth Models: How to Characterize Chick Postnatal Growth." *Zoology* 113(6):326–333.
Tlïg, S., H. Sahli, M. Ouaj, and M. Mzoughi. 2008. "Cenomanian-Turonian Rudist Build-Ups and

Geodynamic Evolution of the Northern Chotts Mountains (Tunisia)." Abstract presented at the Eighth International Congress on Rudists, İzmir.
Tomlinson, P. B. P. 1986. *The Botany of Mangroves*. Cambridge Tropical Biology Series 17. Cambridge: Cambridge University Press.
Tong, T., and E. Buffetaut. 1996. "A New Genus and Species of Pleurodiran Turtle from the Cretaceous of Southern Morocco." *Neues Jahrbuch für Geologie und Paläontologie, Abhandlungen* 199(1):133–150.
Torices, A., F. Barroso-Barcenilla, O. Cambra-Moo, A. Pérez-García, and M. Segura. 2010. "The New Cenomanian Vertebrate Site Algora (Gudalajara, Spain)." Presented at the 70th Anniversary Meeting Society of Vertebrate Paleontology, Pittsburgh.
Trapman, T., F. M. Holwerda, O. Rauhut, J. Reumer, M. Joachimski. 2017. "Stable Isotopes in the Neotheropod Teeth from the Kem Kem Beds, North Africa: Insights in the Palaeoenviroment and Palaeobiology of Large Predatory Dinosaurs." Abstract presented at the 77th Annual Meeting of the Society of Vertebrate Paleontology, Calgary.
Trevisan, L. 1980. "Ultrastructure Notes and Considerations on Ephedripites, Eucommiidites and Monosulcites Pollen Grains from Lower Cretaceous Sediments of Southern Tuscany (Italy)." *Pollen and Spores* 22(1):85–132.
Tristram, H. B. 1865. *The Land of Israel: A Journal of Travels in Palestine, Undertaken with Special Reference to Its Physical Character*. London: Society for Promoting Christian Knowledge.
Tshudy, D. W., S. Donaldson, C. Collom, R. M. Feldmann, and C. E. Schweitzer. 2005. "*Hoploparia albertaensis*, a New Species of Clawed Lobster (Nephropidae) from the Late Coniacean, Shallow-Marine Bad Heart Formation of Northwestern Alberta, Canada." *Journal of Paleontology* 79(5):961–968.
Tucker, M. A., and T. L. Rogers. 2014. "Examining Predator-Prey Body Size, Trophic Level and Body Mass across Marine and Terrestrial Mammals." *Proceedings of the Royal Society B: Biological Sciences* 281(1797):2014–2103.
Tumarkin-Deratzian, A. R., B. S. Grandstaff, M. C. Lamanna, and J. Smith. 2004. "New Material of *Libycosuchus* from the Cenomanian Bahariya Formation of Egypt." Supplement, *Journal of Vertebrate Paleontology* 24(3):123A.
Turner, A. H., D. Poljulia, A. Clark, E. Gregory, M. Erickson and M. A. Norell. 2007. "A Basal Dromaeosaurid and Size Evolution Preceding Avian Flight." *Science* 317(5843):1378–1381.
Tykoski, R. S., and T. Rowe. 2004. "Ceratosauria." In *The Dinosauria*, 2nd ed, edited by D. B. Weishampel, P. Dodson, and H. Osmolska, 47–70, 2nd ed. Oakland: University of California Press.
Underwood, C. J., and S. F. Mitchel. 1999. "Albian and Cenomanian Sleachian Assemblages from North-east England." Special Papers in Palaeontology 60:9–59.
UNESCO. 1970. Convention on the Means of Prohibiting and Preventing the Illicit Import, Export and Transfer of Ownership of Cultural Property, Paris.
Unwin, D. M. 2001. "An Overview of the Pterosaur Assemblage from the Cambridge Greensand (Cretaceous) of Eastern England." *Mitteilungen aus dem Museum für Naturkunde in Berlin, Fossil Record* 4:189–221.
Upchurch, P. 2000. Neck Posture of Sauropod Dinosaurs. *Science* 287:547b.
Upchurch, P., P. M. Barrett, and P. Dodson. 2004. "Sauropoda." In *The Dinosauria*, 2nd ed, edited by D. B. Weishampel, P. Dodson, and H. Osmolska, 259–322. Oakland: University of California Press.
Upchurch, P., and P. D. Mannion. 2009. "The First Diplodocid from Asia and Its Implications for the Evolutionary History of Sauropod Dinosaurs." *Palaeontology* 52(6):1195–1207.
Uthicke, S., B. Schaffelke, and M. Byrne. 2009. "A Boom-Bust Phylum? Ecological and Evolutionary Consequences of Density Variations in Echinoderms." *Ecological Monographs* 79(1):3–24.
Vacelet, J. 1985. "Coralline Sponges and the Evolution of the Porifera." In *The Origins and Relationships of Lower Invertebrates*, edited by S. C. Morris, J. D. George, R. Ginson, and H. H. Platt. Systematics Association. London: Clarendon Press.
Vacelet, J., and E. Duport. 2004. "Prey Capture and Digestion in the Carnivorous Sponge *Asbestopluma hypogea* (Porifera: Demospongiae)." *Zoomorphology* 123:179–190.
Vakhrameev, V. A. and V. A. Krassilov. 1979. "Reproductive Structures of Angiosperms from the Albian of Kazakhstan." *Palaeontol. J. (Moscow)* 1:121–28.
van Helmond, N. A. G. M., A. Sluijs, G-J. Reichart, J. S. S. Damsté; C. P. Slomp, and H. Brinkhuis. 2014. "Chapter 3: A Perturbed Hydrological Cycle during Oceanic Anoxic Event 2." *Geology* 42:123–26.
van Konijnenburg-van Cittert, J. H. A. 1993. "A Review of the Matoniaceae Based on In Situ Spores." *Review of Palaeobotany and Palynology* 78(3-4):235–267.
Varricchio, D. V. 1993. "Bone Microstructure of the Upper Cretaceous Theropod Dinosaur *Troodon formosus*." *Journal of Vertebrate Paleontology* 13(1):99–104.
Vecchia, F. M. D., P. Arduini, and A. W. A. Kellner. 2001. "The First Pterosaur from the Cenomanian

(Late Cretaceous) Lagerstätten of Lebanon." *Cretaceous Research* 22(2):219–225.
Vecchia, F. M. D., and L. M. Chiappe. 2002. "First Avian Skeleton from the Mesozoic of Northern Gondwana." *Journal of Vertebrate Paleontology* 22(4):856–860.
20minutos.es. 2015. "Scientists Involved in the Trafficking of Dinosaur Fossils in Morocco." *20minutos*. June 6, 2015. http://www.20minutos.es/noticia/2482392/0/cientificos-marruecos/trafico-fosiles/dinosaurios/#xtor=AD-15&xts=467263.
Vermeij, G. J. 1977. "The Mesozoic Marine Revolution: Evidence from Snails, Predators and Grazers." *Palaeobiology* 3(3):245–258.
Vernygora, O., and A. M. Murray. 2016. "A New Species of *Armigatus* (Clupeomorpha, Ellimmichthyiformes) from the Late Cretaceous of Morocco, and Its Phylogenetic Relationships." *Journal of Vertebrate Paleontology* 36(1):1–9.
Vidal, N., J-C. Rage, A. Couloux, and S. B. Hedges. 2009. "Snakes (Serpentes)." In *The Timetree of Life* edited by S. B. Hedges and S. Kumar, 390–39. Oxford: Oxford University Press.
Viglietti, P. A., C. R. Penn-Clarke, M. O. Day, and W. L. Taylor. 2014. "Southern African Palaeosciences Documentary Initiative: 3.5 Ga of Life History and Origins, the Responsibility to Showcase It." Proceedings of the 18th Biennial Conference of the Palaeontological Society of Southern Africa, Johannesburg: 60–114.
Villalobos-Segura, E., J. Kriwet, R. Vullo, S. Stumpf, D. J. Ward, and C. J. Underwood. 2021. "The Skeletal Remains of the Euryhaline Sclerorhynchoid *Onchopristis* (Elasmobranchii) from the Mid-Cretaceous and Their Palaeontological Implications." *Zoological Journal of the Linnean Society*, 193(2):1–26.
Vrsanky, P., and D. S. Aristov. 2014. "Termites (Isoptera) from the Jurassic/Cretaceous Boundary: Evidence for the Longevity of Their Earliest Genera." *European Journal of Entomology* 111(1):137–141.
Vullo, R. 2019. "A New Species of *Lapparentophis* from the Mid-Cretaceous Kem Kem Beds, Morocco, with Remarks on the Distribution of Lapparentophiid Snakes." *Comptes Rendus Palevol* 18(7):765–770.
Vullo, R., H. Cappetta, and D. Néraudeau. 2007. "New Sharks and Rays from the Cenomanian and Turonian of Charentes, France." *Acta Palaeontologica Polonica* 51(1):99–116.
Vullo, R., and J-C. Rage. 2018. "The First Gondwanan Borioteiioid Lizard and the Mid-Cretaceous Dispersal Event between North America and Africa." *The Science of Nature*, 105(61):11–12.
Vullo, R., and P. Courville. 2014. "Fish Remains (Elasmobranchii, Actinopterygii) from the Late Cretaceous of the Benue Trough, Nigeria." *Journal of African Earth Sciences* 97:194–206.
Vullo, R., G. Guinot, and G. Barbe. 2016. "The First Articulated Specimen of the Cretaceous Mackerel Shark *Haimirichia amonensis* Gen. Nov. (Haimirichiidae Fam. Nov.) Reveals a Novel Ecomorphological Adaptation within the Lamniformes (Elasmobranchii)." *Journal of Systematic Palaeontology* 14(12):1003–1024.
Vullo, R., and D. Neraudeau. 2008. "When the Primitive Shark *Tribodus* (Hybodontiformes) Meets the Modern Ray *Pseudohypolophus* (Rajiformes): The Unique Co-occurrence of These Two Durophagous Cretaceous Selachians in Charentes (SW France)." *Acta Geologica Polonica* 58(2):249–255.
Wang, H., and D. L. Dilcher. 2006. "Aquatic Angiosperms from the Dakota Formation (Albian, Lower Cretaceous), Hoisington III Locality, Kansas, USA." *International Journal of Plant Sciences* 167(2):385–401.
Wang, S., J. Stiegler, R. Amiot, X. Wang, G-H. Du, J. M. Clark, and X. Xu. 2017. "Extreme Ontogenetic Changes in a Ceratosaurian Theropod." *Cell Biology* 27(1):144–148.
Ward, L. F., W. M. Fontaine, A. Bibbins, and G. R. Wieland. 1905. "Status of the Mesozoic Floras of the United States." *Monographs of the United States Geological Survey* 48, USGS.
Watson, J. 1977. "Some Lower Cretaceous Conifers of the Cheriolepidaceae from the U.S.A. and England." *Palaeontology* 20(4):715–749.
Watson, J., and N. A. Harrison. 1998. "*Abietites linkii* (Römer) and *Pseudotorellia heterophylla* Watson: Coniferous or Ginkgoalean?" *Cretaceous Research* 19:239–78.
Weiler, W. 1930. "VI. Fischreste aus dem Nubischen Sandstein von Mahmid und Edfu aus den Phosphaten OberaÈgyptens und der Oase-Baharõeje. 12–37. In Stromer, E. and Weiler, W. Beschreibung von Wirbeltier-Reste aus dem nubischen Sandstein Obera Egyptens und aus aÈgyptischen Phosphaten nebst Bemerkungen uÈber die Geologie der Umgegend von Mahmid in Obera Ègypten. Ergebnisse der Forschungsreisen Prof. E. Stromers in den Wuesten Egyptens." *Abhandlungen der Bayerischen Akademie der Wissenschaften, Mathematisch-naturwissenschaftliche* Abteilung, Neue Folge 7:1–42.
———. 1935. "Ergebnisse der Forchungsreisen Prof. E. Stromers in den Wasten Agyptens. II. Wirbellierreste der Baharaje- Stufe (unterstes Cenoman). 16. Neue Untersuchungen an den Fischresten." *Abhandlungen der Bayerischen Akademie der*

Wissenschaften, athematischenaturwissenschaftliehe Abteilung.Neue Folge Heft 32:1–57.
Weishampel, D. B., P. M. Barrett, R. A. Coria, J. Le Loeuff, X. Xing, Z. Xijin, A. Sahni, E. M. P. Gomani, and C. R. Noto. 2004. "Dinosaur Distribution." In *The Dinosauria*, 2nd ed, edited by D. B. Weishampel, P. Dodson, and H. Osmolska, 517–606. Oakland: University of California Press.
Wellnhofer, P., and E. Buffetaut. 1999. "Pterosaur Remains from the Cretaceous of Morocco." *Paläontologische Zeitschrif* 73:133–142.
Wells, M. J., J. Wells, and R. K. O'Dor. 2009. "Life at Low Oxygen Tensions: The Behavior and Physiology of *Nautilus pompilius* and the Biology of Extinct Forms." *Journal of the Marine Biological Association of the United Kingdom* 72(2):313–328.
Welten, M., M. M. Smith, C. Underwood, and Z. Johanson. 2015. "Evolutionary Origins and Development of Saw-Teeth on the Sawfish and Sawshark Rostrum (Elasmobranchii; Chondrichthyes)." *Royal Society, Open Science* 2(9):150189.
Wenz, S. 1981. "Un cœlacanthe géant, *Mawsonia lavocati* Tabaste, de l'Albien-base du Cénomanien du sud marocain." *Annales de Paléontologie (vertébrés)* 67:1–20.
Werner, C. 1989. "Die Elasmobranchier-Fauna des Gebel Dist Member der Bahariya Formation (Obercenoman) der Oase Bahariya, Ägypten." *Palaeo Ichthyologica* 5:1–112.
———. 1990. "Biostratigraphical Results of Investigations on the Cenomanian: Elasmobranchian Fauna of Bahariya Oasis, Egypt." *Berliner Geowissenschaftliche Abhandlungen—Reihe* A 120:943–56.
———. 1994. "Die kontinentale Wirbeltierfauna aus der unteren Oberkreide des Sudan (Wadi Milk Formation)." *Berliner Geowissenschaftliche Abhandlungen, E* 13:221–49.
Westermann, G. E. G. 1996. "Ammonoid Life and Habitat." In *Ammonoid Paleobiology*, edited by N. H. Landman, K. Tanabe, and R. A. Davis, 607–707. New York: Plenum Press.
———. 2013. "Hydrostatics, Propulsion and Life-Habits of the Cretaceous Ammonoid *Baculites*." *Revue de Paleobiologie* 32:249–65.
Wheatley, P. V., H. Peckham, S. D. Newsome, and P. I. Kock. 2012. "Estimating Marine Resource Use in the American Crocodile (*Crocodylus acutus*) in Southern Florida." *Marine Ecology Progress Series* 447:211–29.
Wiese, F., and F. Schulze. 2005. "The Upper Cenomanian (Cretaceous) Ammonite *Neolobites vibrayeanus* (d'Orbigny, 1841) in the Middle East: Taxonomic and Palaeocologic Remarks." *Cretaceous Research* 26(6):930–946.
Wiley, E. O. 1998. "Bichirs & Their Allies." In *Encyclopaedia of Fishes*, edited by J. R. Paxton, and W. N. Eschmeyer, 5–76. San Diego: Academic Press.
Wilmers, C. C., R. L. Crabtree, D. W. Smith, K. M. Murphy, W. Getz. 2003. "Trophic Facilitation by Introduced Top Predators: Grey Wolf Subsidies to Scavengers in Yellowstone National Park." *Journal of Animal Ecology* 72(6):909–16.
Wilmsen, M. 2000. "Late Cretaceous Nautilids from Northern Cantabria, Spain." *Acta Geologica Polonica* 50(1):29–43.
Wilmsen, M., and E. Nagm. 2012. "Depositional Environments and Facies Development of the Cenomanian-Turonian Galala and Maghra el Hadida Formations of the Southern Galala Plateau (Upper Cretaceous, Eastern Desert, Egypt)." *Facies* 58:229–247.
———. 2013. "Upper Cenomanian-Lower Turonian Ammonoids from the Saxonian Cretaceous (Lower Elbtal Group, Saxony, Germany)." *Bulletin of Geosciences* 88(3):647–674.
Wilson, J. A., and R. Allain. 2015. "Osteology of *Rebbachisaurus garasbae* Lavocat, 1954, a Diplodocoid (Dinosauria, Sauropoda) from the Early Late Cretaceous-Aged Kem Kem Beds of Southeastern Morocco." *Journal of Vertebrate Paleontology* 35:1–33.
Wilson, J. A., P. C. Sereno, S. Srivastava, D. K. Bhatt, A. Khosla, and A. Sahni. 2003. "A New Abelisaurid (Dinosauria, Theropoda) from the Lameta Formation (Cretaceous, Maastrichtian) of India." *Contributions from the Museum of Paleontology, the University of Michigan* 31(1):1–42.
Wilson, M. A., O. Vinn, and T. J. Palmer. 2014. "Bivalve Borings, Bioclaustrations and Symbiosis in Corals from the Upper Cretaceous (Cenomanian) of Southern Israel." *Palaeogeography, Palaeoclimatology, Palaeoecology* 414:243–245.
Wirsing, A. J., M. R. Heithaus, and L. M. Dill. 2007. "Fear Factor: Do Dugongs (*Dugong dugon*) Trade Food for Safety from Tiger Sharks (*Galeocerdo cuvier*)?" *Oecologia* 153(4):1031–1040.
Wirsing, A. J., and W. J. Ripple. 2011. "A Comparison of Shark and Wolf Research Reveals Similar Behavioral Responses by Prey." *Frontiers in Ecology and the Environment* 9(6):335–341.
Witton, M. P. 2013a. *Pterosaurs*. Princeton, NJ: Princeton University Press.
———. 2013b. "The Solution to Everything: under the (Jurassic) Sea, part 2" *Markwitton.com* (blog), September 26, 2013. http://markwitton-com.blogspot.com/2013/09/the-solution-to-everything-under_26.html.
———. 2014a. "The '*Spinosaurus* Reboot': Sailing in Stormy Waters." Markwitton.com (blog), September 20, 2014. http://markwitton-com.blogspot

.co.uk/2014/09/the-spinosaurus-reboot-sailing-in.html.
———. 2014b. "The *Spinosaurus* Hindlimb Controversy: A Detailed Response from the Authors." Markwitton.com (blog), September 22, 2014. http://markwitton-com.blogspot.de/2014/09/the-spinosaurus-hindlimb-controversy.html.
———. 2020. "*Spinosaurus* 2020: Thoughts for Artists." *Markwitton.com* (blog), May 12, 2020. http://markwitton-com.blogspot.com/2020/05/spinosaurus-2020-thoughts-for-artists.html.
Witton, M. P., and D. Naish. 2008. "A Reappraisal of Azhdarchid Pterosaur Functional Morphology and Paleoecology." *PLoS ONE* 3 (5):e2271.
Worm, B., and K. Lotze. 2006. "Effects of Eutrophication, Grazing, and Algal Blooms on Rocky Shores." *Limnology and Oceanography* 51(2):569–579.
Wright, C. W., and W. J. Kennedy. 1981. *The Ammonoidea of the Plenus Marls and the Middle Chalk*. Palaeontographical Society Monographs. London: Palaeontographical Society.
———. 1984. *The Ammonoidea of the Lower Chalk*. Part I. Monographs of the Palaeontological Society 567. London: Palaeontological Society.
Wright, J. B. 1985. *Geology and Mineral Resources of West Africa*. Crows Nest, George Allen & Unwin.
Wu, W-H., C-F. Chang-Fu Zhou, and B. Andres. 2017. "The Toothless Pterosaur *Jidapterus edentus* (Pterodactyloidea: Azhdarchoidea) from the Early Cretaceous Jehol Biota and Its Paleoecological Implications." *PLoS ONE* 12 (9):e0185486.
Wueringer, B. E., L. Squire Jr., and S. P. Collin. 2009. "The Biology of Extinct and Extant Sawfish (Batoidea: Sclerorhynchidae and Pristidae)." *Reviews in Fish Biology and Fisheries* 19:445–464.
Wueringer, B. E., L. Squire Jr., S. M. Kajiura, N. S. Hart, and S. P. Collin. 2012. "The Function of the Sawfish's Saw." *Current Biology* 22(5):PR150-R151.
Xing, L., E. M. Roberts, J. D. Harris, M. K. Gingras, H. Ran, J. Zhang, X. Xug, M. E. Burns, and Z. Dong. 2013. "Novel Insect Traces on a Dinosaur Skeleton from the Lower Jurassic Lufeng Formation of China." *Palaeogeography, Palaeoclimatology, Palaeoecology* 388:58–68.
Xu, X., J. M. Clark, J. Mo, J. Choiniere, C. A. Forster, G. M. Erickson, D. W. E. Hone, C. Sullivan, D. A. Eberth, S. Nesbitt, Q.Zhao, R. Hernandez, C-K. Jia, F-L. Han, and Y. Guo. 2009. "A Jurassic Ceratosaur from China Helps Clarify Avian Digital Homologies." *Nature* 459:940–944.
Yabumoto, Y., and P. M. Brito. 2013. "The Second Record of a Mawsoniid Coelacanth from the Lower Cretaceous Crato Formation, Araripe Basin, Northeastern Brazil, with Comments on the Development of Coelacanths." In *Mesozoic Fishes 5: Global Diversity and Evolution*, edited by G. Arratia, H. Schultze, and M. Wilson, 489–497. Munich: Dr. Friedrich Pfeil.
Yabumoto, Y., and T. Uyeno. 2005. "New Materials of a Cretaceous Coelacanth, *Mawsonia lavocati* Tabaste from Morocco." *Bulletin of the National Science Museum Tokyo* C, 31.
Yilmaza, I. O., D. Altiner, U. K. Tekin, O. Tuysuz, F. Ocakoglu, and S. Acikalin. 2010. "Cenomanian-Turonian Oceanic Anoxic Event (OAE2) in the Sakarya Zone, Northwestern Turkey: Sedimentological, Cyclostratigraphic, and Geochemical Records." *Cretaceous Research* 31(2):207–226.
Younes, M. A. 2012. "Hydrocarbon Potentials in the Northern Western Desert of Egypt." In *Crude Oil Exploration in the World*, edited by Y. Mohamed, 24–46. London: IntechOpen.
Young, M. T., A. K. Hastings, R. Allain, and T. J. Smith. 2016. "Revision of the Enigmatic Crocodyliform *Elosuchus felixi* de Lapparent de Broin, 2002 from the Lower-Upper Cretaceous Boundary of Niger: Potential Evidence for an Early Origin of the Clade Dyrosauridae." *Zoological Journal of the Linnean Society* 179(2):1–27.
Zaborski, P. M. P. 1995. "The Upper Cretaceous Ammonite *Pseudaspidoceras* Hyatt, 1903, in North-eastern Nigeria." *Bulletin Natural History Museum London* 46:53–72.
———. 1996. "The Upper Cretaceous Ammonite *Vascoceras* Choffat, 1898 in North-eastern Nigeria." *Bulletin of the Natural History Museum London* 52:61–89.
Zimmermann, W. 1959. *Die Phylogenie der Pflanzen*. 2nd ed. Stuttgart: University of Michigan.
Zobaa, M., C. A. S. Botero, C. Browne, F. Oboh-Ikuenobe, and M. I. Ibrahim. 2008. "Kerogen and Palynomorph Analyses of the Mid-Cretaceous Bahariya Formation and Abu Roash "G" Member, North Western Desert, Egypt." *Gulf Coast Association of Geological Societies Transactions* 58. GCAGS national meeting, Houston.
Zomlefer, W. B. 1994. *Guide to Flowering Plant Families*. Chapel Hill: University of North Carolina Press.
Zitouni, S., C. Laurent, G. Dyke, and N-E. Jalil. 2019. "An Abelisaurid (Dinosauria: Theropoda) Ilium from the Upper Cretaceous (Cenomanian) of the Kem Kem Beds, Morocco." *PLoS ONE* 14 (4):e0214055.
Zouhri, S. 2017. "Vertebrate Paleontology of Morocco: The State of Knowledge." Proceedings of the Geologists Association 131(3-4):417–419.

Index

Italic numbers indicate pages with illustrations.

Abu Roash Formation, *13–14*
Abu Qada Formation, 15–*16*
Abyad member (units G). *See* Abu Roash Formation
Actinopterygii, 109
Actinostromarianina, 60
Adrianaichthys, 109–110
Aegisuchus, 165–68, 226–27
Aegyptosaurus, 199, 201
Aegyptosuchus, 165–*166*, 168, 174, 226, 227
Africa, continent, 1–2, 7–8, 12, 20, 28, 32–33, 53, 61, 63, 73, 77, 106, 111, 114, 148, 132, 151–52, 159, 160, 181, 191, 201, *204*, 206–8, 229, 236, 242–43, *244–46*, 252; equatorial Africa, 115, 224; southern Africa, 79, 163; sub–Saharan Africa, 70, 246; west Africa, 75, 219, 242–243
Afropollis, 219–20
Agassizilia, 134, 239
Agathis, 37–38, 218, 220
Agoultichthys, 131, 136, 215
Aidachar, 125–126
Ain Tobi member. *See* Nefusa Formation
Alanqa, 178–180
Akrabou Formation. *See* Kem Kem Formation
Albian, 17, 26–28, 46, 52, 55, 61, 108, 160, 242
algal, 220, 247–48
Algeria, 11, 22, 24–26, 60, 65, 70, 77, 79, 81, 95, 143–44, 160, 165, 193, 199, 208, 211–12, 214, 221, 234, 250; Béchar province, 22–23; Guir Basin, 22, 65; Ksour Mountains, 24; Ouargla province, 26; Tinrhert Plateau, 25
Amazighopsis, 91, *92*
ammonite, 11, 15–*16*, 26, 66–68, 73, 75–78, 80, 212, 231, 250–52
amphibian, 101, 151, 153, 210, 226
angiosperm, 218–19, 220, 257
Angulithes, 68, 80–*18*
Annelida, 63
anoxic, 14, 18, 20, 215, 252
Antaeusuchus, 171
anthozoa, 60, 62
Aoufous Formation. *See* Kem Kem Formation
Apertotemporalis, 154, 156–57
Applinocrinus, 65
Aptian, 15, 38, 46, 108, 207, 220, 243, 246
Aquatifolia, 52–53
Araripesuchus, 171–72
archipelago, 17
Arganodus, *144*
Armigatus, *131*, 137,
Aspidiscus, 60–61
assemblage, 2, 22, 68, 93, 220, 231, 252
Asia, continent, 4, 73, 79, 105, 242–43
Asiatifolium, 46,
Asteracanthus, 107–8
Atractosteus, *110–112*
Australia, continent, 7, 28, 142, 222, 227, 231
Avicennia, 55
Axelrodichthys, 139, *140–42*
Baculites, 68–70
Bahariasaurus, 192–93, 196, 207, 233, 236
Baharipristis, 102
Bahariya Formation, 11–14, 17, 91, 216, 228, 229, 230, 232–34
Bahariya region, 1, 3, *11*–12, 14, 17, 20, 31, 33, 37, 42, 45–47, 50–58, 93, 97, 103, 108, 110, 117, 121, 138, 140, 143, 146, 158–59, 163, 181, *192*, 196, 214, 217–21, 223, 225–26, 229–30, 231–33, 235
Bambusa, 55
Barykovia, 49, 50
basalt, 1
Bawitius, 110, 116–18, 148
Beaconites, 63
Béchar province. *See* Algeria
bed (unit), 10, *11*–14, 17, 19, 20–26, 31, 47, 59, 63, 66, 68, 114, 118, 120, 126, 176, 178, 192, 215–16, 226, 228, 250
beetle, 47, 96, 97, 98, 251
Belonostomus, *131*, *135*, 136
bird, 28, 101, 151, 175, 177, 182, 184, 203, 208, 210, 222–23, 226, 251
Bivalvia, 81
bivalves, 81, 103, 211–12, 214, 241–42, 261
Bordj Omar Driss Formations, 25
Botryococcus, 220
Boulouha region. *See* Tunisia
Brasenites, 53
burrow, 23, 62–63, 66, *94*–95, 98, 129, 179, 212, 214, 224
Burroceras, 72, 73
Calamopleurus, 115–16
Calcaires de Sidi Mohamed Ben Bouziane Formation, 22–23
Calycoceras, 70–71
carbonate platform, 8, 28
Carcharodontosaurus, 187–90, 192, 202, 229, 233, 236
carnivore, 63, 102, 117, 129, 171, 212, 231, 234, 235
Celtis, 55
Cenomanian, age, 1–2, 21, 24–25, 28, 38, 45, 47, 50, 57, 59, 61, 66, 73, 76, 78–81, 90, 120, 130, 146, 153, 158, 160, 165, 197, 207, 211, 242–43, 247, 249; environment, 2, 7, 8, 9, 12, 14, 16–17, *19*, 20, 22, 26, 28, 36, 38, 58, 82, 211, 214, 224, 231, 234, 243–47, 249–50, 251, 253; fossils, 2, 17, 26, 31, 44, 47, 50, 53, 62–65, 73, 75, 82, 93, 101, 108, 113, 151–53, 214–15, 218, 220, 223, 241–42, 245, 250; geology, 2, 11, 14–16, 19, 20–22, 24, 26, 50, 59, 137, 157; migration, 216, *241–42*, *244–45*

Central Gondwanan Desert, 8, 224
Ceratodus, 142–45
Chelicerata, 82
Choffaticeras, 70
Chondrichthyes, 101
Chotts Hanging Reef, 28
Chrysidoidea, 95
Cicatricosisporites, 220
Cladocora, 61–62
Cladocyclus, 126
Cladophlebis, 33–34, 217
Classopollis, 220
climate, 8, 9, 20–21, 36, 57–58, 219, 222, 232, 241
Clupavus, 132, 214
Cocculophyllum, 54–55
Coelodus, 131, 134, 239
coleoptera, 96
Concavotectum. *See Paranogmius*
Continental Intercalaire, 17, 22, 26, 233
continual ecological province, 10, 226, 229
coral, 23, 24, 59, 60–61, 84, 211–12, 214, 247, 249
Corazzatocarcinus, 89–90
crab, 23, 82–33, 85, 91, 93–95, 214, 224, 225, 234
craton 11, 219; metacraton, 11
Cretaceous, 1–2, 22, 41, 47, 48, 55, 57, 60, 97, 101, 105–6, 135, 151, 199
crinoid, 63–65
Cretapenaeus, 83–84
crustacean, 82, 84, 91–95, 103, 169, 179, 213, 221, 249
Cretagalathea, 88–89
Cretodus, 105, 107
Cretolamna, 105–7
Crocodylomorpha, 164, 238; crocodile, 22, 180, 163, 166, 185–86, 227–28, 237
Cribroperidinium, 220
Cunningtoniceras, 79
Crybelosprites, 220
Cyathidites, 220
Cyclonephelium, 220
Cyperites, 56
Cytherella, 93

Dadoxylon, 38–39
Davichthys, 132
Dekkar Group, 20
Dekkar Formation. *See* Dekkar Group
delta, 9, 16, 19, 22, 26, 116, 118, 156, 215, 217, 219
Deltadromeus, 192–96, 206, 207, 236
Dentilepisosteus, 113
Dercetis, 131, 135
detritivore, 88, 97, 224
Dicksonia, 35, 218
Dictyophyllidites, 220
dinoflagellate, 220
Dinopterygium, 220
dinosaur, 1–4, 7–8, 12, 27, 60, 67, 98, 108, 153, 161, 172, 174–75, 181, 184, 190, 201, 208–9, 222, 224–25, 229–32, 232–35, 237, 245, 247
Diplocraterion, 95
Diplomystus, 137–38
Diplospondichthys, 128–29, 219
Dipterocarpophyllum, 50
Dirqadim, 156–57
dolostone, 24
Douira Formation. *See* Hamadian Supergroup
Douira region. *See* Morocco
Draa Ubari. *See* Libya
Dryophyllum, 49–50

East Themed area. *See* Egypt
Ecology, 97, 105, 176, 206–7, 211, 214, 222; paleoecology, 31
ecosystem, 9, 12, 31, 50, 97, 101, 211, 222–23, 225, 235, 251; ancient, 2, 11, 13, 17, 21–22, 27, 58, 81, 96, 120, 144, 213–16, 219, 220–21, 227–28, 232–38, 247, 252, 253
Echinodermata, 63, 65
Egypt, 1, 3, 11–12, 22, 31, 34–35, 47, 51, 60–61, 63, 69, 70, 72, 75, 78–9, 91, 94, 101–3, 105, 116, 129, 134, 145–46, 157, 174, 189, 206, 216, 221, 223–24, 236–38; ancient, 38, 44, 50, 76, 77, 81–82, 95, 182, 214, 217, 226–27, 231, 250; Eastern Desert, 15; East Themed area, 14–16, 212–13, 250; Gebel Areif El-Naqa, 212; Sinai region, 14–15, 59, 93, 250; Wadi Quseib, 15, 212–13, 250; Western Desert, 1, 14, 65, 219
Elaterosporites, 219–20
El Heiz member. *See* Bahariya Formation
Ellimmichthyiformes, 118, 120, 138, 148
Elopopsis, 130, 214
Elosuchus, 164–65, 237–38, 253
Enchodus, 113–14, 132, 221–22
endemism, 2, 229
Ephedripites, 220
era, 1–2, 4, 48, 50, 68, 92, 146, 247, 254
Erfoudichthys, 127–28
Errachidia, 130–31
Eucalycoceras, 79
Euomphaloceras, 70–72, 79
Europe, continent, 4, 7, 28, 34, 56, 65, 70, 73, 75, 79, 80, 105, 110–111, 132, 137, 146, 159, 213, 221, 241, 244, 245, 246
Eutrephoceras, 81
Exochosphaeridium, 220
Exogyra, 82
extinction, 2, 26, 60, 63, 130, 216, 247, 249, 250, 251, 252–53; extinct, 1–2, 36, 61–62, 76, 81, 84, 105, 115, 151, 161, 211, 215, 253; mass extinction, 26, 60, 247

fauna, 2, 7–8, 10, 15, 17, 23, 26, 28, 59, 101, 151, 161, 211, 215–16, 222–23, 226, 229, 230, 241, 243, 245–46, 252; infauna, 212–14; epifauna, 212
Ficus, 52
Fikaites, 79–80
Filograna, 63
fish, 23, 101, 109, 114–15, 118, 120–22, 128, 131–32, 134–35, 137–38, 140, 145, 146, 165–67, 210, 212, 214–15, 221, 226, 230, 234–36, 239, 251
flora, 12, 20, 28, 31, 53, 217, 219, 221, 226, 232, 241
Florentinia, 220
flower, 31, 43, 47, 53, 217
Forbesiceras, 78–79
forest, 1, 8–9, 12, 20, 26, 33, 37–38, 50, 58, 216, 219, 220–22, 228
Frenelopsis, 41–42, 47

Gabel el Dist member. *See* Bahariya Formation
Gabal el Magrafa member. *See* Naqb el Sellem Formation
Gafsa region. *See* Tunisia
Galianemys, 155–57
Galala Formation, 12, 14–16, 212
Galathea, 87–88
Gara Sbaa Formation. *See* Hamadian Supergroup
Gara Sbaa region. *See* Morocco
Gargal (Gargaf), 8, 28
Garasbahia, 53

Garian member. *See* Nefusa Formation
Gattar member. *See* Zebbag Formation
gastropod, 23, 65–66, 169, 211, 213, 225, 234, 250
Gebel Areif El-Naqa. *See* Egypt
Gebel Ghorabi member. *See* Bahariya Formation
genus, 35, 37, 39, 40–42, 68, 223, 229, 230; amphibian, 151; archosaur, 165, 172, 174–78, 180, 183, 187–89, 193, 196, 202–3, 206, 232; fish, 101–3, 105–6, 108–15, 117, 120, 126, 128, 129, 132–37, 140–45, 148, 234; invertebrate, 66, 70, 73–75, 78–789, 81, 83–5, 87–88, 91, 93; mammal, 154; plant, 32–35, 37–38, 40–47, 50–53, 55, 57; reptile, 154–55, 158–59, 160, 162, 163, 244
Ginkgo, 40, 42–43, 218, 220
global warming, 2, 224, 253
Glomerula, 63
Glossifungites 13
Glyphea, 84–5
Gondwana (Gondwanaland), 7–8, 16, 28, 37, 60, 151, 160, 211, 218, 224, 226, 241, 251
Great Britain, 41, 105, 146, 153, 163, 177, 245, 248
Gres rouges Formation, 22–24
Guir Basin. *See* Algeria
Gymnura, 103
gypsum, 15, 18, 21–5

habitat, 9–10, 26, 33, 42, 61–63, 65, 83, 105, 108, 113, 128, 151, 208, 213, 217–18, 222–23, 225, 227–28, 235, 251
Haimirichia, 107–8
Halal Formation, 15
Hamadachelys, 156, 157
Hamadasuchus, 169–73, 238
Hamadian Supergroup, 18–19
Heckelichthys, 126
herbivore, 55, 63, 124, 180, 198, 199, 200, 212, 218, 223–24, 231–32, 237, 251
Heterodiadema, 63
Hibiscoxylon, 47–48
High Atlas region. *See* Morocco
Homalopagus, 133
Homoptera, 95

Ichnofacies, 13
Ichthyotringa, 114–15, 132, 214
Idrissides islands, 20
Ifezouane Formation. *See* Kem Kem Formation
index fossil, 68, 213
insect, 43, 54, 95, 96–98, 110, 169, 222, 224, 226
invertebrate, 59, 81–82, 110, 120, 122, 143, 177, 179, 224, 226, 235, 237, 247
Isidobatus, 103
island, 8, 16–17, 20, 28, 224, 235, 245
isopod, 70, 92
Isoptera, 96–97

Jbel Tselfat: region. *See* Morocco
Jeddaherdan, 161
jungle, 9

Kababisha, 152–53
Kem Kem Formation, 17–20, 22, 116, 215, 218–19, 229, 232, 234
Kemkemia, 174, 175
Kerker member. *See* Zebbag Formation
Kermichthys 133
Konservat Lagerstätte, 18, 20
Koreanaornis, 208–9
Ksar Ayaat: region. *See* Tunisia
Ksour Mountains. *See* Algeria

Laganosuchus, 168, 226, 227, 239
lagoon, 8, 14, 16, 20, 22, 24–26, 28, 42, 65, 83, 93, 103, 120, 142, 217
Laurasia, 7, 28, 160
Laurophyllum, 48–50
Lavocatchampsa, 169–70, 172, 238
leaf, 32–33, 35–36, 40–43, 46–47, 50–51, 53–57, 62, 97, 222, 224
Lepidotes, 109–10, 113
Libya, 11, 21–22, 153, 158, 160, 221, 242–43; Draa Ubari region, 21
Libycosuchus, 169, 172
limestone, 13–14, 16, 19, 21–27
Liriodendrites, 43–44
lizard, 48, 158, 161–5
Lusitanichthys, 124–25
lungfish, 63, 138, 142–45, 215, 234

madtsoiid, 162
Maghrabi Formation, 14, 157
Magnolia, 43–49, 217, 228
Magnoliaephyllum, 47
Magrebichthys 131, 123
mammal, 1, 101, 151, 153, *154*, 207, 223, 229
mangrove, 8, 12–14, 32–34, 37, 55, 58, 82, 111, 203, 211, 217–18, 221–23, 224–25, 227–28, 233, 235, 25
mansour member (units F). *See* Abu Roash Formation
Marocarcinus, 91
Marckgrafia, 102
marine, 2, 12–13, 15–16, 18–20, 25–27, 33, 42, 63, 80–81, 108, 116, 118, 132, 142, 160, 163, 211–12, 214–17, 219–21, 235, 239, 241, 243, 247, 249, 252
Marnes à gypse inférieures Formation, 22, 23
Marl, 15–16, 19, 21–25, 27
Marsilea, 36
Mawsonia, 101, 121, 138–39, 140–42, 148, 215
Mazoula Formation, 25
Mazuza member. *See* Mizdha Formation
Mdaouer Formation, 24–25
Metoicoceras, 76
Metengonoceras, 78
Mesolimulus, 82
Mesozoic, 1–2, 21, 42, 50, 53, 63, 67, 108, 111, 152–53, 163, 218, 222, 235–36, 247, 251
microconglomerates 18, 24
microcrinoids, 65
miospores, 58, 219, 257
Mizdah Formation, 21–22
Morocco, 3–4, 11, 17, 22–3, 31, 35, 40, 50, 59, 63–64, 66, 78, 83–84, 89, 91–94, 97–98, 101, 107–9, 110–12, 114, 116–18, 120–21, *127*, 129, 130, 133–38, *141*, 144–46, 148, 152–53, 155–56, 158, 160, *162*, 168, 171, 173–76, 179, 182, 187–89, 197, 201, 204–5, 207–9, 213, *221*, 224, 236, 238; ancient, 20, 223, 220, 226–27, 231, 235, 250; Douira region, 18–19; Gara Sbaa region, 12, 18–19, 126, 162–63, 215–16; High Atlas region, 20; Jbel Tselfat region, 58, 216, 277; southwestern Morocco, 242, 252; Tinghir region, 19
mountain, 11, 19, 24, 38, 109, 211, 219, 220, 228
Mudrongia, 220

mudstone, 12–15, 19, 23–26
Muelleristhes, 85–87

Naqb el Sellem Formation, 14
Nefusa Formation, 21
Nehvizdya, 42
Nelumbites, 51–52, 56, 217
Nelumbo, 56
Neoceratodus, 142–44
Neocallichirus, 83
Neolobites, 67–68, 80–81, 213
Neoproscinetes, 134, 239
Neovermilia, 63
Nigericeras, 75, 79
Nigerophiid, 162
niche partitioning, 105, 180, 226, 234, 236–39
Noasauridae, 194–96, 206–7
Norisophis, 160
North Africa, 2–4, 7, 9, 11–12, 17, 37, 41, 48, 50, 54, 58–59, 61–62, 66–67, 69, 70, 72–73, 75–78, 83, 92–93, 96–97, 102–8, 110, 114, 118, *131*, 137–38, 142, 151, 153–54, *157*, 162, 164, 168, 175, 177, 180, 196, 203, 208, 229, 235–39, 246; ancient, 1, 6, 8, 15, 27, 31, 33, 36, 38, 49, 53, 63, 65, 80, 82, 95, 101, 120, 125, 132, 134, 144, 158, 165, 174, 182, 188, 211, 214–15, 218–19, *221–23*, 226–27, 231–32, 234, 241–43, 247, 252
North America, continent, 56, 65, 68, 70, 78–80, 105, 114, 132, 134, 146, 208, 241, 244–45
nutrient, 33, 111, 124, 212, 218–19, 221–22, 224–26, 247, 251
nymphaea, 52

OAE 2 event, 252
oasis, 1, 11, 14, 144, 209
ocean, 2, 7, 9, 23, 26, 65, 75, 103, 132, 140, 142, 211–12, 216, 218, 227, 235, 241–42, 249, 250, 252; currents, 2, 7–8, 241–42, 244, 249, 250; oceanic dispersal, 111, 226, 241–45
Odonata, 97
Omosoma, 130
Omosomopsis, 123, 129–30
Omphalopus, 56
Onchopristis, *100–1*, 103, 106, 121
Oniichthys, *112*
ontogeny, 103, 124, 132, 151, 188, 197, 208
Ophiomorpha, 95
ornithopod, 190, 207–9, 231, 236
Orthogonocrinus, 65
osteoderm, 165
ostracod, 83, 93, *242–43*, 262
Ouargla province. *See* Algeria
Oumrkoutia, 150–51

packstone, 15, 23, 25
Palaeonotopterus, 129
palaeontology, 102–3, 164
palynology, 58; palynomorph, 9, 16, 58, 220
Pangaea, 7
Paradoxopteris. *See Weichselia*
Paralititan, 199, *200–1*, 224
Paranogmius, 121–23
Paranursallia, *131*, 137
Parataxodium, 56
Paravinciguerria, 132, 137
Parasitism, 60
Parnassia, 57
Pchelinsevia, 66, 214
Pediastrum, 220
period: geological, 1–4, 7, 9, 11, 23–26, 37, 46, 55, 57, 63, 65, 81, 97, 105, 160, 247
Peseudoceratium, 220
Peyeria, *100*, 103
piscivore, 176, 186, 225, 236–37
Placenticeras, 79
plankton, 26, 70, 215, 217, 220, 247, 249; bacterioplankton, 235; phytoplankton, 220, 222, 235; zooplankton, 222, 235
plant, 8–9, 12, 18, 20, 22, 27, 31–36, 39, 41–42, 46–47, 53–54, 56–58, 62, 64, 97, 124, 155, 161, 198, 211, 214, 216, 218–22, 224–25, 228, 235, 247, 251
Platanus, 57
Plesiosauria, 163
Plethodus, 129
Podozamites, 42
Polypterus, 117
Polytremacis, 61
Populus, 50
Porifera, 59
Pseudotorellia, 40–42
Pseudaspidoceras, 74–75, 251
pterosaur, 175–78, 180–81, 226, 235, 238
Ptychotrygon, 104
Raha Formation, 15, 59, 93
Rebbachisaurus, *197–98*, 202, 231, 246
reef, 8, 16, 24–28, 60–62, 65, 79–82, 84, 93, 120, 134, 160, 211–14, 216, 231, 242, 250, 252
Renpetia, 103
reptile, 101, 151, 163, 210, 212, 214
Retodus, *143–44*
Rhadouane member. *See* Zebbag Formation
Rharbichthys, 136
El Rhelida Formation, 24–25
Rhinoptera, 103
Rhoundjaïa Formation, 24–25
Rhynchodercetis, *131*, 134–35
river, 8–9, 12, 14, 19–20, 22, 26, 33–34, 41, 47, 84, 103, 110–11, 118, 156, 167, 174, 177, 185–86, 211, 216–22, 224, 227, 229, 233, 249
Rhombichthys, *131*, 136
Rogersia, 46
Roveacrinus, 65
Rubroceras, 79
rudists, 60, 81, 211, 242

Sabkha, 24–25, 211
Salix, 46
sandstone, 12–15, 18–19, 23–24, 218
Sauroniops, *190–191*, 236
Saurorhamphus, *131*, 136
sauropod, 190, 197–99, 200–1, 209, 224, 226, 229, 231, 236, 246, 251
Schizorhiza, *102*
sea, 1, 2, 8, 15–19, 20, 23–34, 26, 60, 80, 85, 103, 111, 125, 132, 137, 211, 214–15, 219, 221, 227, 230, 238, 243, 250; seabed, 59–60, 64, 106, 212; sea levels, 7, 12–13, 17, 20, 22–25, 28, 58, 243–44, 250, 253
sea anemone, 60, 62–63, 66
sediment, 2, 8, 11, 17–19, 22, 26–28, 32–3, 60–61, 63, 81, 108, 118, 160, 179, 209, 212, 214, 216–20, 224, 229, 238, 247, 249, 250
Serenoichthys, *118*, 219
shark, 101, *104–5*, 107–8, 146, 219, 231
shell, 13, 60, 63, 66–70, 72, 74, 76, 78, 80–81, 93, 108, 110, 154–56, 179, 214, 234, 249, 250
Simoliophis, 158, *159–60*
Sinai: region. *See* Egypt
Sigilmassasaurus, *182–83*, 186–88, 225, 236, 236–37
Siroccopteryx, 175–76

skeleton, 2, 4, 12, 26, 59, 60–62, 101, 104, 106, 128, 154, 162, 171, 185, 188–89, 193, 197, 199; exoskeleton, 213, 249
skull, 101, 112, 114, 124, 128–30, 132–33, 136, 141, 151–52, 154–56, *165*, 166–69, 172, 173, *183*, 185–86, *189–91*, 238
South America, continent, 7, 48, 115, 141, 151, 201, 241–46
Southern Meander System, 9, 221
Sorbinichthys, 118–19
Species, 3, 8, 31, 42, 63, 82, 84, 85, 103, 114, 210–13, *215–17*, 222–25, 228–31, 238, 247, 250–52; amphibian, 151–53; archosaur, 164–69, 171–74, 176–77, 179, 182–83, 186–88, 190, 195–97, 199, 201–3, 206, 208–9, 225–26, 229, 231, 233–34, 236–38, 246, 253; fish, 101, 103–6, 108–9, 110–19, 120–*21*, 123–26, 129, 130–39, 140–46, 221, 234; invertebrate, 60–63, 65–70, 72–79, 80–83, 85, 87, 90–98, 211–12, 214, 225, 243, 251; mammal, 153, 229, 251; plant, 31–38, 40–48, 50–51, 54, 56–57, 218, 220–221, 234–35, 241, 251; reptile, 155–59, 160, 162–63, 244–45;
Spinocaudichthys, 122, *124*, 219
Spinosaurus, 3, 173, *181–88*, 190, 225, 229, 233, 236–39, 253
sponge, 26, 59–60, 211–12, 214
Squalicorax, *104–5*, 107
Squamata, 158, 162
Squatina, 104–5, *107*
Steinbachodus, 107, 109
Stomatosuchus, 166–67, 174, 226–27, 238
Subtilisphaera, 220
supergroup, 10, 18
swamp, 12, 14, 26, 37–38, 41, 58, 111, 211, 217, 228–29, 247
taxon, 38, 50, 106, 138, 140, 153, 159, 164, 171, 177, 180–81, 197, 204, 230, 246; archosaur, 165–66, 169, 171–73, 176, 181–82, 185, 187–88, 190, 193–96, 207, 233–34, 238; fish, 102–3, 105–6, 108–9, 110, 113, 116, 128, 130–37, 140, 145–46, 148, 239; invertebrate, 65–68, 70, 72, 78–80, 83, 90, 93; plant, 31–32, 35, 38, 40–2, 44, 46–48, 50–53, 55–57, 61; reptile, 156, 161–62, 244
taxonomy, 40, 47, 61, 83, 88, 138, 140, 153, 164, 172, 176, 187, 190, 193, 197, 204, 234
Telamonocarcinus, 90
temperature, 8–9, 44, 228
Terminalioxylon, 44–45, 217
termite, 97–98, 224
Testudine, 154–58
Tethys, 7–9, 16, 27, 60, 111, 218, 241, 243, 247, 250, 252
Tetragramma, 63–64
Thala member. *See* Mizdah Formation
Thalassinoides, 23, 94–95
theropod, 118, 174, 182, 185–6, 188, 193, 195–96, 202, 204, 206, 209, 224, 226, 229, 231, 233–34, 236–39
Thomelites, 77–78
Thorectichthys, *120–21*
Tigrinna member. *See* Mizdah Formation
Tinghir: region. *See* Morocco
Tingitanichthys 130
Tinrhert Plateau. *See* Algeria
titanosaur, 201, 223, 231–32, 246, 251
tooth plate, 129, 142–45
Tortoflabellum, 61
trace fossil, 62–63, 66, 94, 97, 154, 158, 174, 208–9
Trans-Saharan Seaway, 8, 16, 27, 140, 226–27, 229, 241–44, 252
tree, 22, 26, 31, 37–38, 42–43, 46, 48–49, 50–51, 55, 57–58, 123, 166, 181, 216–18, 220, 251
Tribodus, *107*, 108, 239
Triplomystus 133
Tristemma, 57
trophic cascade, 210, 251
Tselfatia, 132, 214
Tunisia, 22, 26, 28, 38, 60, 65, 70, 78–79, 108, 137, 154, 213, 221, 223, 250, 252; Boulouha region, 27; Gafsa region, 27; Ksar Ayaat region, 27
Turonian, 14, 16, 18, 21, 24–26, 46–47, 59, 73, 79, 114, 120, 130, 211, 242, 249, 250, 252
Typhaephyllum, 54

Unusuropode, 92–93

Vascoceras, 76–77, 79
vertebrae, 101, 113, 118, 124, 127, 133, 146, 152–54, 159, 160, 162–66, 178, 186–87, 189, 192, 193–94, 198–99, 207
Vitiphyllum, 55

wackestone, 15, 23, 25
Wadi Quseib. *See* Egypt
Weichselia, 30–33, 42, 216–18
Western Desert. *See* Egypt
wetland, 9, 14, 26, 41, 51, 57, 59, 222, 231, 247, 251, 253
World War II, 3, 103, 114, 117, 139, 166, 168

xerophyte, 9, 32
Xiphophoridium, 220

Yafran member. *See* Nefusa Formation

Zebbag Formation, 22, 26–27
Zizyphoides, 57

A graduate of Liverpool John Moores University with a degree in Palaeobiology and Evolution, **Jamale Ijouiher** is an expert on Mesozoic African biota. Ijouiher has participated in fieldwork in Britain and Morocco.